SpringerWienNewYork

Michael Hesse
Heidemarie Halbritter
Reinhard Zetter
Martina Weber
Ralf Buchner
Andrea Frosch-Radivo
Silvia Ulrich

Pollen Terminology

An illustrated handbook

SpringerWienNewYork

Univ.-Prof. Dr. Michael Hesse
DDr. Heidemarie Halbritter
Ao.Univ.-Prof. Dr. Reinhard Zetter
Ao.Univ.-Prof. Dr. Martina Weber
Dipl.-Biol. Dr. Ralf Buchner
Andrea Frosch-Radivo
Mag. Silvia Ulrich
University of Vienna, Austria

Printed in Austria

SpringerWienNewYork is part of
Springer Science+Business Media
springer.at

Cover Illustrations: *Agrostemma githago* (SEM, HH), *Pinus* sp. fossil (LM, RZ), *Ruellia graecicans* (SEM, HH), *Phyllanthus x elongatus* (SEM, HH), *Argyranthemum* sp. (TEM, AF-R), *Acacia myrtifolia* (SEM, HH), *Leontodon saxatilis* (SEM, HH)
Layout: Ralf Buchner, Barbara Hermanowski
Drawings: Heidemarie Halbritter, Ralf Buchner
Printing: Holzhausen Druck und Neue Medien GmbH, 1140 Wien, Austria

Printed on acid-free and chlorine-free bleached paper
SPIN: 12045303

With numerous (partly coloured) Figures

Library of Congress Control Number: 2008941153

ISBN 978-3-211-79893-5 SpringerWienNewYork

In memory
of

Jan MULLER and Wilhelm KLAUS,

who played a prominent role
in the study of fossil and extant pollen.

GENERAL CHAPTER

ILLUSTRATED GLOSSARY

ALPHABETIC GLOSSARY

ANNEX

There are more things in heaven and earth,
than are dreamt of in our philosophy.

Preface

The principal aim in compiling this book was to provide the reader with first-hand information about the structure and outlook of the extremely manifold pollen in seed plants.

Pollen Terminology. An illustrated Handbook should not be seen as a mere collection of striking and/or informative light and electron micrographs. Each of the micrographs is intended to convey a specific message related to properties and functions of the pollen grains shown. The authors hope that the book will be useful for experienced researchers as well as for beginners in palynology, but also for medicine, biochemistry, or even for lawyers and artists as an aid and guide for the evaluation and interpretation of pollen features.

Acknowledgements

Many people were involved in various aspects of this book project and the authors would like to acknowledge and thank them for their time, suggestions and encouragement during the various developmental phases of this book and/or for providing plant material.

Among the many other colleagues, who contributed substantially, the authors want to thank (in alphabetical order):

Wolfram ADLASSNIG
Elisabeth ANGER
Josef BOGNER
Paula BOMBOSI
Perica BRODARIC
Thomas CROAT
Bernadette DIETHART
David Kay FERGUSON
Manfred A. FISCHER
Alfred GLASER
Lynn HANNON
Barbara HERMANOWSKI
Anton IGERSHEIM
Irmgard JÄGER-ZÜRN
Christel KASSELMANN
Nadja KAVCIK
Alexander KOCYAN
Wolfgang KOLLER
Thomas LENDL
Claudia LOOS
Božena MITIĆ
Wolfgang OBERSCHNEIDER
Marianne PEROUTKA
Paul RADIVO
Saskia SAM-RICHTARZ
Johannes SCHACHNER
Ursula SCHACHNER
Barbara SIMON
Susanne SONTAG
Robert STANGL
Rupert STINGL
Matthias SVOJTKA
Angelika SYROVATKA
Enikö TWERASER
Walter TILL
Stefan VOGEL
Bruno WALLNÖFER
Arabella WURZINGER
Klaudia ZETTER
The staff members of the Botanical Garden of the University of Vienna (HBV)
The staff members of the Bundesgärten Wien_Innsbruck

A special note of thanks is due to Mag. Franziska BRUGGER and to Mrs. Silvia SCHILGERIUS of the team at Springer Vienna for their great support and patience during the preparation of this book.

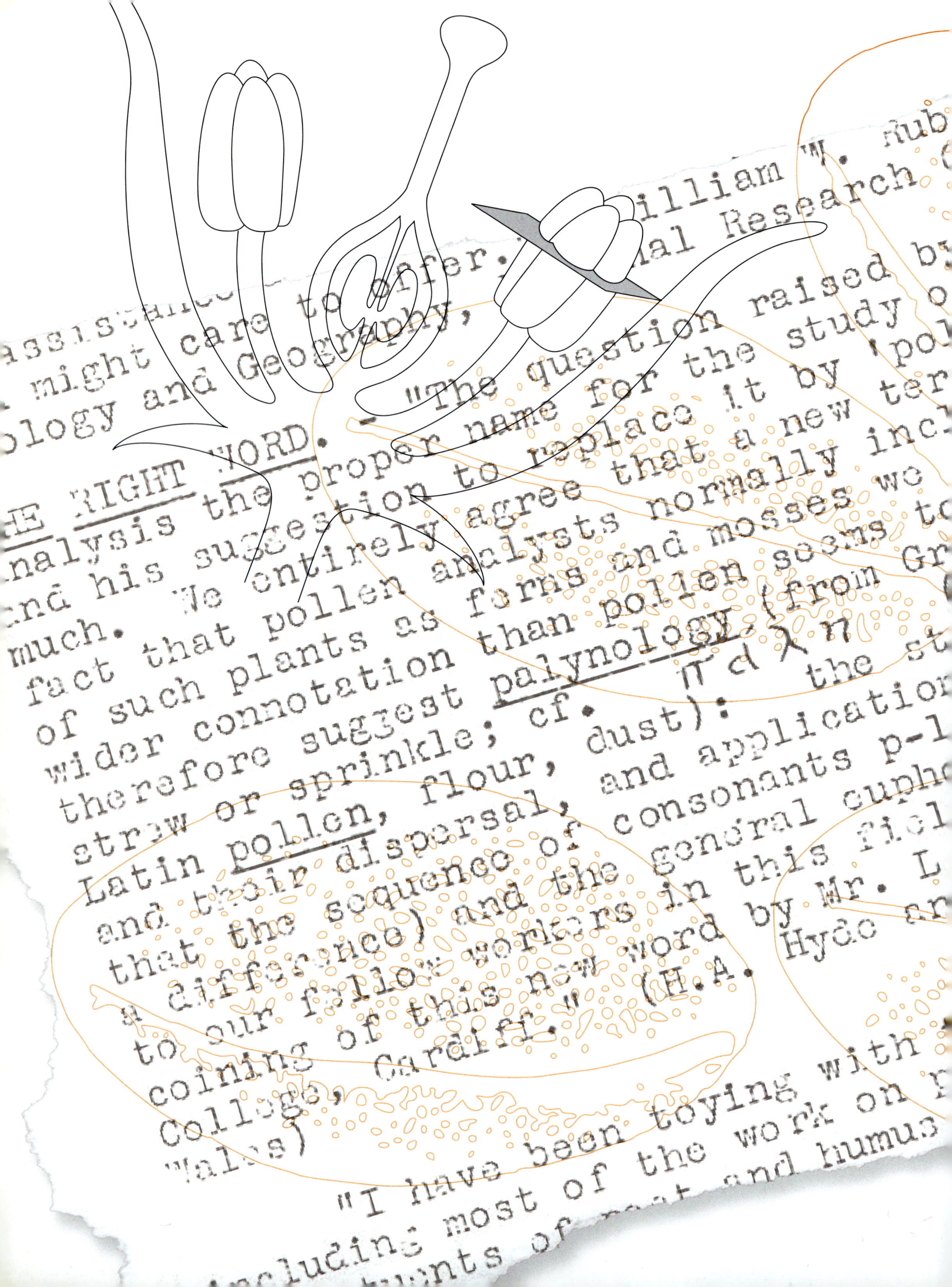
assistance
might care to offer. William W. Rub
ology and Geography, al Research C
HE RIGHT WORD. — "The question raised b
nalysis the proper name for the study o
nd his suggestion to replace it by 'pol
much. We entirely agree that a new ter
fact that pollen analysts normally inc
of such plants as ferns and mosses wo
wider connotation than pollen seems to
therefore suggest palynology (from Gr
strew or sprinkle; cf. παλη
Latin pollen, flour, dust): the st
and their dispersal, and applicatio
that the sequence of consonants p-l
a difference) and the general eupho
to our fellow workers in this fiel
coining of this new word by Mr. L
College, Cardiff." (H.A. Hyde an
Wales)
"I have been toying with a
including most of the work on p
unts of peat and humus

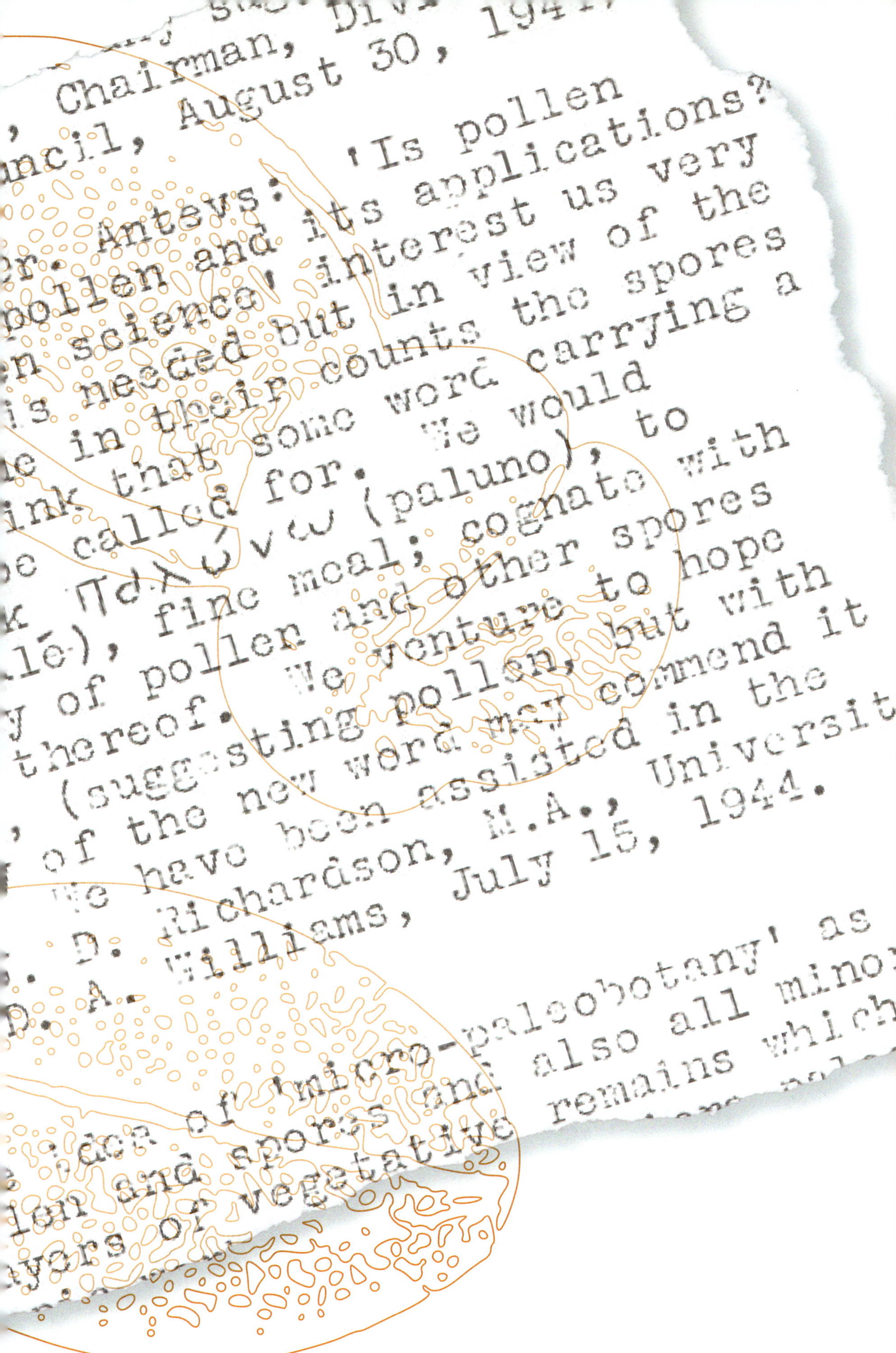

GENERAL CHAPTER

GENERAL CHAPTER

CONTENT

Introduction

Pollen Terminology. An illustrated Handbook is a collection of useful terms in palynology, well illustrated with light (LM) and electron microscope (EM) micrographs. The reader will not find an encyclopedic compilation of terms; in that respect see KREMP (1968). The focus is on the pollen of seed plants, predominantly angiosperms, while spores are considered only exceptionally. Therefore the terminology rarely includes spore or gymnosperm characteristics (e.g., leptoma, trilete mark).

Since 1994, the Glossary of Pollen and Spore Terminology, co-authored by Wim PUNT, Stephen BLACKMORE, Siwert NILSSON and Annick LE THOMAS, was the standard reference publication in palynological terminology. Then, in 1999 the online version by Peter HOEN (*http://www.bio.uu.nl/~palaeo/glossary/glos-int.htm*) appeared, with several additions. The online version was published by W. PUNT, P.P. HOEN, S. BLACKMORE, S. NILSSON and A. LE THOMAS in 2007 and provides informative schematic drawings containing the essentials of each term and colored to indicate the wall and aperture components, mostly using LM findings.

Although extremely useful for overview purposes, drawings cannot show the full range of features. This can be achieved only with micrographs, which demonstrate – a picture is telling more than thousand words – the often stunning diversity of features. For that reason, the explanatory power of micrographs produced with scanning electron microscopy (SEM) and transmission electron microscopy (TEM) is used in the present volume. The numerous SEM micrographs illustrating the astonishing diversity of pollen ornamentation. Where important terms have appeared ambiguous or have been hitherto underrated, the term has been reviewed and brought into focus (e.g. harmomegathy, or pollen class versus pollen type).

It is self-evident that such a book cannot renounce the basics of palynology. In this context please consult standard textbooks in palynology, e.g., ERDTMAN (1952), FÆGRI and IVERSEN (1989) or BEUG (2004). The principles of pollen development and morphology are incorporated as separate chapters for purposes of clarity and in order to correctly interpret the detailed structures of the pollen wall and the full range of ornamentation.

Guidelines

The aim of this book is to provide a fully illustrated terminology and glossary of the most **important** palynological terms, including a substantial standardization of definitions. If terms are not considered here, then they appeared to us as insignificant, or they belong to the terminology of fern spores, which is not considered here. A comprehensive description of pollen grains with terms mentioned in "*Pollen Terminology. An illustrated Handbook*" is easily accomplishable.

A strict rationalization of terms on the basis of practical criteria has been attempted. For consistency, phrases are standardized as far as possible; for example, features of ornamentation are stereotypically defined as "**pollen wall with** ....", and pollen wall features (or pollen shape and size) as "**pollen grain with** ....".

Where it was necessary, definitions have been reworded, newly circumscribed, or brought into focus. In addition, consistent application of EM techniques and the nowadays better understanding of pollen features have made redefinition of some terms necessary. Moreover, we have classified terms according to applied techniques (LM, SEM, TEM) and their usage in morphological, anatomical and/or functional context. In chapter "Alphabetic Glossary" the entries are arranged alphabetically. The definitions are provided with numbers in bold referring to the respective page in chapter "Illustrated Glossary"[1] and numbers in square brackets referring to important literature (see chapter "Bibliography").

Emphasis is given to the numerous illustrations. The worldwide largest database on pollen, PalDat (*http://www.paldat.org/*) is the main source of pictures. Each term is illustrated with LM or EM pictures in order to point out the **character range** of a term (or, more precisely, to show the full range of a single character). Brief information on the method of preparation is often provided. In preparing pollen for SEM micrographs, acetolysis was avoided as far as possible.

Underrated pollen conditions, e.g., the physical condition of the turgescent, life-like pollen, are considered. The SEM micrographs usually represent the turgescent condition, without further notice. Consequently, pollen grains are often shown in dehydrated stage, marked as "dry pollen". The deviating characters in turgescent and dry pollen grains are designated by descriptive pictorial terms such as cup-shaped, boat-shaped and aperture sunken.

Comments are provided where this may help in the application of a term or to qualify the circumstances in which it is used.

Self-explanatory general terms are usually not defined; in such cases the context is noted (e.g., circular, see **outline**). For more information on these see the appropriate page(s) in chapter "Illustrated Glossary".

Three categories of terms are used: important terms are printed in **bold** and are usually illustrated; terms of minor importance are printed in regular script, usually without illustrations (if necessary, terms in chapter "Alphabetic Glossary" are sometimes also illustrated in a footnote); terms printed in italics are not recommended and often provided with an explanatory comment.

The chapter "Illustrated Glossary" is subdivided into larger topics, e.g., "Shape and Size" or "Ornamentation". The terms themselves are listed according to their resemblance in order to provide the user with a side-by-side spectrum of similar characters. For a quick orientation please use the last page of "*Pollen Terminology. An illustrated Handbook*". It is a fold-out page with terms alphabetically arranged. Numbers indicate the page in chapter "Illustrated Glossary".

In contrast to chapter "Illustrated Glossary" the terms in chapter "Alphabetic Glossary" are throughout arranged alphabetically as the noun and the corresponding adjectival

[1] Please note: literature references are not necessarily the earliest publication in which the term was used. The comprehensive literature list (see chapter "Bibliography") includes beside the references more and other (and preferably recent) publications which have been selected as sources of further information.

form, if appropriate. Few terms are used exclusively as nouns or exclusively as adjectives. Sometimes two adjectival variants (-ate, -ar) are used but, if so, in two different meanings. For example: from the noun granulum (sculptural or structural element of differing size and shape, less than 1 µm in diameter) derive the two adjectival forms granular and granulate (both meaning "with granules"); these are corresponding terms used in two quite different contexts: **granular** describes a distinct type of infratectum hence a structural feature whereas **granulate** refers to an ornamentation feature – a sculpturing element.

Both the singular and the plural are given consistently for Latin terms. The English spelling of the Latin term is added (porus, pl. pori, engl. pore) if the English form is preferable.

Cross-references are given to terms that are **synonyms** (the preferable one is printed in bold) or that indicate the opposite condition (**antonyms**), e.g., homo- and heterobrochate.

Numbered literature references are given for each term in chapter "Alphabetic Glossary" and are not necessarily the earliest publication in which the term was used.

PUNT et al. (2007) provide the basis of the present terminology. Many terms in palynology were coined at a time when only LM observations were available. Mainly for historical reasons, inconsequent nomenclatural applications, enumerations of synonyms, and even differing definitions have been found for one and the same term.

During the 20th century questions of terminology became more and more problematic. The main reasons were the greatly increasing number of publications in palynology, dealing with sometimes insufficiently described or "uncommon" pollen features, and simultaneously the advent of manifold applied fields of palynology. For various reasons, nearly all authors used their own terminology. The situation became worse in the 1970s and 1980s, leading to a variety of terminological "schools".

Nonetheless, in the 1950s attempts were made to restrict the wording and to state the definitions of terms more precisely. A deserving, widely accepted but all-too restricted list of pollen morphological terms and definitions was published as early as 1950 by IVERSEN and TROELS-SMITH. Later, KREMP (1968), in his famous encyclopedia, provided a monumental enumeration of all known terms .

Being aware of the danger that pollen terminology tends to become foggy, REITSMA (1970) took the first resolute step to overcome this problem. A concise terminology now became available, though unfortunately not taking account of the range of variation of most of the palynological features, and without drawings or micrographs. FÆGRI and IVERSEN (1989, 4th ed.) restricted their glossary to terms exclusively used in their book. MOORE et al. (1991, 2nd ed.) provided a glossary of selected terms used in their pollen and spore keys. Standardization came with the glossary by PUNT et al. (1994), updated in 2007. The main advance of their concise and comprehensive terminology is the consistent usage of drawings and the critical comments on terms and usage.

Rules for Using Prefixes

If both a Greek and a corresponding Latin form exist for a prefix, then the Greek form is used consistently: panto- (not peri-), ekto- (not ecto-), or the Greek di- (dis-), and not the Latin bi- (bis-). There are few exceptions from this rule. If the Latin form is more widely used, then the term is treated as a nomen conservandum; for example, bisaccate is found exclusively in the literature and not the Greek form disaccate.

Some prefixes need a comment. **Micro**- is used to denote features <1 µm: microreticulate, -echinate, -verrucate, -baculate, -clavate, -gemmate, -rugulate. However, some possible combinations are not applicable; for example, micro-striate or microperforate. Striae are not known to be <1 µm in length, and perforate by definition describes a feature <1 µm.

Terms not listed in the glossary belong to fern or moss spores, or are considered as obsolete, diffuse or redundant (e.g., multiplanar tetrad), superfluous (e.g., polyplicate, because plicate pollen grains are always equipped with several to many plicae), or may be a permanent source of confusion (zon-, zona-, zoni-, zono-).

"*Pollen Terminology. An illustrated Handbook*" aims to clearly separate the types and classes of pollen. **Pollen type** is a general term categorizing pollen grains by distinct combinations of characters and is often used in connection with a distinct taxon (e.g., *Polygonum aviculare* type).

Pollen class[2] is an artificial grouping of pollen grains that share a single, distinctive character. Pollen classes refer to pollen units, to aperture form and location, or to an extremely distinctive ornamentation character. Classes include the terms polyads, tetrads, dyads, saccate, inaperturate, sulcate, ulcerate, colpate, colporate, porate, synaperturate, spiraperturate, lophate, clypeate and plicate. These classes are useful in identification keys as they have a good diagnostic, although mostly no systematic, value. In general, a pollen grain may belong to more than one pollen class; in such cases the more significant feature should be ranked first (e.g., *Pistia*: plicate - inaperturate, *Hemigraphis*: plicate - colporate, *Typha*: tetrads - ulcerate, *Rhododendron*: tetrads - colporate).

[2] "Pollen type" is sometimes (colloquially) misused; for example, *Croton* type, which is a distinct feature of ornamentation and is correctly termed *Croton* pattern.

Palynology

The Science of Pollen and Spores

The term palynology was coined after a written discussion with Ernst ANTEVS and A. Orville DAHL in the Pollen Analysis Circular no. 8 by HYDE and WILLIAMS (1944) and is a combination of the Greek verb paluno (παλύνω, "I strew or sprinkle"), palunein (παλύνειν, "to strew or sprinkle"), the Greek noun pale (παλη, in the sense of "dust, fine meal", and very close to the Latin word pollen, meaning "fine flour, dust") and the Greek noun logos (λογος, "word, speech").

Palynology is the science of palynomorphs, a general term for all entities found in palynological samples. A dominating object of the palynomorph spectrum is the pollen grain, the point of origin and the carrier for the male gametes (sperm cells).

What makes pollen grains so unique? Pollen grains represent an extra generation in seed plants, the highly reduced male gametophyte (the enclosing sporoderm and the cellular content, consisting of two or three cells, and the pollen tube). Pollen grains are therefore not simply parts of a plant, such as leaves or seeds, but are the haploid counterpart of the much larger diploid plant body "as we see it in nature". During transport pollen grains are completely separated from the parent plant and perfectly adapted for their role – the transfer of male genetic material – and are able to resist hostile environmental stress on their way to the female flower parts. These tiny (male haploid) organisms usually have as variable parameters: the pollen shape and size, the number, type and position of apertures and the pollen wall with its extremely diverse structure and sculpture. The characters of these parameters in comparative pollen (and spore) morphology and plant systematics are at least as important as any other morphological character of the diploid generation.

The pollen grains of seed plants and the spores of mosses and ferns share many homologies. However, although probably equivalent, the terminology of spore wall strata differs, mainly for historical reasons, from the terms used for pollen grains. Some elements and/or features of spores are unknown in pollen grains, e.g., the outermost wall layer in many fern spores, called the perine or perispore.

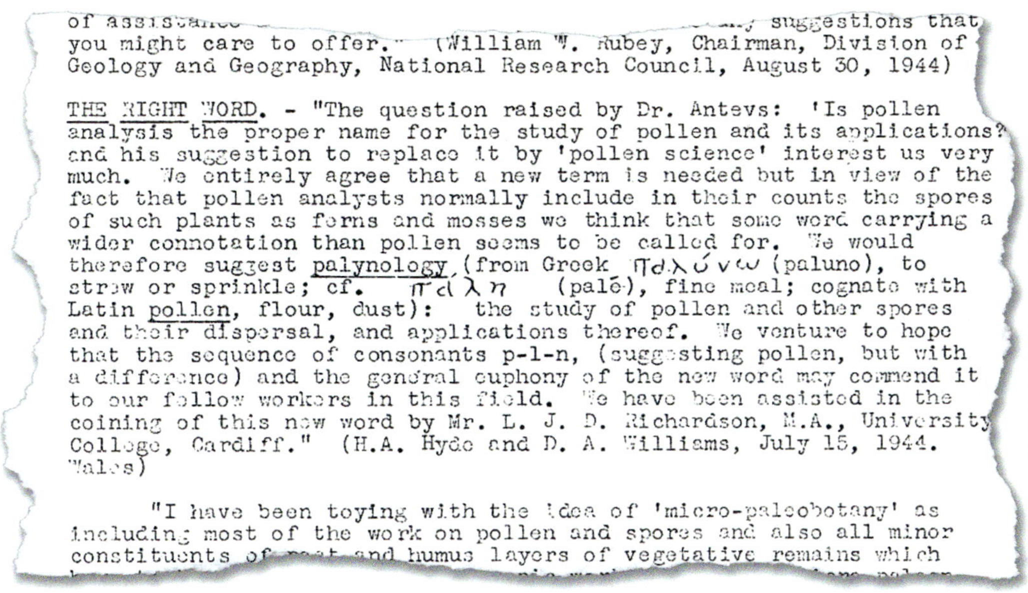
of assistance ... suggestions that you might care to offer." (William W. Rubey, Chairman, Division of Geology and Geography, National Research Council, August 30, 1944)

THE RIGHT WORD. - "The question raised by Dr. Antevs: 'Is pollen analysis the proper name for the study of pollen and its applications?' and his suggestion to replace it by 'pollen science' interest us very much. We entirely agree that a new term is needed but in view of the fact that pollen analysts normally include in their counts the spores of such plants as ferns and mosses we think that some word carrying a wider connotation than pollen seems to be called for. We would therefore suggest palynology (from Greek παλύνω (paluno), to strew or sprinkle; cf. παλη (palē), fine meal; cognate with Latin pollen, flour, dust): the study of pollen and other spores and their dispersal, and applications thereof. We venture to hope that the sequence of consonants p-l-n, (suggesting pollen, but with a difference) and the general euphony of the new word may commend it to our fellow workers in this field. We have been assisted in the coining of this new word by Mr. L. J. D. Richardson, M.A., University College, Cardiff." (H.A. Hyde and D. A. Williams, July 15, 1944. Wales)

"I have been toying with the idea of 'micro-paleobotany' as including most of the work on pollen and spores and also all minor constituents of peat and humus layers of vegetative remains which

HYDE and WILLIAMS (1944)
The right word.
Pollen Analysis Circular 8: p. 6

A Brief History of Palynology

The Very Early Beginnings

Assyrians are said to have known the principles of pollination, but it is unclear if they recognized the nature and power of pollen itself. Greeks and Romans, and the Middle Ages up to the 16th century did not contribute substantially, as far as is known.

The Era of the Light Microscope

A comprehensive historical survey is found in WODEHOUSE (1935) and especially in DUCKER and KNOX (1985). Only the most important scientists can be mentioned here; the list is not exhaustive.

It was Nehemiah GREW who as early as 1662 in his famous work "*The Anatomy of Plants*" described the constancy of pollen form within the same species; in other words, he founded pollen morphology and was the first to recognize that all plants have "their" pollen. Carl von LINNÉ (1751) first used the term pollen (in Latin). During the 18th and the early 19th centuries there was considerable progress on pollen and the understanding of pollination. For example, Joseph Gottlieb KOELREUTER (1766), together with Christian Konrad SPRENGEL, the founder of flower ecology, perceived the importance of insects in flower pollination and found for the first time that the pollen grain has an important part in determining the characters of the offspring.

SPRENGEL (1793) was the first to recognize pores and furrows in the pollen wall; he also demonstrated the effects of cross pollination, of dichogamy, and distinguished between entomo- and anemophily.

Johannes PURKINJE (1830) and Franz Andreas (Francis) BAUER, among others, also made substantial contributions. BAUER is famous for his fine and exact drawings and watercolors of pollen, now held in the Botanical Library of the Natural History Museum, London. Only a few facsimiles have been published, e.g., in KESSELER and HARLEY (2004). Robert BROWN (1828, 1833) reported and confirmed BAUER's earlier work, and gave the first description of the origin of the pollen tube.

New and better microscopes enabled Hugo von MOHL (1834) and Carl Julius FRITZSCHE (1837) to separate clearly the principal layers of the pollen wall and to publish surveys on pollen morphology of many angiosperm families. The terms pollenin, exine and intine go back to FRITZSCHE. Johann Heinrich Robert GÖPPERT (1837) and Christian Gottfried EHRENBERG (1838) were the first to describe and depict fossil pollen grains. Eduard STRASBURGER (1882) achieved ground-breaking insights into the development and internal structure of pollen. Hugo FISCHER (1890) was the first to summarize the arguments for the phylogenetic value of pollen characters. Pollen statistics represented a first step towards an applied field of science and in 1916 Lennart von POST published the first pollen diagram (pollen profile).

The 20th century up to ca 1960 was dominated by the skilful use of the LM, with many new findings; for example, the LO-analysis, a method for analyzing patterns of exine organization by light microscopy: focusing at different levels distinct features appear bright (L = Lux) or dark (O = Obscuritas). Textbooks by Roger WODEHOUSE (1935), Gunnar ERDTMAN (1943, 1952, 1969), or Knut FÆGRI and Johannes IVERSEN (1950) summarized the knowledge on pollen at that time and to a great extent have maintained their value.

In the first half of the 20th century palynology as a predominantly basic science "went applied", giving rise to a series of diversifications. Applied fields, worldwide in use, include aeropalynology, biostratigraphy, copropalynology, cryopalynology, forensic palynology, iatropalynology, melissopalynology, paleopalynology, pharmacopalynology, among others.

The Era of the Electron Microscope

As pointed out by KNOX (1984, p. 204): "*The terminology applied to the pollen wall is daunting, especially as it has been developed from early light microscopy work, and then transposed to the images seen in the transmission and scanning electron microscopes*".

Electron Microscopy with its two most important types, TEM and SEM, facilitated the major breakthrough in palynology: the ultrastructure of developing and mature pollen and the stunning visualization of pollen morphological characters.

During the 1950s and early 1960s considerable progress in TEM preparation methods (from fixation to microtomy and staining) took place. The resolving power of the TEM was the basis for new information on pollen grain ultrastructure and pollen development. Nevertheless, EM-based information on ornamentation details of pollen grains was rare up to the mid-1960s. Only TEM-based casts or replica methods were available, all of them with limited resolution and depth of focus (e.g., the single-stage carbon replica technique; ROWLEY and FLYNN 1966, FLYNN and ROWLEY 1967). The time-consuming and laborious TEM replica procedures were an obstacle to extensive surveys of pollen morphology and have now been successfully replaced by SEM (HARLEY and FERGUSON 1990).

Today barely conceivable, the introduction of SEM in palynology in the second half of the 1970s was a key innovation in the study of the fine relief of pollen surfaces. Advantages of SEM include the relatively simple and rapid preparation methods, the unsurpassed depth of focus revealing an overwhelming vividness and power. SEM was accepted in the very first moment as the quantum leap in EM (HAY and SANDBERG 1967). The first SEMs of pollen grains were published by THORNHILL et al. (1965) and ERDTMAN and DUNBAR (1966).

Since then palynologists have been provided with a plethora of beautiful micrographs. "*The scanning electron microscope has provided a greater impetus to palynology than any other technical development during the history of the subject.*" (BLACKMORE 1992).

Nowadays the LM (with basic and advanced equipment) and the two main types of EM form an expedient combination of imaging techniques. The LM remains the workhorse method (TRAVERSE 2007; see the compendia by REILLE 1992, 1995 and 1998) but is limiting insofar as morphological and structural features at species level, not observable by LM but of diagnostic value, are routinely determinable only by SEM. The role of SEM as an essential part in illustrating exine sculpture and ornamentation cannot be overrated (HARLEY and FERGUSON 1990).

A Tentative Outlook

Nowadays, palynology, as an organismic-based science, can serve as an indispensable tool for various applied sciences, but clearly also can stand alone as one of the most developed basic sciences.

In general, compared to the diplont the male gametophyte in seed plants is yet poorly investigated. From at least 250.000 plant species onlyca 10 percent have been studied with respect to pollen grain morphology, and regarding pollen grain anatomy it is much less.

In the 21st century, no matter what role palynology will play, being a basic field of science or more probably a bundle of applied fields, a vital issue will be the increase of our knowledge of pollen grains and in this context the enhancement of pollen terminology. Modern palynologists, making use of LM as well as EM, need for descriptive matters a clearly defined and pictorial pollen terminology, covering the richness of features and the enormous spectrum of characters.

Pollen Morphology

A diagrammatic representation of the main morphological features of a palynomorph (preferably pollen grains or spores) is called **palynogram**. It includes parameters of symmetry, shape and size, aperture number and location, ornamentation and stratification.

Polarity and Symmetry

Mature pollen is shed in **dispersal units**. The post-meiotic products either remain permanently united or become partly or usually completely disintegrated. In the latter case the dispersal unit is a single pollen grain, a **monad**; if the post-meiotic products remain united, **dyads** (a rare combination), **tetrads** or **polyads** (**massulae**, **pollinia**) are the result. **Pollinaria** are dispersal units of two pollinia including the sterile, interconnecting appendage.

Tetrad stage orientation of microspores

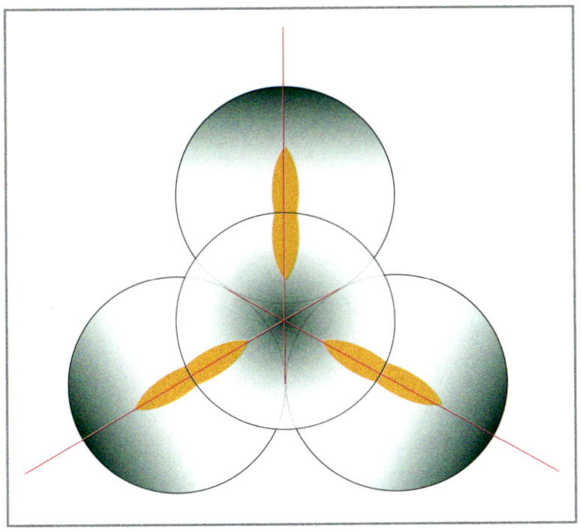

distal poles shaded green

Pollen shape and aperture location directly relate to pollen **polarity**, which is determined by the spatial orientation of the microspore in the meiotic tetrad and can be examined only in the tetrad stage. The **polar axis** of each microspore runs from the **proximal pole**, orientated towards the tetrad center, to the **distal pole** at the outer tetrad side. The **equatorial plane** is located at the microspore's center, perpendicular to the polar axis. Therefore, the **equatorial plane** divides the pollen grain into a proximal and a distal half.

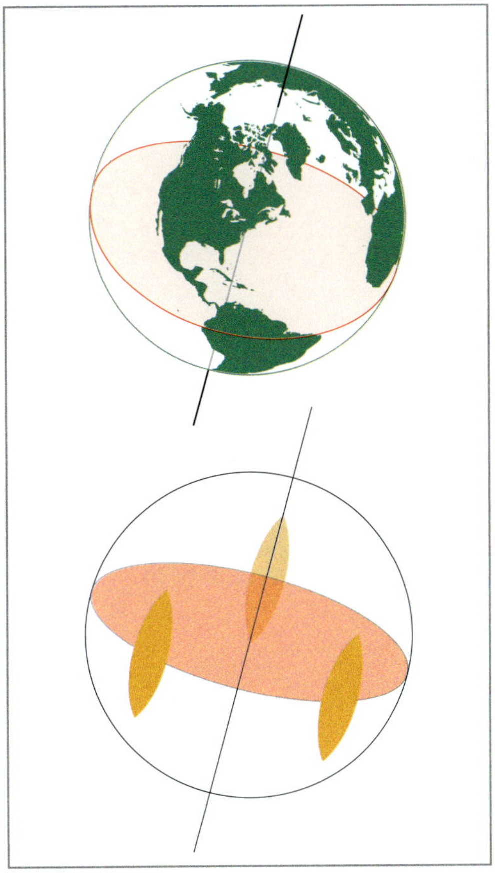

Polar axis and equatorial plane

Isopolar pollen grains have identical proximal and distal poles, thus the equatorial plane is a symmetry plane. In **heteropolar** pollen grains the proximal and distal halves are different.

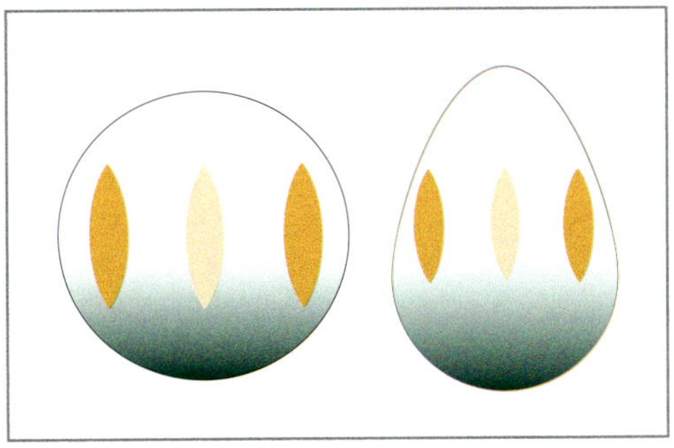

Polarity

left: isopolar

right: heteropolar

The various arrangements of the four microspores within permanent or disintegrating **tetrads** depend on the simultaneous or successive type of cytokinesis and on the type of intersporal wall formation. The spatial arrangement of microspores after **simultaneous** cytokinesis is usually a **tetrahedral tetrad**. This arrangement is of systematic relevance. The spatial arrangement of microspores after **successive** cytokinesis leads to different tetrad types without any systematic relevance: **planar** (tetragonal, linear, T-shaped) or non-planar (decussate or tetrahedral).

Tetrad arrangement

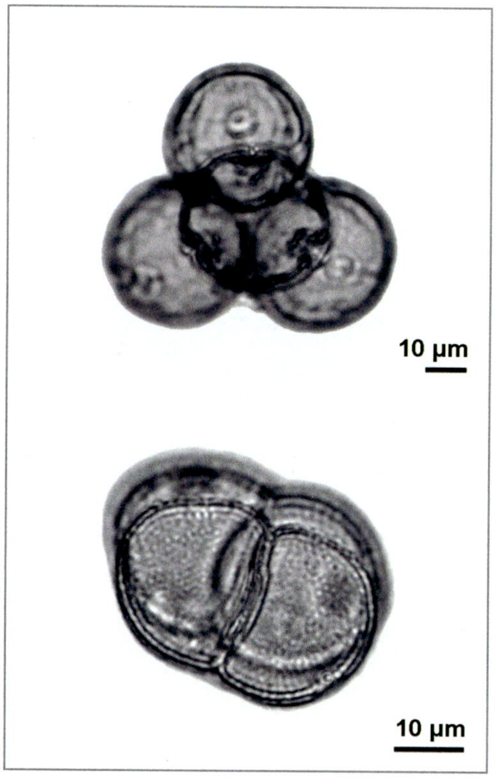

tetrad tetrahedral

Fagus sp.
Fagaceae, fossil (exceptional finding)

tetrad planar

Typha latifolia
Typhaceae

In pollen grains with three apertures, two types of aperture arrangement occur after simultaneous cytokinesis (disintegrating or permanent tetrahedral tetrads). **Fischer's law** refers to the most frequent arrangement where the apertures form pairs at six points in the tetrad (e.g., Ericaceae, permanent tetrads). **Garside's law** refers to the unusual arrangement of apertures where they form groups of three at four points in the tetrad (probably restricted to Proteaceae, no permanent tetrads).

Aperture arrangement

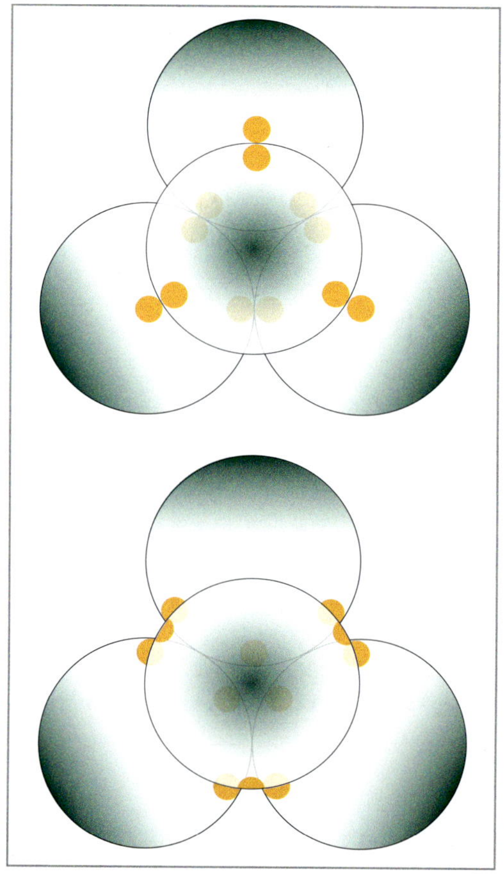

Fischer's law

Garside's law

Pollen **shape** refers to the **P/E-ratio**: the ratio of the length of the polar axis (P) to the equatorial diameter (E). In **spheroidal** (or **isodiametric**) pollen grains the polar axis is ± equal to the equatorial diameter. Pollen grains with a polar axis longer than the equatorial diameter are called **prolate**; grains where the polar axis is shorter than the equatorial diameter are described as **oblate**.

Pollen shape

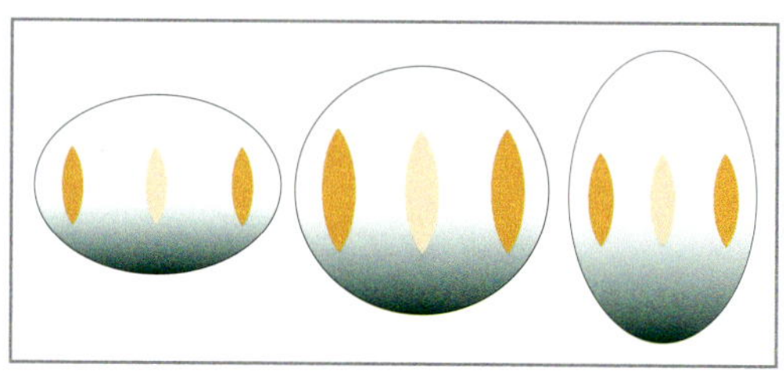

left: oblate

mid: spheroidal

right: prolate

Pollen **size** varies from less than 10 µm to more than 100 µm. To indicate the pollen size the largest diameter is used. It also depends on the degree of hydration and the preparation method. Because of this and natural variation, a bandwidth designation is recommended. A diameter indication in the range of, e.g., less than 1 µm is not recommended.

The use of the following size categories may be helpful: very small (<10 µm), small (10–25 µm), medium (26–50 µm), large (51–100 µm) and very large (>100 µm).

Apertures

The many facets of an allegedly simple character

Nomenclature and Typology

An **aperture** is a region of the pollen wall that differs significantly from the rest of the wall in its morphology and/or anatomy, and is presumed to function usually as the site of germination and to play a role in harmomegathy. Pollen grains lacking an aperture are called **inaperturate**. The aperture definition fits both angiosperm and gymnosperm pollen, but in gymnosperms the type of aperture usually differs from that in angiosperms, since often a **leptoma** is present. Note: unless stated otherwise, the following sections deal with angiosperm aperture constructs only.

The polarity of the pollen determines the aperture terminology. A circular aperture is called a **porus** if situated equatorially or globally; if situated distally it is called an **ulcus**. An elongated aperture is called a **colpus** if situated equatorially or globally; if situated distally it is called a **sulcus**. A combination of porus and colpus is termed a **colporus**; colpori are situated only equatorially or globally. Colpi and colpori (colpi and pori) may be present simultaneously in some taxa; this condition is called **heteroaperturate**. A circular or elliptic aperture with indistinct margins is a **poroid**.

The number of equatorial apertures (pori, colpi, colpori) is indicated by the prefixes di- or tri-. However, tetra-, penta- or

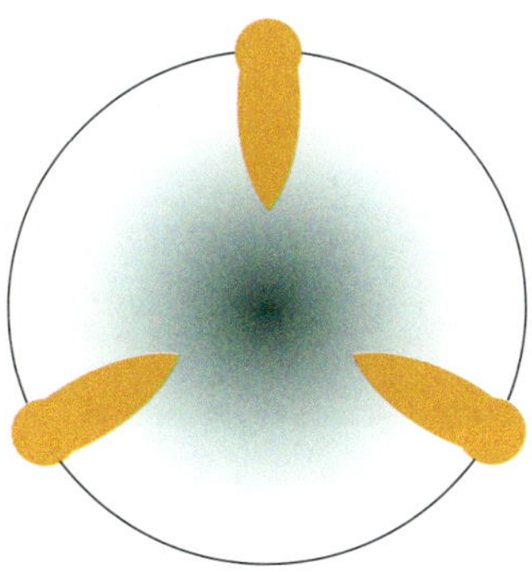

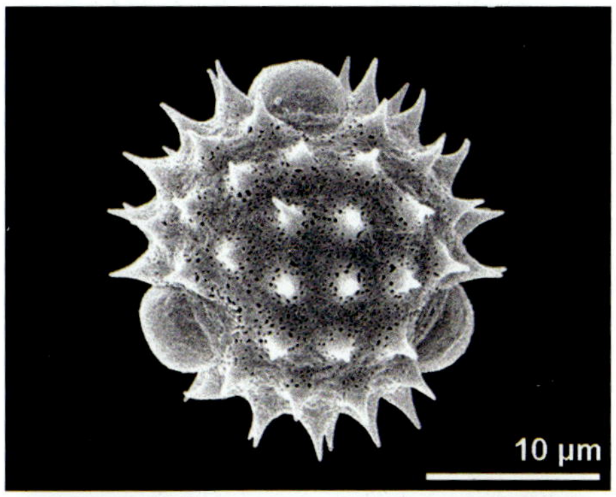

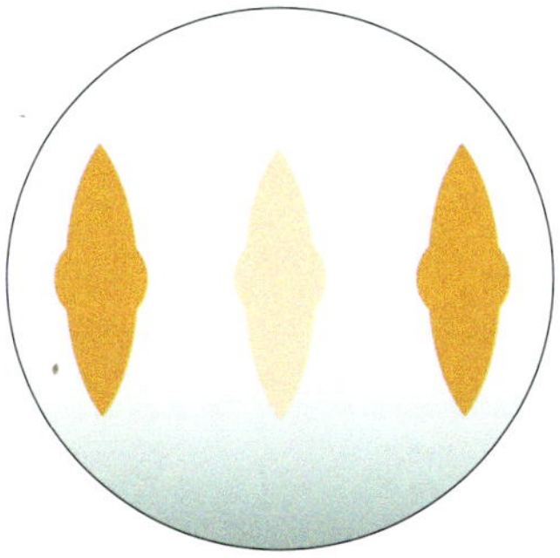

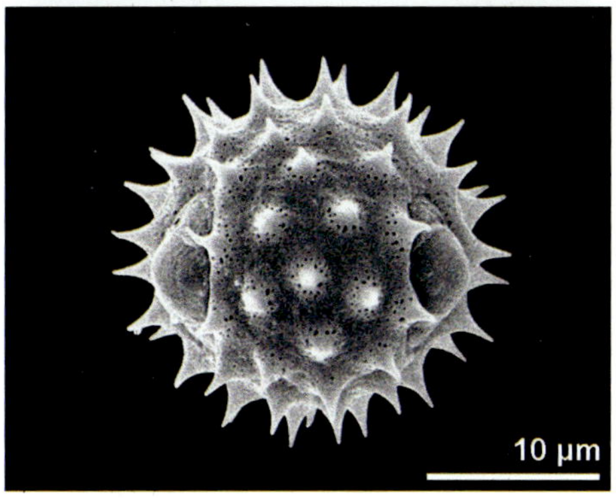

Pollen grain polarity dicots

Bellis perennis
Asteraceae

polar view

equatorial view

Pollen grain polarity monocots

Allium paradoxum
Alliaceae

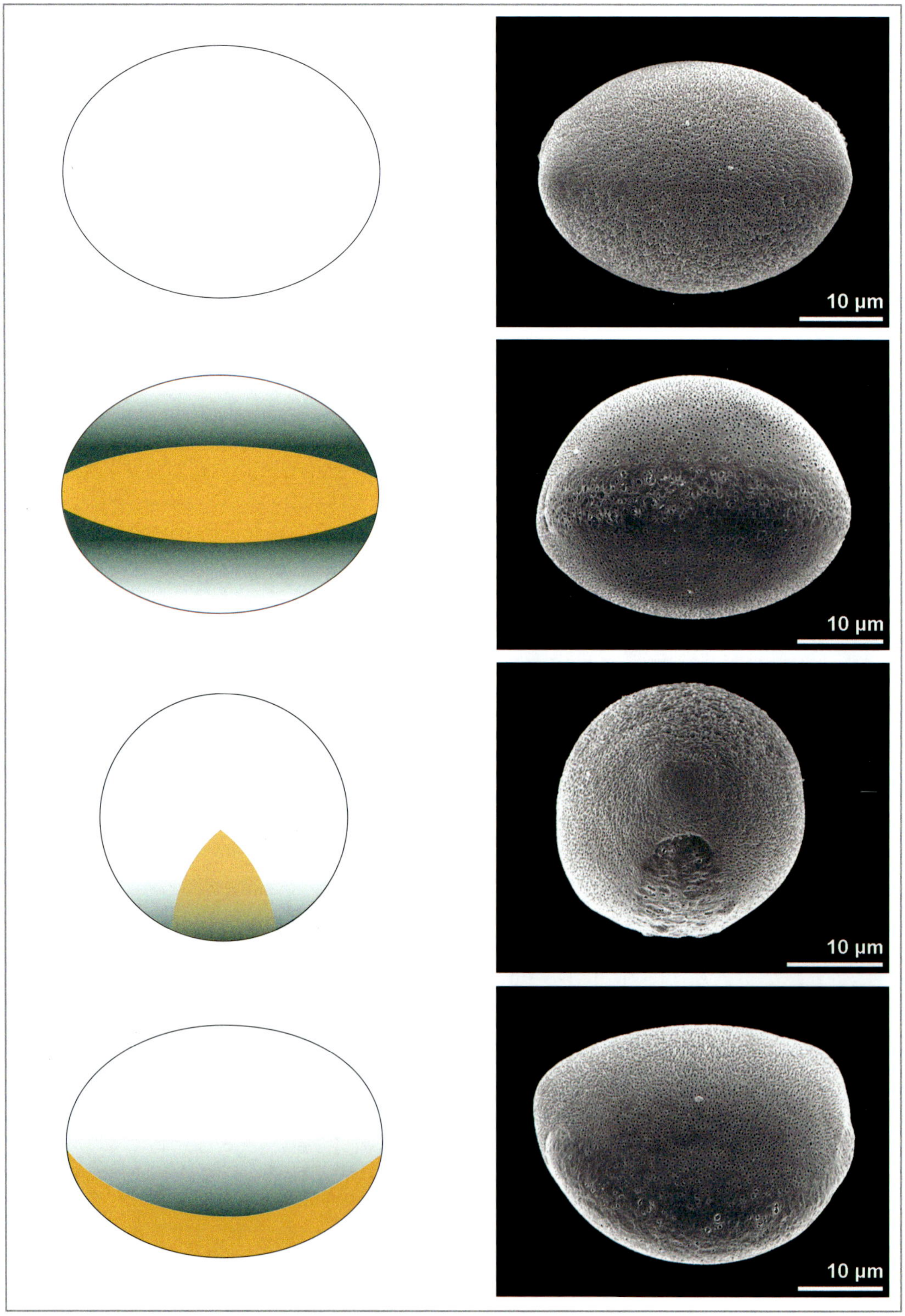

proximal polar view

distal polar view

equatorial view

equatorial view

hexa- are sometimes used. (Writing numbers instead of prefixes is in common use, e.g., 4-porate or tetraporate, 6-colpate or hexacolpate. "*Pollen Terminology. An illustrated Handbook*" prefers the use of prefixes.) Any pollen grain with more than three apertures at the equator is also called **stephanoaperturate** (**stephanoporate**, **stephanocolpate**, **stephanocolporate**). Pollen grains with globally distributed apertures are called **pantoaperturate**.

The polarity gives rise to the polar and the equatorial view. In dicots there is usually one polar and one equatorial view. In monocots, due to the mostly distal aperture, there are four views: a proximal polar, a distal polar, and two different equatorial views.

Proximal germination is unknown in seed plants and is restricted to **spores**, which germinate at the **tetrad mark**, the so-called **laesura** (extensive overview: TRYON and LUGARDON 1991).

Pre-(prae-)pollen (microspores of certain extinct seed plants) is characterized by proximal and distal apertures, and by presumed proximal germination, producing motile spermatozoids.

Pre-pollen

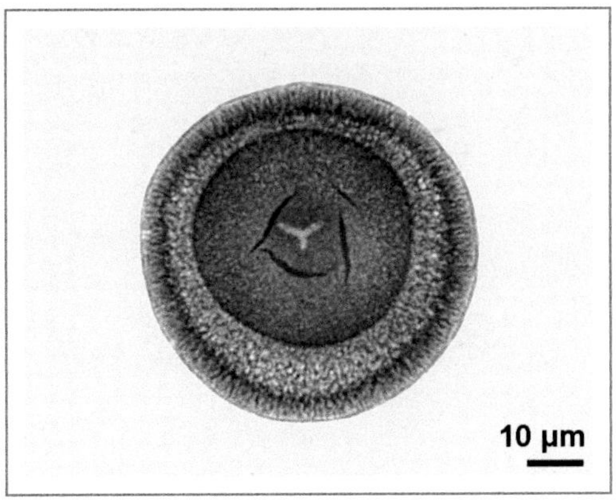

polar view

Apertures are normally covered by an exinous layer, the **aperture membrane**. **Aperture membranes** can be **ornamented**, e.g., covered with various exine elements, or can be **smooth**. In contrast, an **operculum** is a thick, coherent exine shield and covers the aperture like a lid.

In general, aperture membranes are infolded in dry pollen state; after acetolysis the aperture membrane may be lost.

Tetrad mark in spores

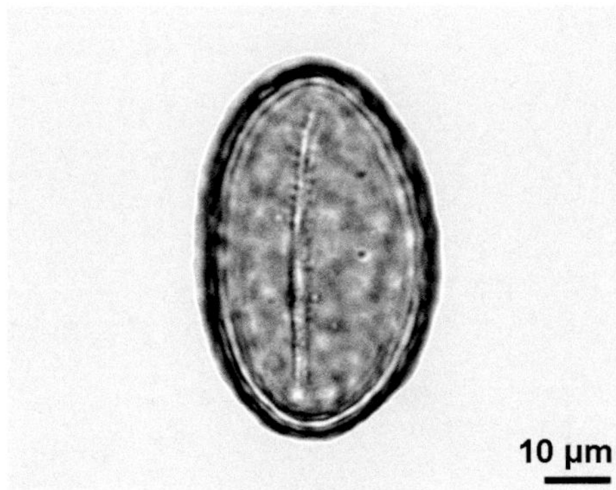

Polypodium sp.
Polypodiaceae, fossil

monolete tetrad mark
polar view

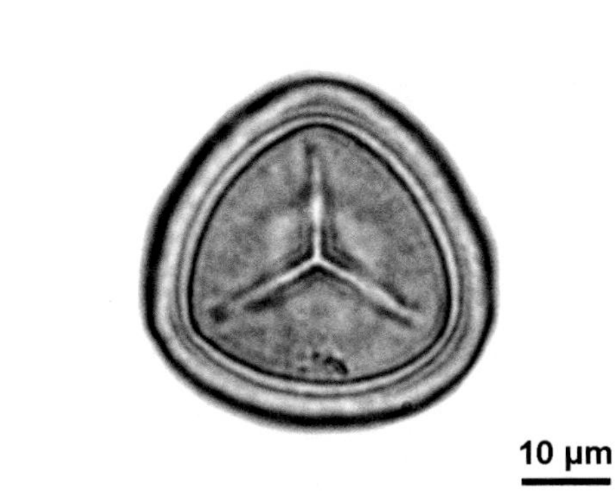

Sphagnum sp.
Sphagnaceae, fossil

trilete tetrad mark
polar view

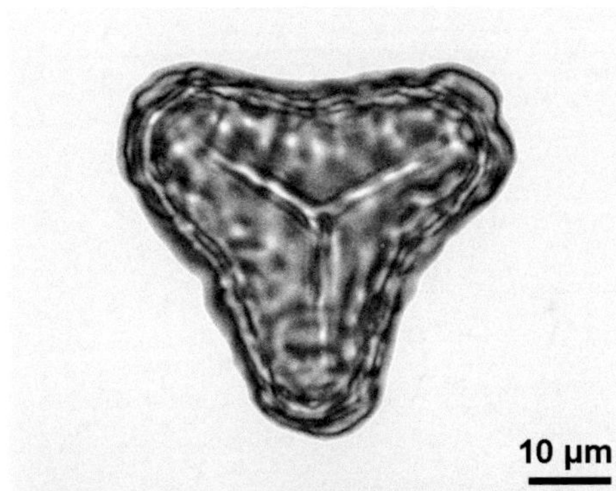

indet.
Pteridaceae, fossil

trilete tetrad mark
polar view

Cryptogramma crispa
Pteridaceae

trilete tetrad mark

Taxoid pollen germination

Cephalotaxus sp.
Cephalotaxaceae

exine shedding prior to pollen tube formation

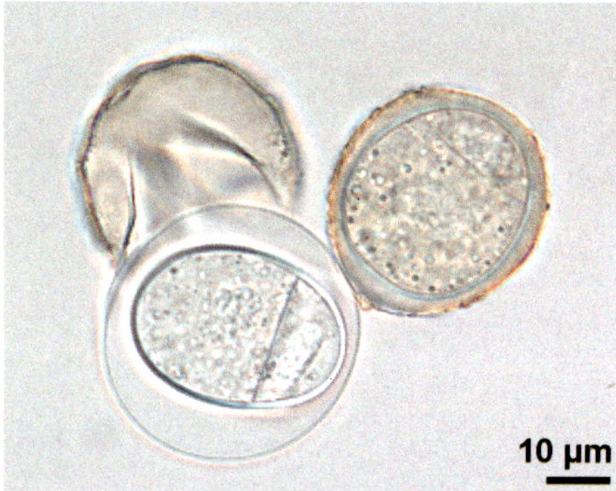

fresh pollen in water

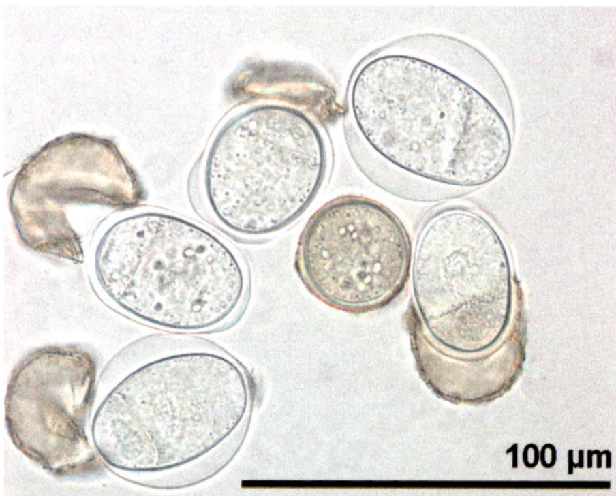

Instant pollen tubes

Scabiosa caucasica
Dipsacaceae

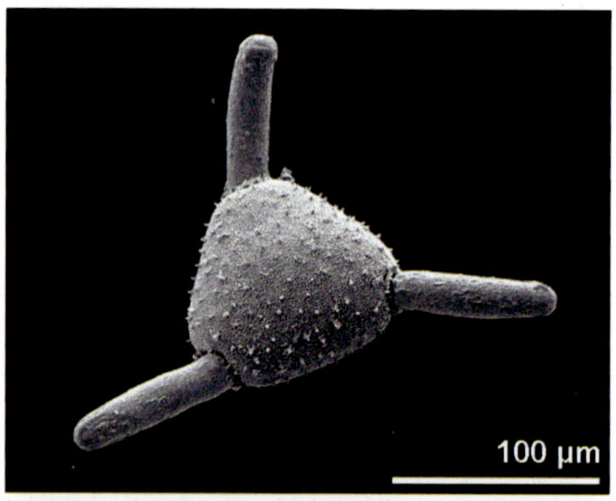

Morina longifolia
Morinaceae

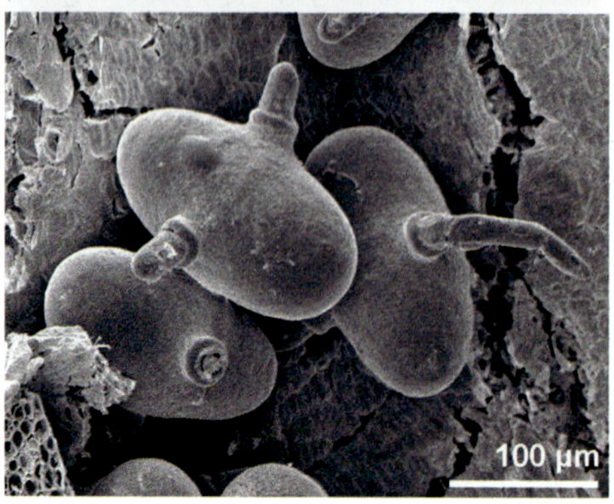

Number, type and position of apertures are genetically determined and usually fixed within a species; however, it may sometimes vary (e.g., number of apertures in stephanoaperturate pollen grains).

Structure and Function

The aperture usually acts as the (exclusive) **germination** site. Pollen tubes in inaperturate angiosperm pollen are produced without a preformed exit zone. In **taxoid** pollen the exine ruptures during hydration at a specialized region, the tenuitas, ulcus, or papilla in the center of a circular leptoma and is subsequently shed.

The intine including the protoplast is released and a pollen tube can be formed anywhere (resembling functionally an inaperturate pollen grain). Furthermore some angiosperm taxa shed the exine before pollen tube formation, e.g., in some Annonaceae.

During germination, usually a single pollen tube is formed. However, sometimes tube-like structures ("instant pollen tubes") are simultaneously formed in the anther or very quickly in shed pollen immediately after water contact. Their production is interpreted as a pre-germinative process (BLACKMORE and CANNON 1983).

Pollen Wall

In general, **the pollen wall** (**sporoderm**) of seed plants consists of two main layers: the outer **exine** and the inner **intine**. The exine consists mainly of **sporopollenins**, which are acetolysis- and decay-resistant biopolymers. The intine is mainly composed of cellulose and pectin. Commonly, the pollen wall in apertural regions is characterized by the reduction of exinous structures or by a deviant exine, and a thick, often bilayered intine.

Two layers within the exine are distinguished: an inner endexine and an outer ektexine. The ektexine consists of a basal

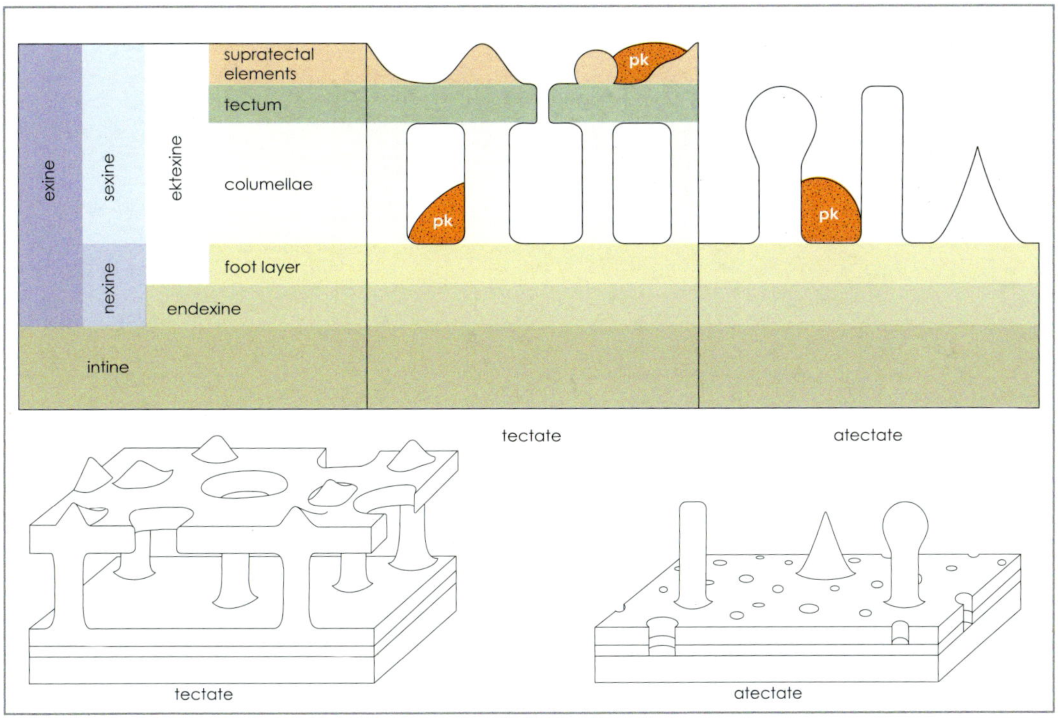

Pollen wall stratification

pk: pollenkitt

foot layer, an **infratectum** and a **tectum**, the endexine is a mainly unstructured, single layer. There are many deviations from this principal construction: layers may be thickened, variably structured, or lacking. In apertural regions the pollen wall is characterized by a different exine construction.

The terms **sexine** for the outer, structured, and **nexine** for the inner, unstructured exine layer are widely used in light microscopy, but do not fully correspond to ekt- and endexine, respectively.

The angiosperm pollen wall

The **ektexine** consists in general of **tectum**, **infratectum** and **foot layer**. The outer layer, the more-or-less continuous tectum, can be covered by supratectal elements. The infratectum beneath is **columellate** or **granular** (a second layer of columellae may form an internal tectum). The foot layer may be either continuous, discontinuous or absent. The **endexine** may be characterized as continuous or discontinuous, spongy or compact, is present overall, only in apertures, or even completely absent. Some typical deviations of the wall thickness are named with special terms: **arcus**, **annulus**, **tenuitas** (see "Illustrated Glossary") and **costa** (a thickening of the nexine/endexine bordering an endoaperture).

Costa

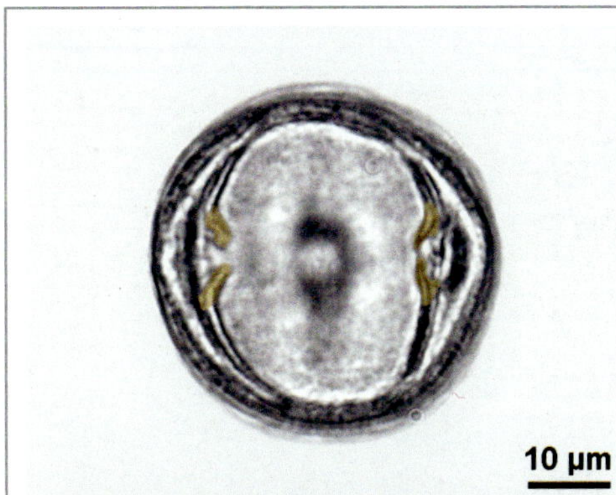

Nyssa sp.
Nyssaceae, fossil

equatorial view

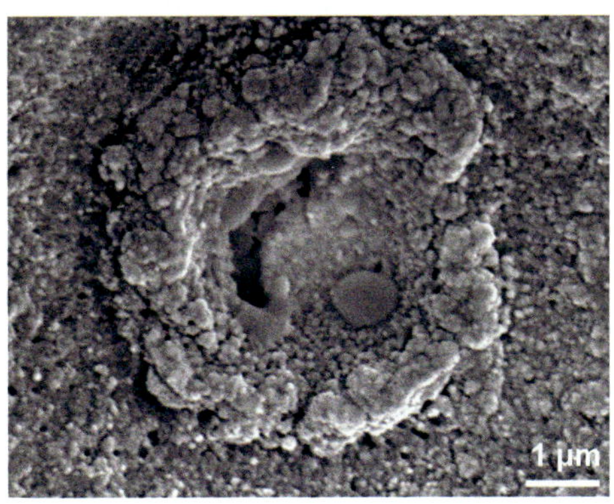

Austrobuxus nitidus
Picrodendraceae, fossil

broken grain, thickening around the endoaperture

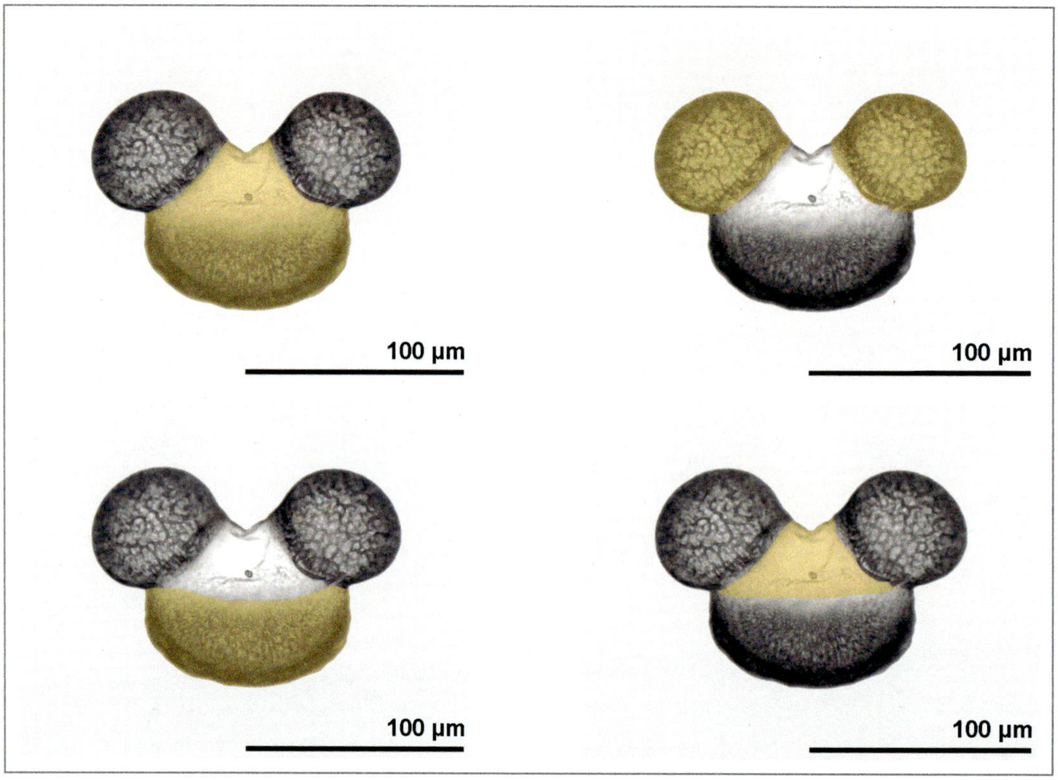

Pollen terminology in saccate gymnosperm pollen

Abies sp.
Pinaceae, fossil

equatorial view

left: corpus

right: sacci

left: cappa

right: leptoma

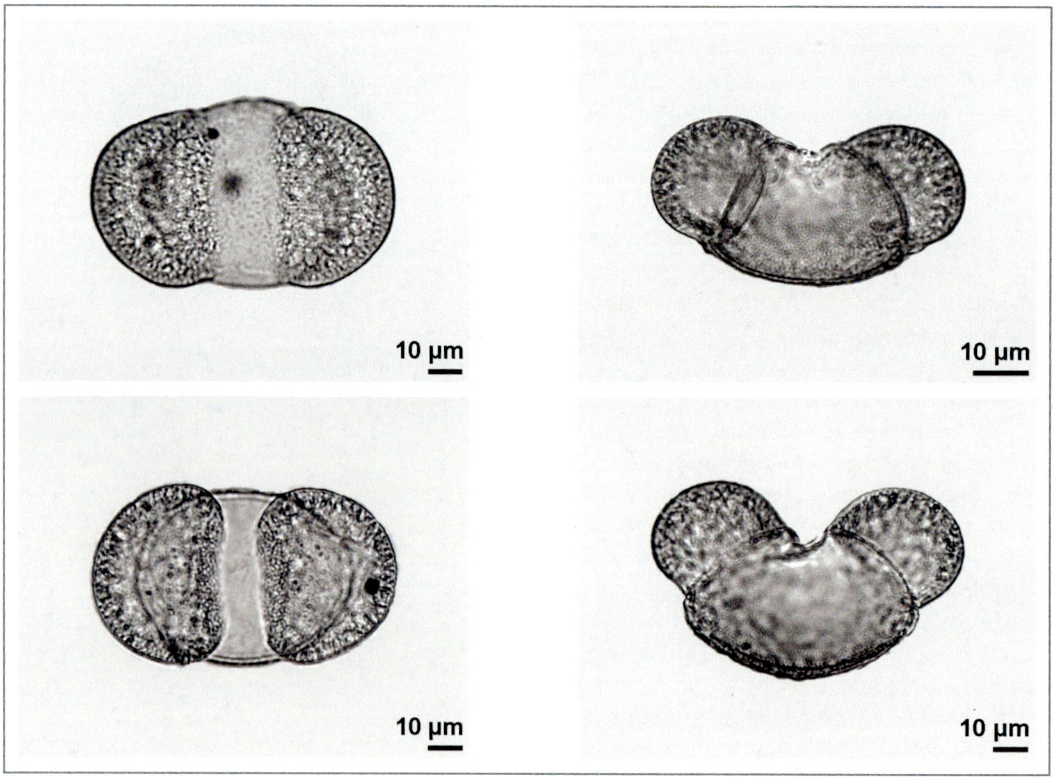

Pollen types in saccate *Pinus* pollen

Pinus sp.
Pinaceae, fossil

left: polar view

right: equatorial view

Haploxylon-pollen-type

Diploxylon-pollen-type

Extreme examples of variable ektexine design include massive forms lacking almost any subdivision or stratification, extremely reduced forms, or even their complete absence.

The typical angiosperm aperture shows a thick, bilayered **intine**.

The Gymnosperm Pollen Wall

The "Gymnosperms" comprise cycads, *Ginkgo*, conifers and Gnetales. The gymnosperm pollen wall differs from that in angiosperms in two characters: 1. the endexine is always **lamellate** in mature pollen stages. 2. the infratectum is never columellate. The four gymnosperm classes exhibit diverse, special constructions of the apertures.

The principal stratification (ektexine, endexine and intine) of the gymnosperm pollen wall is identical to that of angiosperms. A tectum is present in all cycads, in *Ginkgo*, in all Gnetales, but not in all conifers: in some taxa the tectum is completely lacking (sculpture elements are situated on the foot layer). The infratectum is either **alveolate** or **granular** but never columellate.

A special terminology is applied to saccate pollen, i.e., Pinaceae and Podocarpaceae. The **saccus** is a large hollow projection from the **corpus**, the central body of saccate pollen grains. It is a typical deviation of the pollen wall conformation, composed only by the exine with an alveolate infrastructure. Most frequently, two sacci are present, in some taxa even three, or only a single one. Saccate pollen grains show on the proximal side of the corpus a region termed **cappa**, and on the distal side a thinned region, the **leptoma**.

In *Pinus* two pollen types are recognized as of systematic value. The **Haploxylon-pollen-type** is characterized by pollen grains with broadly attached half-spherical air sacs – in LM the leptoma shows remarkable thickenings (black spots). The **Diploxylon-pollen-type** is characterized by pollen grains with narrowly attached, spherical air sacs - the leptoma does not show any thickenings.

Structure and Sculpture

The internal construction of the pollen wall is its **structure**; ornamenting elements on the pollen surface (ornamentation) are summarized under the term **sculpture** or sculpturing. However, it is not always possible to distinguish between structure and sculpture (e.g., free-standing columellae).

Ornamentation

This general term in palynology is applied to surface features. All the ornamenting features (**areola**, **clava**, **echinus**, **foveola**, **fossula**, **granulum**, **gemma**, **plicae**, **reticulum**, **rugulae**, **striae**, **verruca**) are artificial categories that mark extremes within a broad morphological series and are therefore regarded as extremely variable; nevertheless, they are important in pollen description.

For practical purposes a distinct feature can be subdivided into ornamenting elements extending 1 µm in diameter, or if smaller then marked with the prefix **micro**-.

Combinations of sculptural elements are common. Frequently a specific ornamentation is difficult to describe only in words because of the high plasticity of its ornamenting elements. A typical micrograph characterizes sculptural elements to a much higher degree.

The arrangement of ornamenting elements on the pollen surface is very often disparate, particularly in apertural regions. Pollen coatings like **pollenkitt** or **tryphine** may obscure the ornamentation.

Harmomegathy

Harmomegathic Effect (Wodehouse Effect)

All living pollen grains are able to absorb and release water; thus, each living grain exists in two morphologically different states: the **dry** and the **hydrated** condition. Harmomegathic mechanisms, e.g., infolding of the pollen wall, accommodate the change of the osmotic pressure in the cytoplasm during hydration or dehydration.

Harmomegathic effect

Cistus creticus
Cistaceae

left: spheroidal

right: dry pollen
prolate, lobate

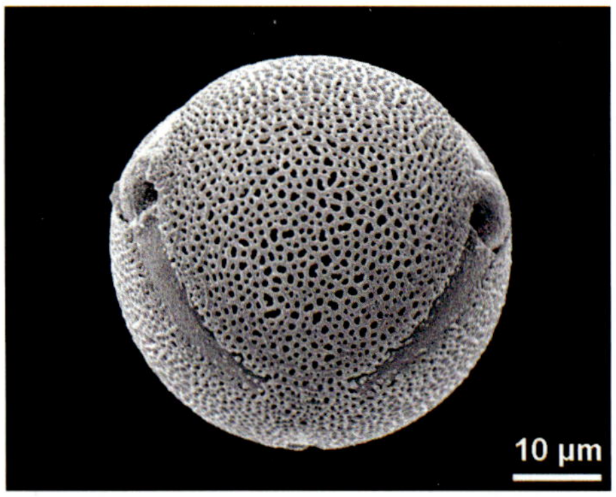

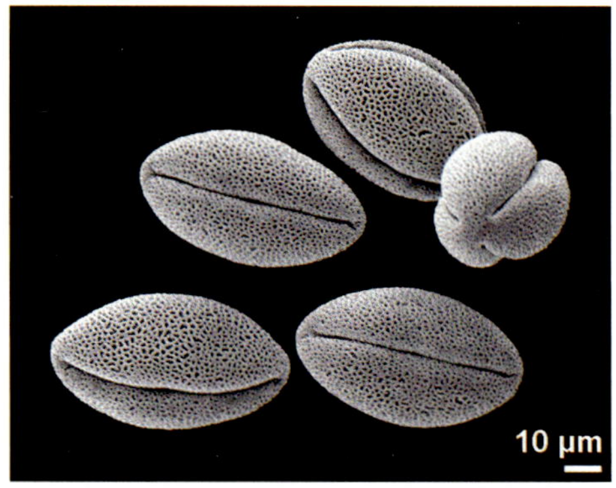

Galium rotundifolium
Rubiaceae

left: oblate

right: dry pollen
prolate, lobate

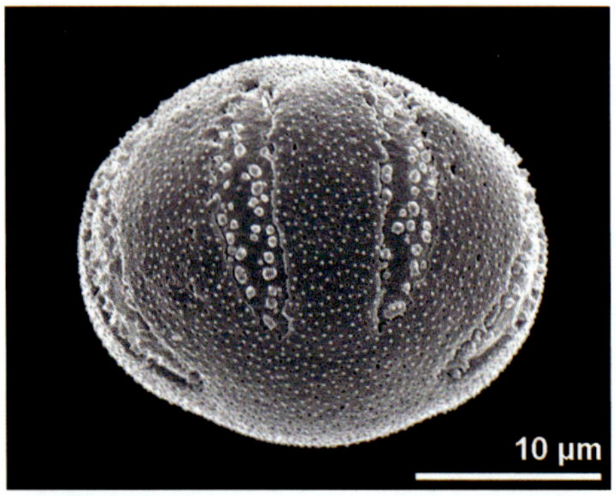

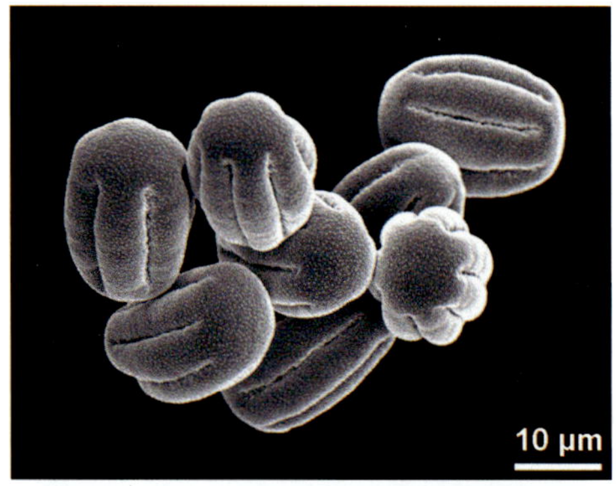

Vriesea pabstii
Bromeliaceae

left: oblate

right: dry pollen
boat-shaped

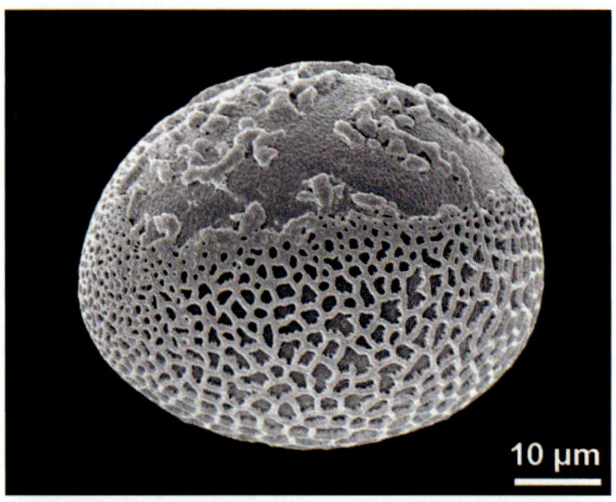

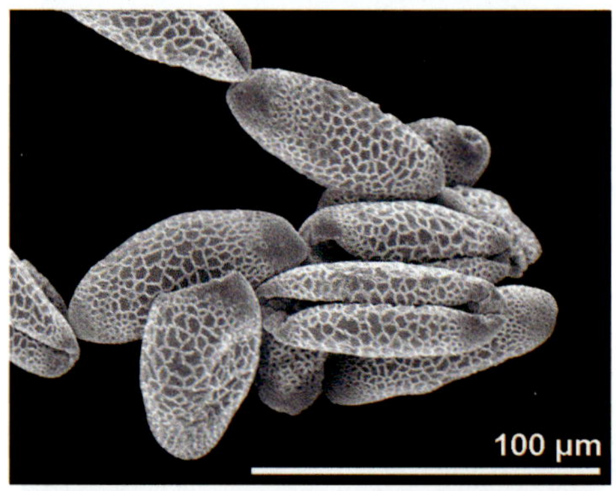

Lamiastrum montanum
Lamiaceae

left: spheroidal

right: dry pollen
prolate, outline elliptic

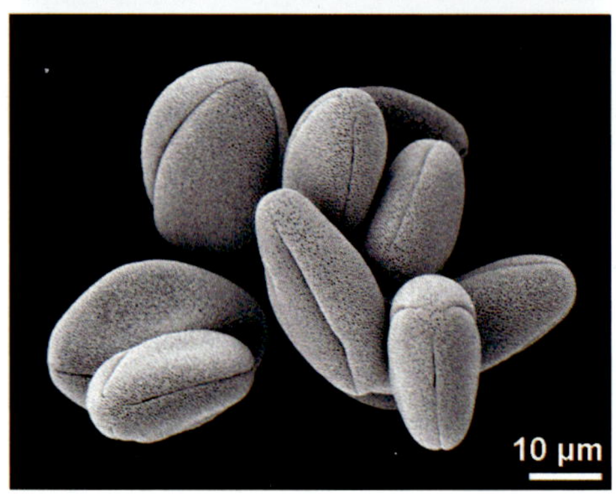

The main purpose of the **harmomegathic effect** is to protect the male gametophyte against desiccation during pollen presentation and dispersal, and is often related to pollination biology.

In mature anthers, pollen is turgescent before shedding. After anther dehiscence and during pollen presentation, water loss takes place and the pollen grain becomes typically infolded, depending on aperture form and/or number, and specific wall thinnings or thickenings. The pollen grain in proper dry state represents the genuine harmomegathic effect and its shape is very often typical for a family and/or genus and is therefore of systematic relevance.

The harmomegathic effect is to some degree reversible. Rehydrated pollen with water uptake at the stigma, or under laboratory conditions, is again turgescent and largely recalls the shape before shedding. A second dehydration does not necessarily result in the typical dry shape but, if pollen walls are sufficiently stable, the harmomegathic effect can be induced several times in the same way. In the case of thin walls, the susceptible internal structure may become irreversibly damaged, and the harmomegathic effect may result in differing shapes, often randomly.

The harmomegathic effect is also observed in pollen taken from herbarium material, and to some degree in fossil material (HALBRITTER and HESSE 2004).

Infolding of the pollen wall after acetolysis is mostly not comparable with that in dry state.

The harmomegathic effect depends predominantly on **the various characters of the pollen wall**. Several pollen features (harmomegathic factors) collectively influence the mode of infolding and cannot be considered separately:

- **apertures** (the most important character): their position, number and form.
- **pollen wall structure**: thinned or thickened regions; in particular, internal girdles or **endoapertures**. If the ektexine is considerably reduced, its role is taken over by other wall strata, namely, by a thick endexine or intine. On the other hand, if the exine is extremely rigid, then the harmomegathic effect is only marginal.
- **ornamentation** type.
- **pollen size**: small pollen grains with thin walls exhibit a lesser degree of infolding.
- **pollen coatings**: if abundant, pollen coatings act as an insulating layer or sheath against desiccation.

Terms used for common phenotypes of dry pollen include: apertures sunken, boat-shaped, cup-shaped, interapertural area infolded, irregularly infolded, not infolded. In addition, technical terms such as, e.g., barrel-like, disk-like, or kidney-like might be helpful for an adequate description.

Why Do We Need Categories?

Nature itself neither needs categorization nor has any knowledge of categories. However, for the scientist, categories are essential for classifying natural characters in their diversity, for defining their range and for placing them in systematic order. Nevertheless, **categories are artificial and always delimited by an individual or collective convention**, mostly not by nature.

In addition to the theoretical concept, categorization always depends on the manner in which a character is perceived: i.e. on the **visibility** of a character, and/or their specific value. Categorization also greatly depends on the technical equipment and method(s) used, as well as on the **subjective interpretation** of character(s)[1]. Thus, categorization of features is difficult to standardize. A well known example is pollen size[2]. However, depending on the preparation method(s), the pollen sample may show pollen grains of one and the same plant species fitting into more than one size category (pollen size categories: see "Pollen Morphology"). Moreover, sometimes the size of pollen grains is found just at the boundary between two adjacent pollen size categories. Placing the pollen grain in one of the size categories therefore depends entirely on the material, the preparation method(s) and the observer's evaluation.

Characterization of pollen ornamentation is even more complex, modifications of basic ornamentation characters or combinations of different characters usually giving rise to a seamless transition between neighboring characters or to a combination of characters.

Seamless transitions between related but clearly defined features exist, e.g., in gemmate pollen and its "neighbor" clavate pollen. Both types of ornamentation are very variable in shape and size and rather rare in their typical form.

Combination of ornamenting characters is very commón. Often, the ornamentation is composed of two or more characters, such as reticulate and foveolate, or a combination of echinate and perforate (for examples see Illustrated Glossary). From the observer's viewpoint it is desirable to name the ornamentation characters in a defined order: in the case of two or more combined characters, the most eye-catching, prominent character (the "leading term") should be mentioned first.

For example, in *Aristolochia*, the pollen grain surface bears very prominent verrucae

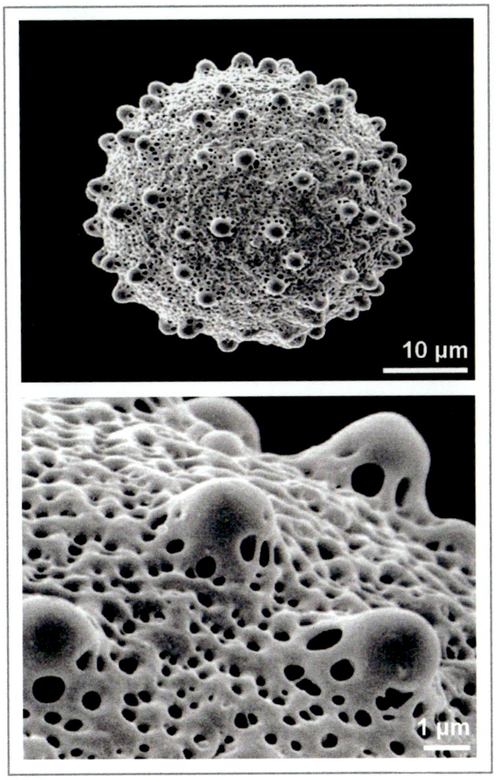

Combination of ornamenting characters

Aristolochia arborea
Aristolochiaceae

inaperturate, spheroidal
verrucate, perforate

surface detail
verrucae and perforations

[1] To be successful in characterization consider the following hints: be familiar with good microscope practice. The microscope, LM or EM, should be in good order. Primary magnification should be adequately high, but any enlarging of details beyond a beneficial magnification is counterproductive. For high magnifications in SEM, the best point resolution should be achieved. Quality of sample preparation is an all-too-often underrated item.

[2] The importance for dimension measurements is acknowledged but there is no need for decimal places, since dimensions vary considerably according to different treatments, as already shown by REITSMA (1969).

(the "leading term") combined with a great number of small perforations. Such ornamentation therefore should be called verrucate, perforate.

Sometimes it is debatable which feature represents the "leading term". As a sample, in Caryophyllaceae, there are numerous, more-or-less regularly arranged microechini and perforations. In some taxa the microechini are more prominent (microechinate, perforate), in others the perforations (perforate, microechinate). There are also taxa, where the two features are on a par (microechinate and perforate). Micrographs elucidate the actual situation at a glance.

Combination of ornamenting characters

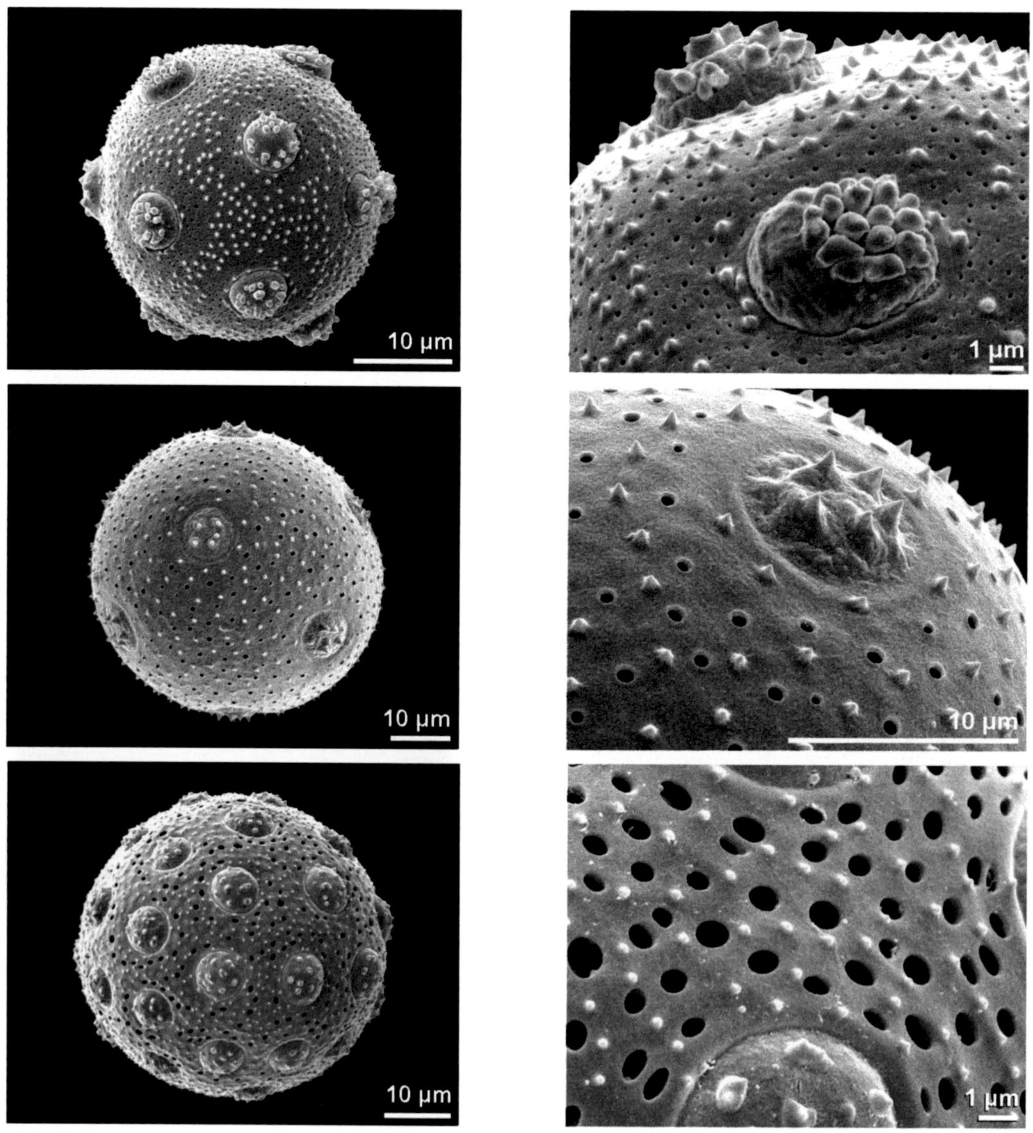

Stellaria media
Caryophyllaceae

microechinate, perforate

Saponaria officinalis
Caryophyllaceae

microechinate and perforate

Silene succulenta
Caryophyllaceae

perforate, microechinate

Distinct areas of the pollen grain surface may show different ornamentation types.

The type of ornamentation may be irregularly distributed over the pollen surface, or restricted to distinct surface regions.

Some examples may elucidate this feature:

- The polar region of *Fallopia convolvulus* is psilate to perforate, apertural regions are microechinate.
- In *Sideritis montana* polar and interapertural areas are perforate to foveolate, apertural regions are psilate.
- In *Salvia austriaca* the polar area is psilate to perforate, all other areas being bireticulate.
- *Solandra longiflora* is an example where the polar areas are reticulate, while in equatorial view the ornamentation is striato-reticulate.

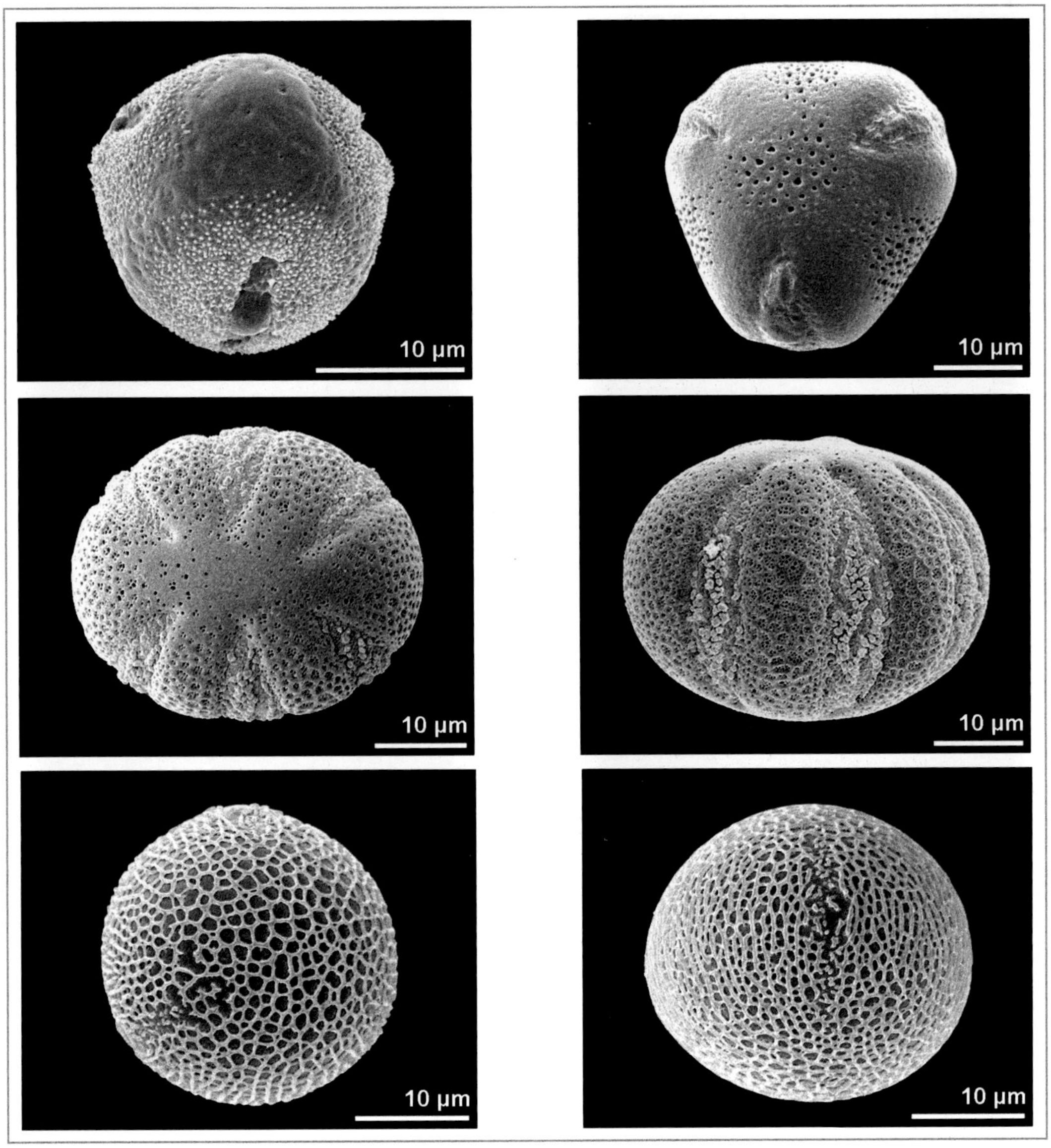

Combination of ornamenting characters

left:
Fallopia convolvulus
Polygonaceae

polar view

right:
Sideritis montana
Lamiaceae

polar view

Salvia austriaca
Lamiaceae

left: polar view

right: equatorial view

Solandra longiflora
Solanaceae

left: polar view

right: equatorial view

Interpretation of ornamenting characters

Sanchezia nobilis
Acanthaceae

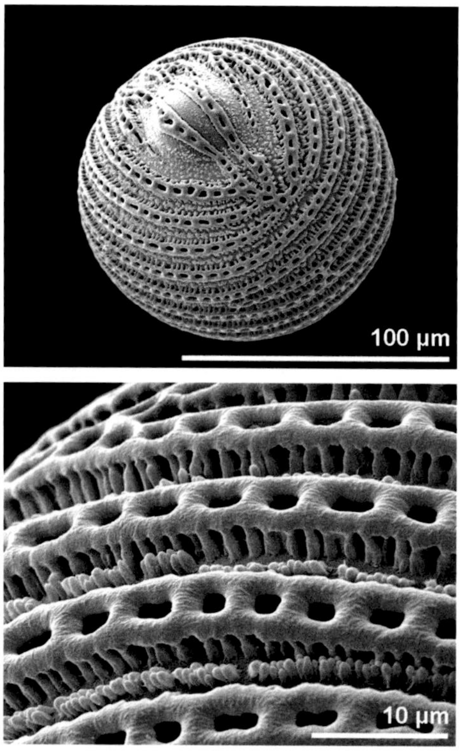

oblique equatorial view

surface detail

Sometimes it depends on the individual researcher to interpret ornamenting features: for example, to call *Sanchezia nobilis* (Acanthaceae) plicate and striate, but also reticulate? And should the rod-like elements be termed clavae, or free-standing columellae? Moreover, is the aperture to be interpreted as a porus or a colporus?

A special case deserves attention. In **heterostylous** species two different pollen types occur. Size and number of apertures, e.g., in *Primula*, or the ornamentation e.g., in *Linum*, may differ.

For better illustration *Linum flavum* and *Primula* pollen of the two floral types (long-styled and short-styled, pin and thrum morphs) is shown here.

In *Linum flavum* the short-styled-morph pollen is baculate, and the long-styled-morph clavate.

In *Primula veris* the pollen of the short-styled morph (thrum) is larger and has more apertures than the pollen of the long-styled morph (pin).

Heterostyly

Linum flavum
Linaceae

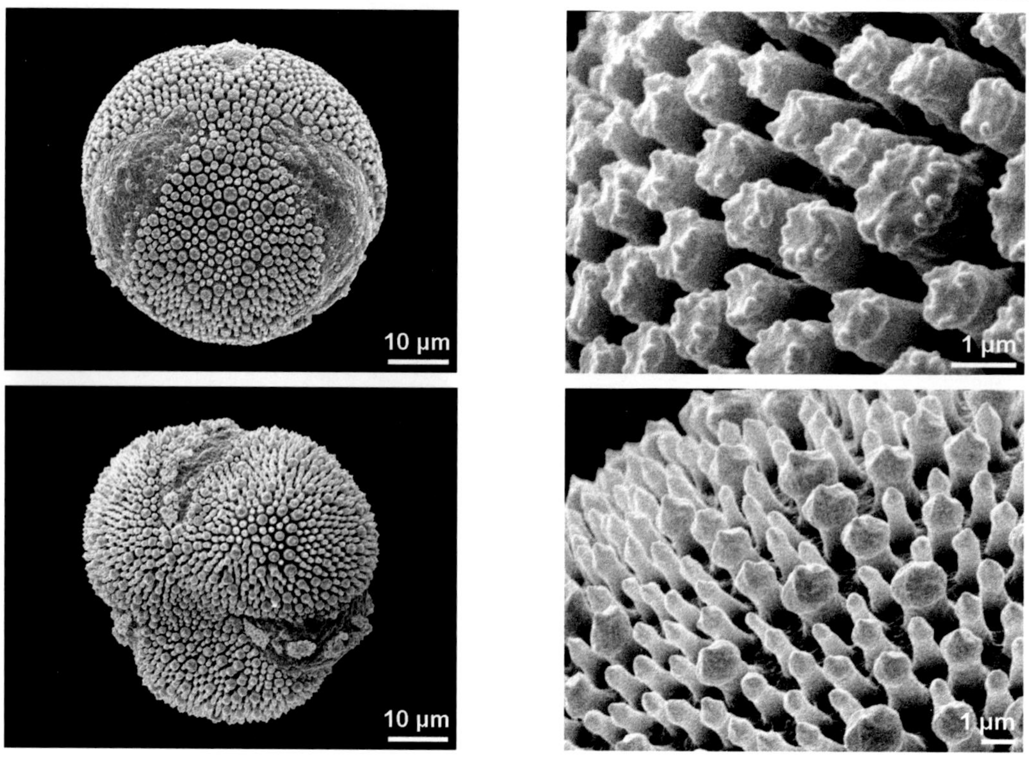

short-styled morph

baculate

long-styled morph

clavate

Heterostyly

Primula veris
Primulaceae

left:
short-styled
morph

right:
long-styled
morph

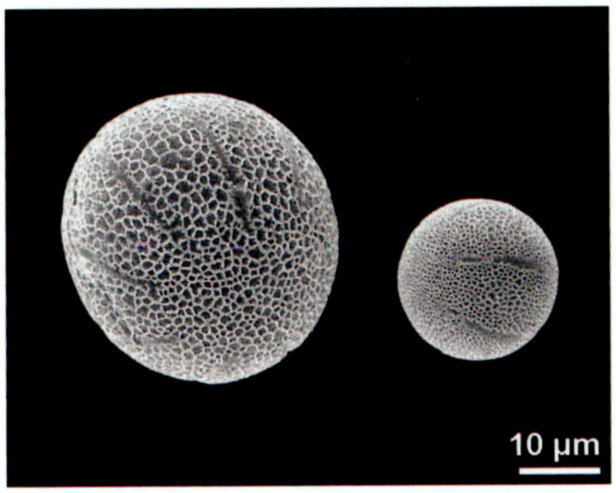

Terms derived from LM level cannot always be extended and applied to SEM level.

A classical example: *Ulmus* pollen at LM level was described as **rugulate** (rugulae: elongated exine elements longer than 1 µm; irregularly arranged). In low SEM magnifications the term **verrucate** (verrucae: wart-like element more than 1 µm, broader than high) would describe the ornamentation in a better manner. High SEM magnifications show additional **granules** (structure or sculpture elements of different size and shape; smaller than 1 µm). A typical rugulate ornamentation at SEM level is present in, e.g., *Sanicula*, which is quite dissimilar to the ornamentation seen in *Ulmus* at SEM high resolution level.

This is a good place to mention interpretative pitfalls. The denotation of ornamentation frequently depends on the optical magnification used and particularly on the point resolution. Very many (paleo-)palynologists have relied on LM only. Even low-power SEM may not be sufficient to distinguish pollen grains unequivocally which are in LM very similar (for examples and discussion see FERGUSON et al. 2007).

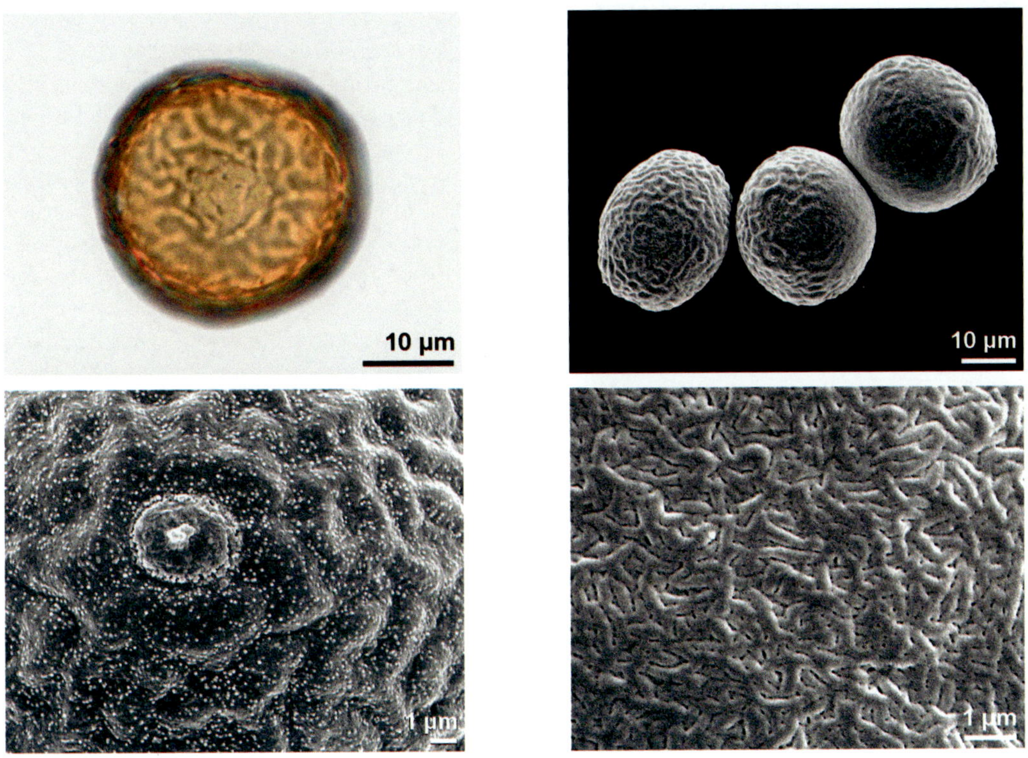

Ornamentation in LM and SEM view

Ulmus laevis
Ulmaceae

left: rugulate (LM)

right: verrucate (SEM)

left:
Ulmus laevis
Ulmaceae

surface detail
verrucate, granulate

right:
Sanicula europaea
Apiaceae

surface detail
rugulate

A second example is **scabrate**, a term used for light microscopy only, describing minute sculpture elements of undefined shape and of a size close to the resolution limit of the light microscope. As an example, *Juglans* pollen is scabrate in LM and (with some reservation) under low power SEM, but microechinate at high resolution SEM.

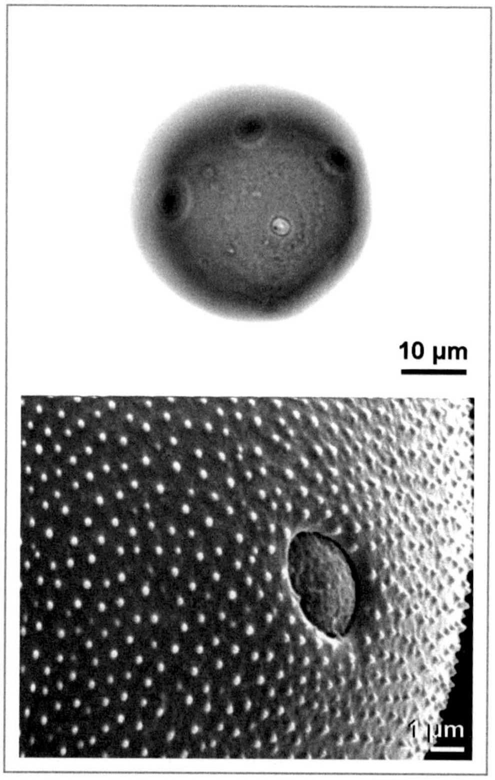

Ornamentation in LM and SEM view

Juglans sp.
Juglandaceae

polar view
scabrate to psilate (LM)

microechinate (SEM)

Another example for different interpretations in LM and SEM is the term **psilate**. Many pollen grains are psilate in LM view, but show a distinct ornamentation at SEM level. For example, in LM view pollen of *Allium* is psilate (see "Illustrated Glossary" – psilate), in SEM view it is striate and perforate (see "Pollen Morphology").

The term **granulate** (describing minute sculptural elements of undefined shape and of a size close to the resolution limit of the LM) is adequate for features at low SEM magnification. At significantly higher resolution a more adequate description is often possible. The actual shape of such "granules" depends on the much better resolution of details at high SEM magnification, where a "granulate ornamentation" emerges as, for example, a great number of very small spines (microechini), the pointed ends seen best in profile, not from top view. The allegedly granulate ornamentation of many Poaceae is in fact microechinate; see "Illustrated Glossary".

Another interpretative pitfall does not depend on optical magnification. Ornamentation sometimes depends entirely or to a high degree on the **preparation method**. A striking example is the presence or complete absence of distinct echini on pollen of many Araceae/Aroideae: fresh or dry material exhibits a distinct echinate ornamentation, whereas after acetolysis the echini are completely removed. These echini are composed of polysaccharides (singular exception) and lack sporopollenin completely. The pollen is then – correctly – called psilate (WEBER et al. 1999).

An example for **different possible interpretations** in relation with a differing degree of hydration is *Trichosanthes anguina* (Cucurbitaceae), where the ornamentation reflects the differing degree of hydration. The overview micrograph on the left shows a fully turgescent pollen, and on the right a less turgescent one. The ornamentation can be described as either areolate, or verrucate or even fossulate. Perforations are clearly visible in fully turgescent pollen only. So ornamentation should better be called verrucate and perforate.

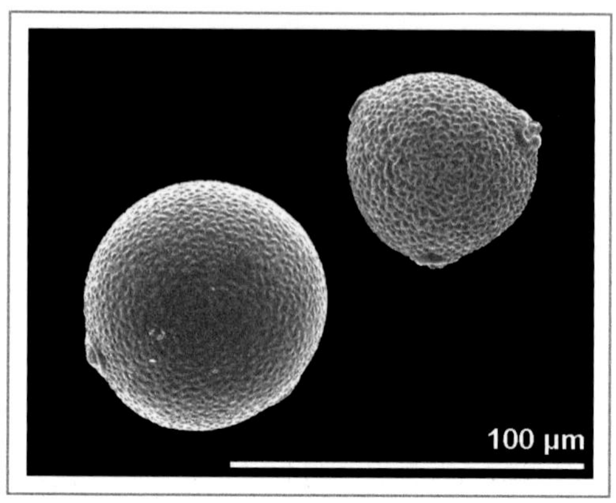

Hydration

Trichosanthes anguina
Cucurbitaceae
pollen grains of different state of hydration

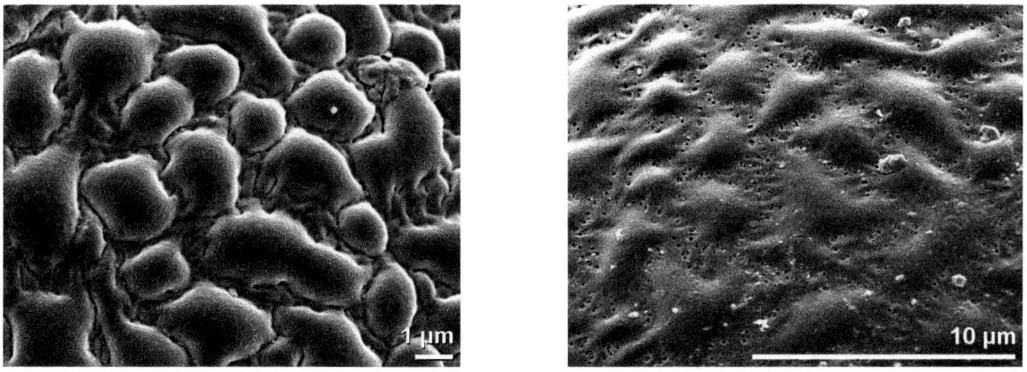

Hydration

Trichosanthes anguina
Cucurbitaceae

left: surface detail areolate

right: surface detail verrucate, perforate

A specific ornamentation sometimes even depends on peculiarities during pollen development. **Ubisch bodies** are usually found as isolated particles between pollen grains, or lining the mature locular wall (HUYSMANS et al. 1998, HALBRITTER and HESSE 2005, VINCKIER et al. 2005; equivalents are found in ferns: LUGARDON 1981). Pollen grains of Cupressaceae and Taxaceae are often equipped with adhering (adnate) Ubisch bodies, which are - strictly speaking - part of the specific pollen ornamentation (for example *Chamaecyparis* or *Juniperus*, see "Illustrated Glossary").

A complex category issue in (Paleo-) Palynology is the nomenclature question.

In **Paleopalynology**, for morphotaxa often form-generic names are used.

The nomenclature of form-genera is either "artificial" when the relationship is not known at all (e.g., *Oculopollis* and *Trudopollis* from the Normapolles group), or "half-natural", when reference to an extant taxon is suspected but not proven (e.g., *Liliacidites*). However, if reference to extant taxa is certain, then a "natural" nomenclature is possible (e.g., *Quercus* sp.).

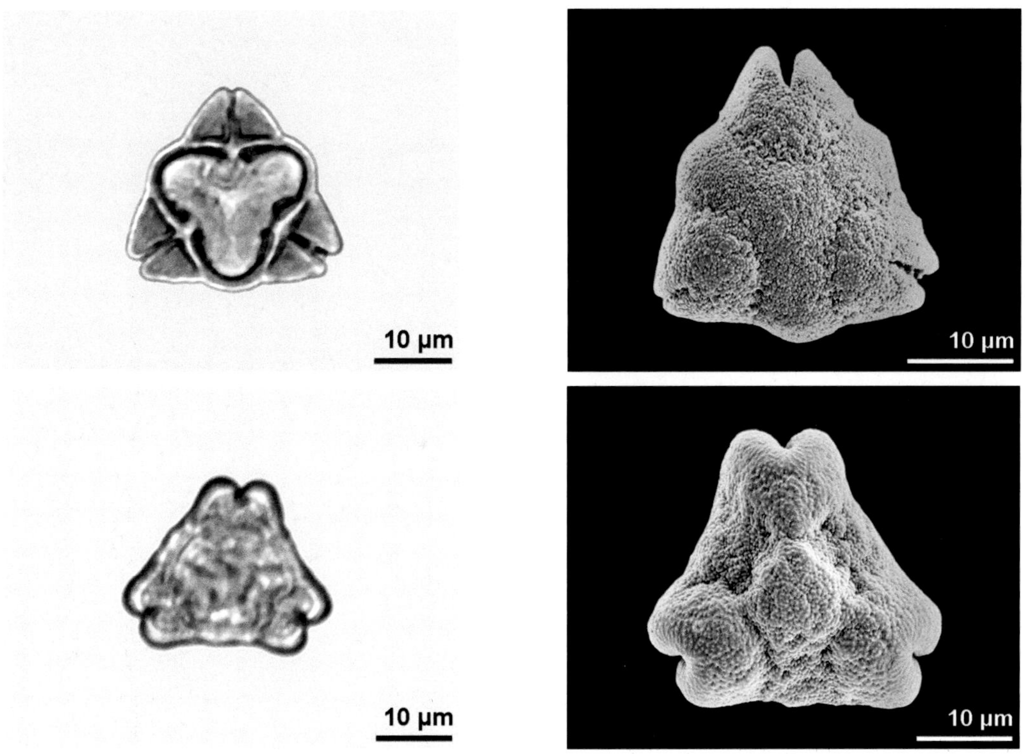

Nomenclature in Paleopalynology

Oculopollis sp.

Trudopollis sp.

Pollen Development

Microsporogenesis and Microgametogenesis

The unicellular pollen grain represents the **microspore** of seed plants, the multicellular pollen grain the male **gametophytic generation** of seed plants and is source and transport unit for the male gametes (or their progenitor cell). The development of a pollen grain includes (micro)sporogenesis [1-4] and (micro)gametogenesis [5-9]. **Microsporogenesis** starts with the differentiation of microspore mother cells (MMC) resp. **pollen mother cells** (**PMC**) [1]. These diploid cells become enclosed by a thick callose wall and undergo **meiosis**, forming a **tetrad** of four haploid **microspores**, each encased in a second callose wall insulating them from each other and from the surrounding diploid tapetal cells [2].

Cytokinesis following meiotic nuclear divisions is accompanied by the formation of cleavage planes determined by the configuration and orientation of the meiotic spindle axes. In the case of **successive cytokinesis**, planes are formed after the first and second meiotic divisions leading to the formation of various tetrad types (see "Pollen Morphology"). During **simultaneous cytokinesis** the cleavage planes are formed concurrently after the second meiotic division; in this case microspores are arranged in a **tetrahedral tetrad**.

Pollen wall formation starts when the microspores are still arranged in tetrads and encapsuled by callose [3]. The first step consists of the deposition of the **primexine**, a fibrillar polysaccharidic material, on the surface of the microspores. The primexine forms a template where sporopollenin precursors and finally **sporopollenin** are subsequently deposited, building the final pollen wall. Apertures are developed where the endoplasmic reticulum has prevented the deposition of primexine.

During pollen formation and maturation the **tapetum** plays an important role, usually forming a single layer of cells circumscribing the loculus. Tapetal cells are specialized and have a short lifespan; they finally lose their cellular organization and are reabsorbed. Two types of tapetum are known: the **secretory** (or glandular or parietal) and the **amoeboid** (or periplasmodial). In the secretory type (e.g., in Apiaceae) the tapetal

Microsporogenesis tetrads

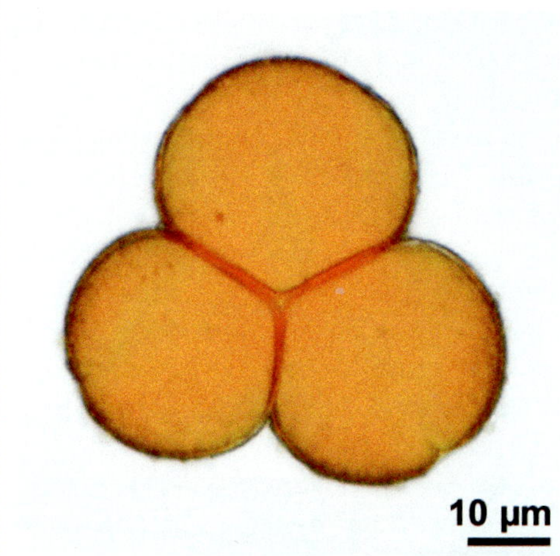

Scrophularia nodosa
Scrophulariaceae

tetrad tetrahedral
iodid

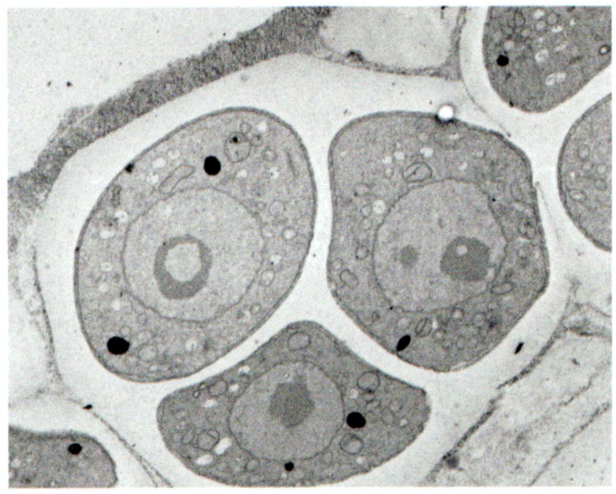

Spiraea sp.
Rosaceae

tetrad tetrahedral
PA+TCH+SP

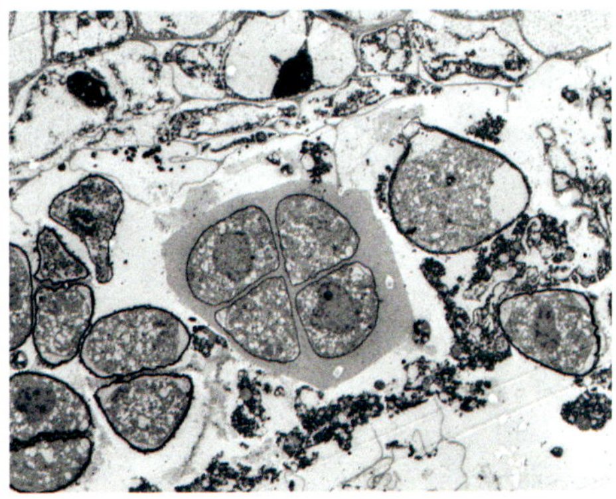

Orobanche hederae
Orobanchaceae

tetrad planar
$KMnO_4$

cells remain stationary until they finish their physiological functions. In the amoeboid tapetum type (e.g., in Araceae) cells lose their individuality in an early developmental stage by degeneration of the cell walls. The protoplasts then fuse and intrude into the locule where they enclose the pollen grains.

The tapetum plays an important role during several stages of pollen development. Its main function is the nourishment of the microspores but it also synthesizes enzymes (e.g., callase), exine precursors, pollen coatings, forms Ubisch bodies and viscin threads (both equivalents to the ektexine). The most striking material produced by the tapetum is **pollenkitt** (and **tryphine** in Brassicaceae, **elastoviscin** in Orchidaceae), a sticky, heterogeneous material composed of neutral lipids, flavonoids, carotenoids, proteins and polysaccharides. Pollenkitt serves numerous functions: for example, keeping pollen grains together during transport; protecting pollen from water loss, ultraviolet radiation, hydrolysis and exocellular enzymes; maintaining sporophytic proteins inside exine cavities.

Microgametogenesis in angiosperms includes first and second pollen mitosis, leading to the formation of the male gametes (sperm cells). Gametogenesis starts with formation of a central vacuole within the uninucleate microspore, pushing the nucleus towards the wall [5]. As long as the nucleus is in a central position within the cytoplasm, the cell is called a **microspore** [4]. With the dislocation of the microspore nucleus the cell becomes the young **pollen grain**.

The **first pollen mitosis** is followed by an asymmetric cell division, leading to the formation of a smaller **generative cell** and a larger **vegetative cell** [6]. When the generative cell is formed it is pressed against the pollen wall; it later separates and is then located within the cytoplasm of the vegetative cell [7]. After detachment, the generative cell, which is sparse in organelles, becomes modified in shape from spherical to **spindle-shaped** (the shape of the generative nucleus changes correspondingly). During the **second pollen mitosis**, which is followed by a symmetric cell division, the generative cell is divided into two sperm cells, the final stage of gametophytic development [8]. In about 25 % of flowering plants investigated, the pollen grains are **three-celled** at the time of anthesis [8]; in 75 % of flowering plants, pollen grains are shed from the anther at a **two-celled** stage. In the latter case the second pollen mitosis takes place in the **pollen tube**, after germination of the pollen grain onto a stigma or a corresponding structure [9].

Microgametogenesis in gymnosperms includes several mitotic divisions. Normally, pollen grains of gymnosperms are multicelled at anthesis, and comprise prothallial cell(s), a large tube cell and a small antheridial cell. The tube cell becomes a pollen tube; the antheridial cell undergoes division into the stalk cell and the spermatogenous cell, the latter finally dividing into the male gametes (sperm cells or spermatozoids).

Pollen development in angiosperms

Inherence of Misinterpretation

Investigation of recent and fossil pollen material often reveals interesting features that in some cases may be misinterpreted. Selected examples are various tripartite surface features that may actually be or only resemble apertures. Other examples are conspicuous, even eye-catching ornamentation features that are potentially misinterpreted as apertures, while the genuine, very inconspicuous apertures might be overlooked. The study of a morphological series can be of help clarifying ambiguous features.

Tripartite Features

Mature pollen of conifers, such as *Abies*, *Larix* and *Pseudotsuga*, often shows proximally a Y-shaped bulge, comparable to a tetrad mark, which is called an **impression mark** (HARLEY 1999). The mark results from the close proximity of the four pollen grains at the post-meiotic tetrad phase and is retained afterwards. Impression marks are also found in palm pollen. Note: the term tetrad mark is restricted to spores, where it is the germination feature, the impression mark of pollen grains is no germination feature.

Superficially similar features in angiosperms are not comparable to that in gymnosperms. In recent and fossil Sapindaceae a **three-armed feature** (more precisely a triangle) is found. *Cardiospermum* has a narrow triangle (tenuitas) proximally, whereas other

Tripartite features

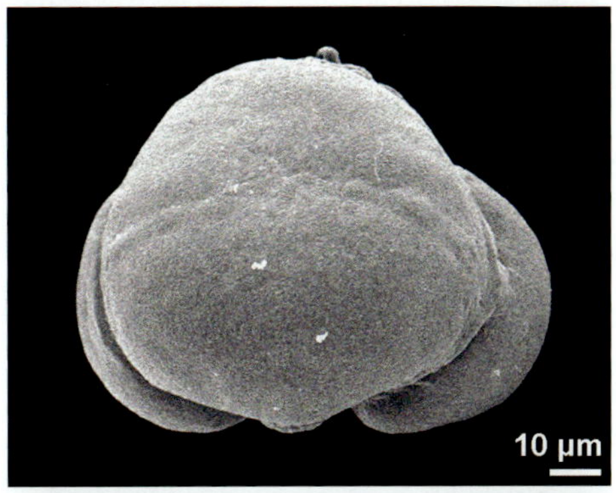

Abies cephalonica
Pinaceae

proximal polar view
indistinct impression mark

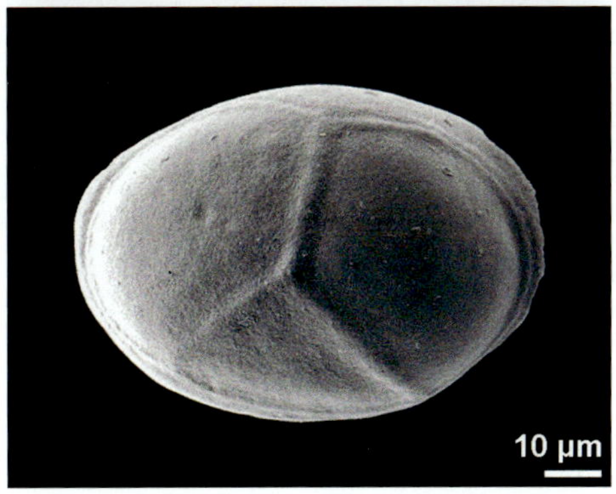

Larix sp.
Pinaceae, fossil

proximal polar view
Y-shaped impression mark

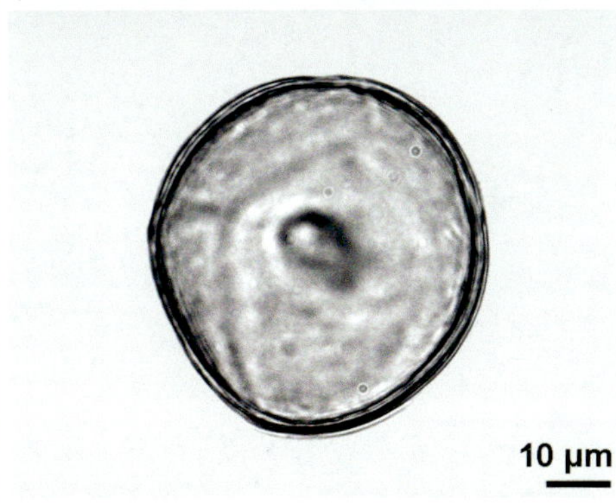

Larix sp.
Pinaceae, fossil

Y-shaped impression mark

Tripartite features

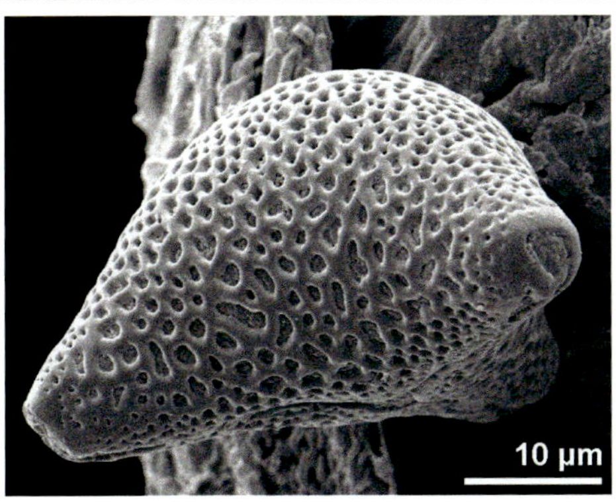

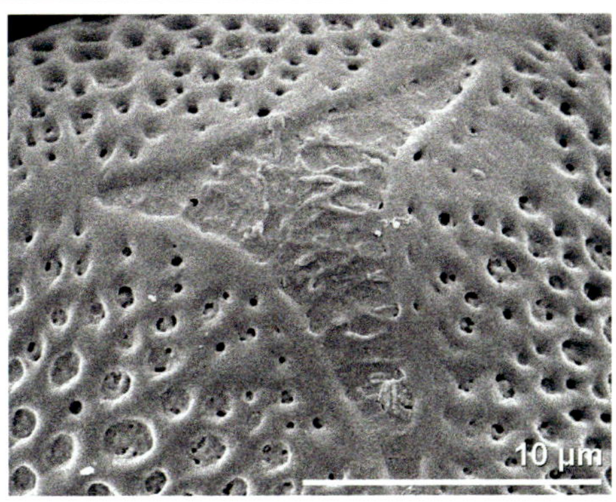

Cardiospermum corindum
Sapindaceae
tricolporate

left: equatorial view

right: proximal pole with triangular area

Tripartite features
synaperturate pollen

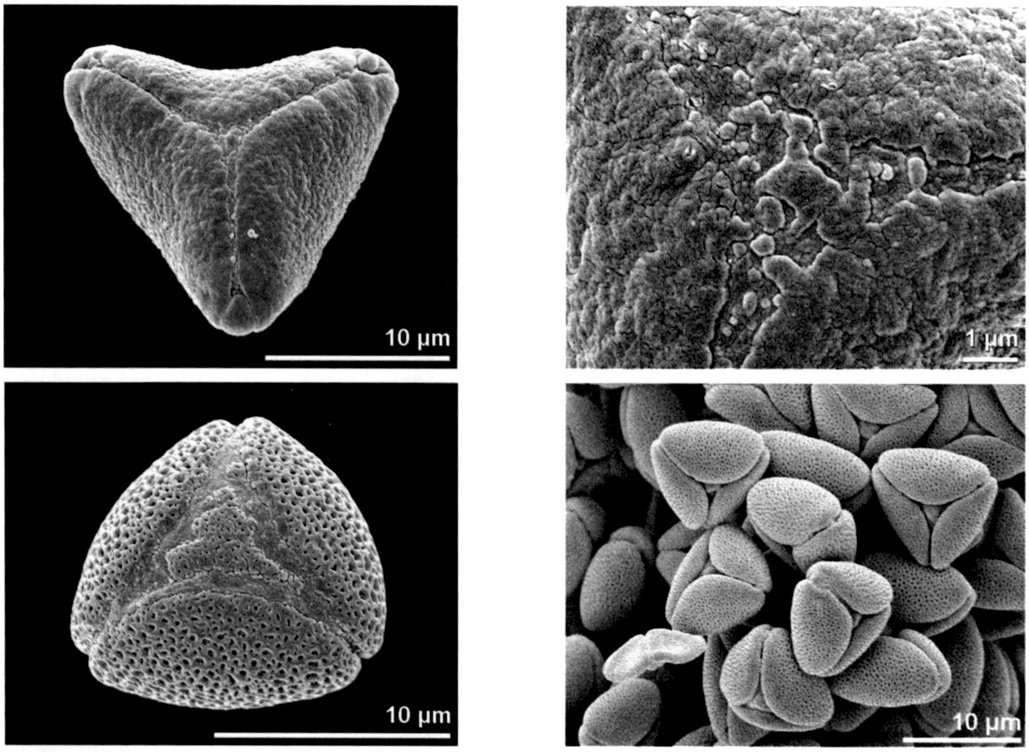

Melaleuca armillaris
Myrtaceae
syncolporate

left:
polar view

right:
polar area

syncolpate

left:
Primula denticulata
Primulaceae
polar view

right:
Primula farinosa
Primulaceae
dry pollen

Tripartite features
trichotomosulcus

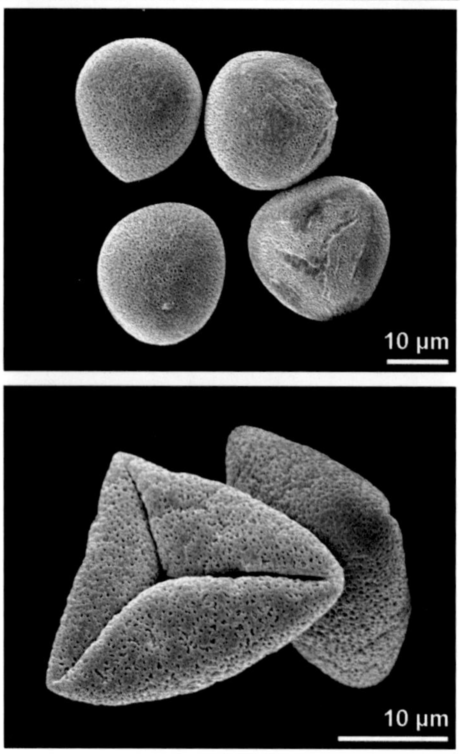

Dianella tasmanica
Phormiaceae

dry pollen

recent and fossil Sapindaceae show such a feature at both poles.

The triangular pollen as found in Myrtaceae, some Primulaceae (*Primula farinosa* or *P. denticulata*) or in some Loranthaceae is characterized by a **triangular field** in both polar areas. The angles elongate to meridional rays (colpi) directed to the roundish pollen tips, the rays crossing the equator and forming an equivalent triangle on the antipodal polar area. Pollen is synaperturate (syncolpate, syncolporate).

Another tripartite feature is the **trichotomosulcus** (HARLEY 2004), a three-armed sulcus found exclusively distally, as, e.g., in *Dianella* and Cretaceous fossils. Trichotomosulcate pollen has been discussed in relation to the evolution of the tricolpate dicot condition, but so far without success.

In contrast to the trichotomosulcus also trisulcate pollen is found. The angiosperm-like pollen of the fossil genus *Eucommiidites* has a distal sulcus and, at angles of ca 120° seen from the sulcus, two folds on the proximal side of the pollen. This feature was erroneously misinterpreted as more-or-less tricolpate pollen (with "colpi" equatorially situated, which has proved to be incorrect). A similar arrangement of a distal sulcus and two small additional "sulci" on the proximal face was described, for example, in some species of *Tulipa* (Liliaceae) and *Tinantia* (formerly *Commelinantia*, Commelinaceae), but these cases were never interpreted as equivalent to a tricolpate condition (HARLEY 2004). In some cases the three sulci are of quite equal size. The aperture condition is very similar to a tricolpate one. The interpretation "trisulcate" is possible and of use only in context with a morphological series.

Tripartite features

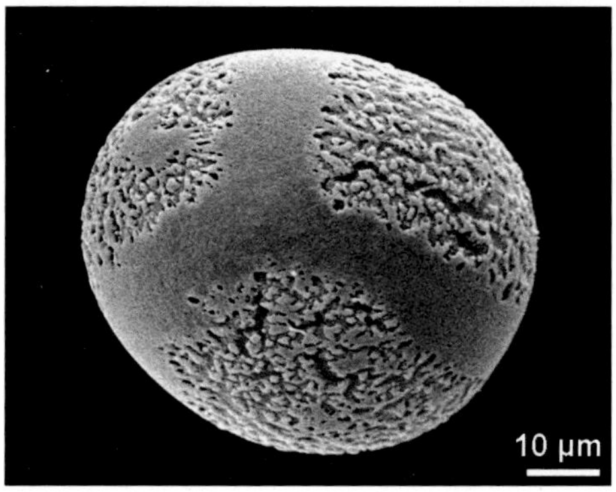

Tulipa kaufmanniana
Liliaceae
trisulcate

equatorial (!) view

Another three-armed feature is the triradiate aperture in *Thesium alpinum* (Santalaceae) pollen (an additional suggestive feature is the pollen shape forming a tetrahedron, the four triangular faces showing conspicuous reticulate areas that might be misinterpreted at first sight as four apertures). In fact *Thesium alpinum* pollen is three-aperturate, the apertures placed in the three tapered edges of the tetrahedron. Each aperture forms a very inconspicuous triradiate figure, which is situated equatorially. Two of the arms point towards the neighboring tetrahedron edge and are rather short; the third, elongated arm is directed to the rounded edge, which is probably the proximal pole.

Tripartite features

Thesium alpinum
Santalaceae

tricolpate, heteropolar

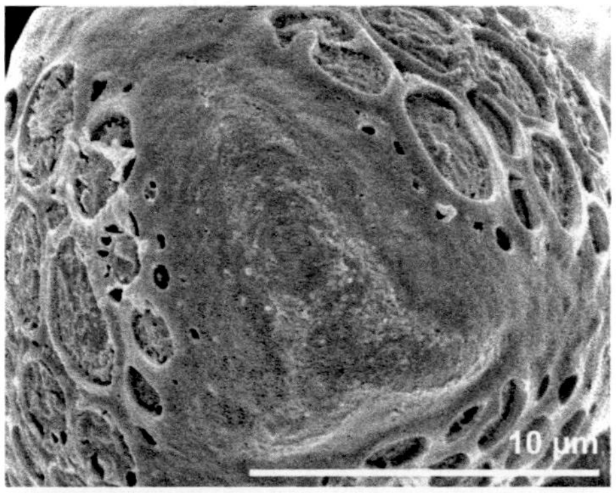

triradiate colpus

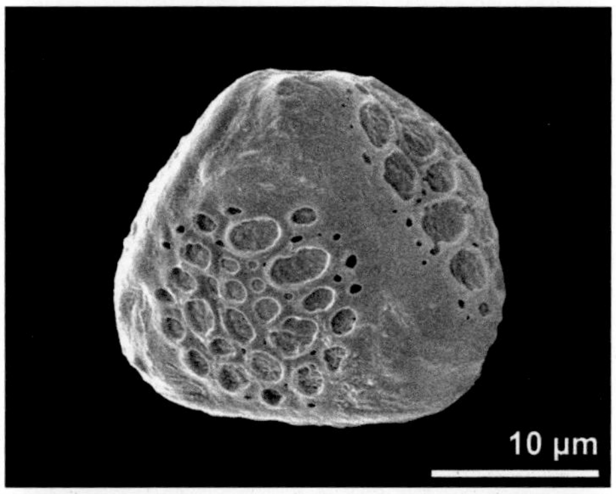

polar view

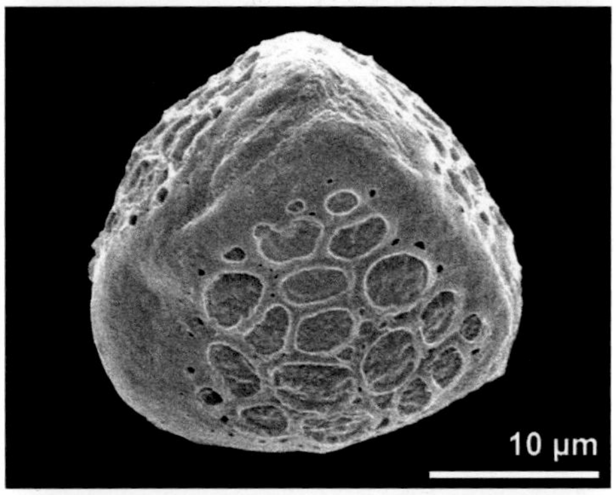

equatorial view

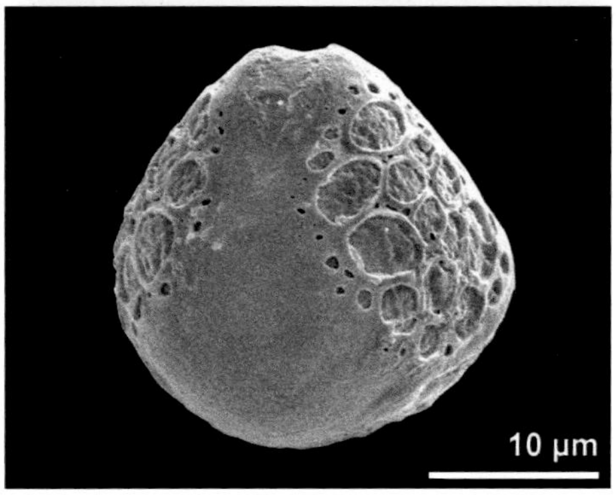

polar view

Apertures as Pitfalls

Sometimes the apertures are inconspicuous and not discernible at first sight. In *Pachira aquatica* (Bombacaceae) pollen three large, more-or-less hemispherical areas are seen equatorially, which may at first sight be interpreted as three pores. However, a detailed observation reveals planaperturate pollen grains with three short colpi.

Apertures as pitfalls

Pachira aquatica
Bombacaceae

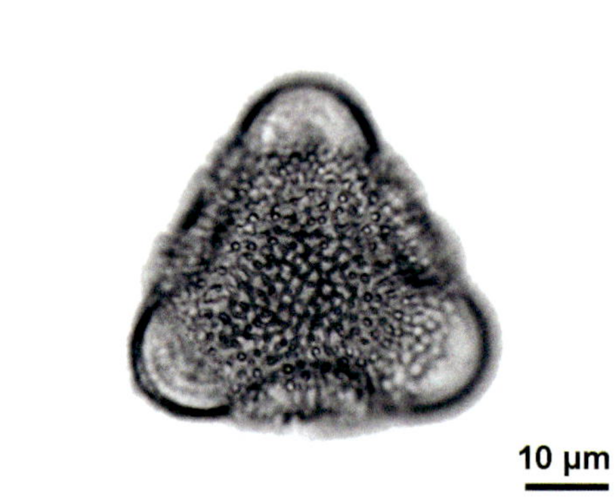

polar view

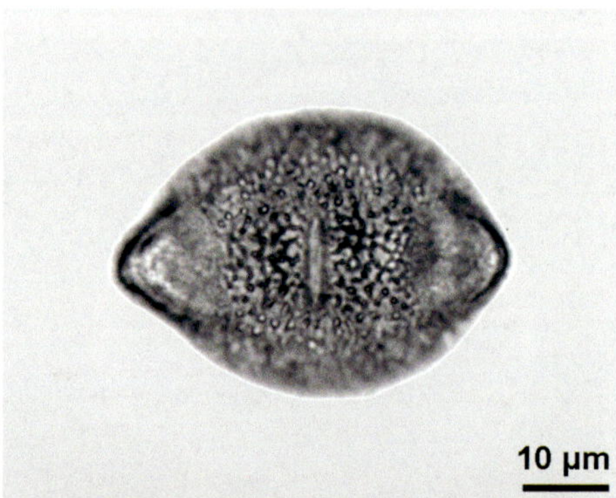

equatorial view

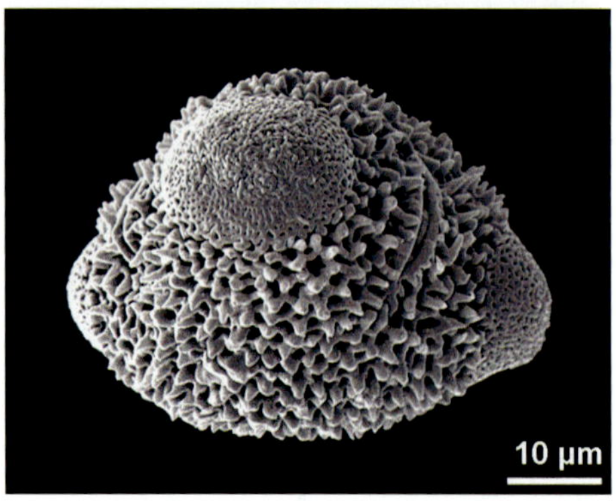

oblique
equatorial view (SEM)

The monads of the *Calliandra emarginata* (Mimosaceae) polyad are separated by narrow groove-like depressions. At low magnification the presence and localization of the apertures remain indistinct; high SEM magnification reveals that the apertures are very inconspicuous pores, situated equatorially, usually at the conjunction of three or four monads.

Apertures as pitfalls

Calliandra emarginata
Mimosaceae

massula
dry state

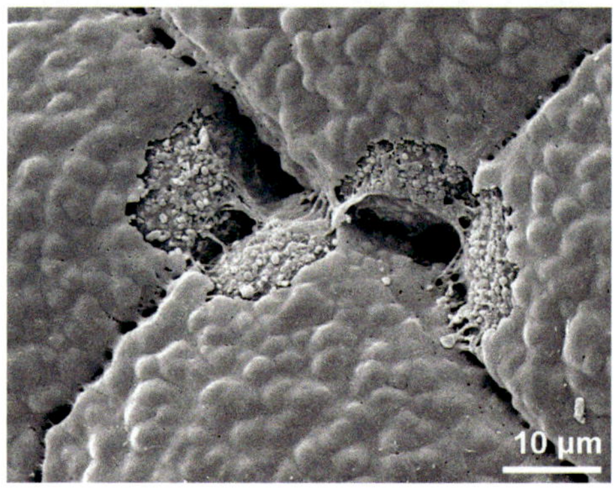

apertures (pores) at the junction of four monads

The apertural condition may be hidden by prominent features. The clypeate pollen of *Phyllanthus* × *elongatus* (Euphorbiaceae) seems to be inaperturate. Only close-ups reveal the inconspicuous pores between the exine shields.

The disk-like pollen of *Oryctanthus* sp. (Loranthaceae) shows at both poles conspicuous circular depressions that are not apertures (FEUER and KUIJT 1985). The pollen is tricolpate with inconspicuous short slit-like colpi between the polar depressions on

Apertures as pitfalls

Phyllanthus × elongatus
Euphorbiaceae

left:
clypeate pollen seemingly inaperturate

right:
inconspicuous pores between the exine shields

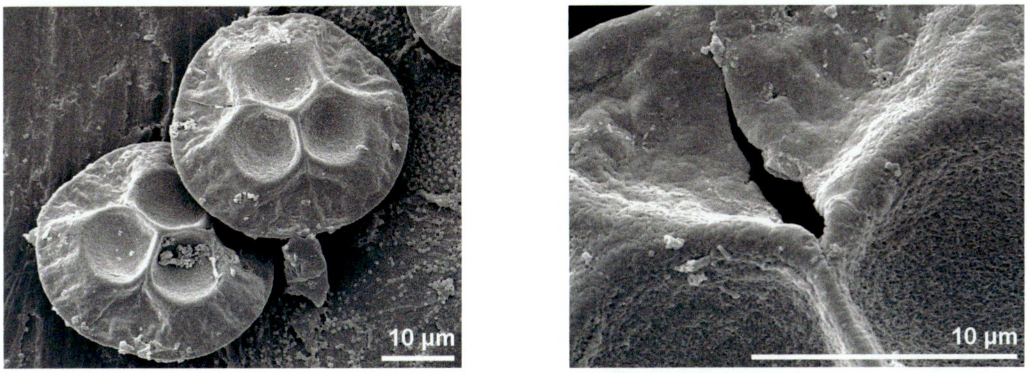

Apertures as pitfalls

Oryctanthus alveolatus
Loranthaceae
acetolyzed

left:
tricolpate (short colpi on both sides, with bridge), pollen grains in polar view

right:
colpus

both sides. The colpi are interrupted by a broad exine bridge at the equator plane; thus calling the aperture condition "hexacolpate" would be a possible interpretation.

Trapa (Trapaceae) pollen is triaperturate (colpate or porate). Recent and fossil pollen grains are distinguished by unique meridional exine features (crests); these cover the apertures so that only after removal of the crest the aperture is visible.

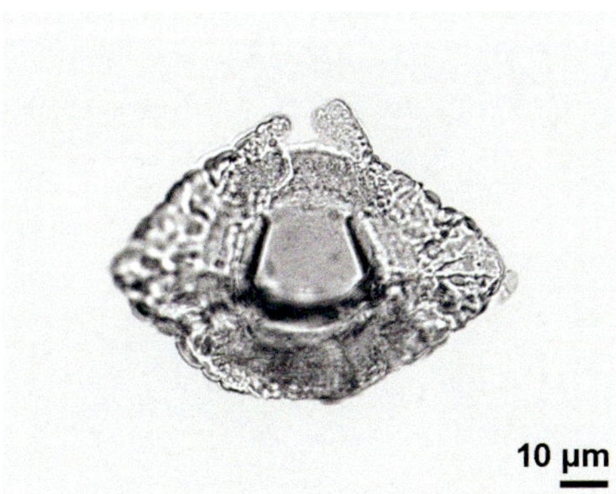

Apertures as pitfalls

Trapa sp.
Trapaceae, fossil

equatorial view
crest broken

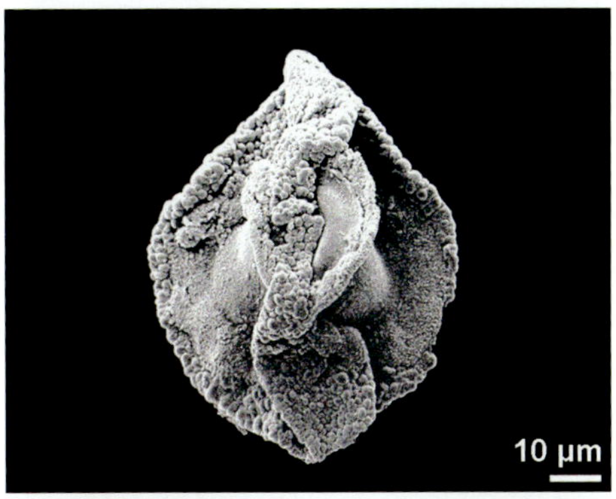

Apertures as pitfalls

Trapa sp.
Trapaceae, fossil

equatorial view
crest in part broken
colpus visible

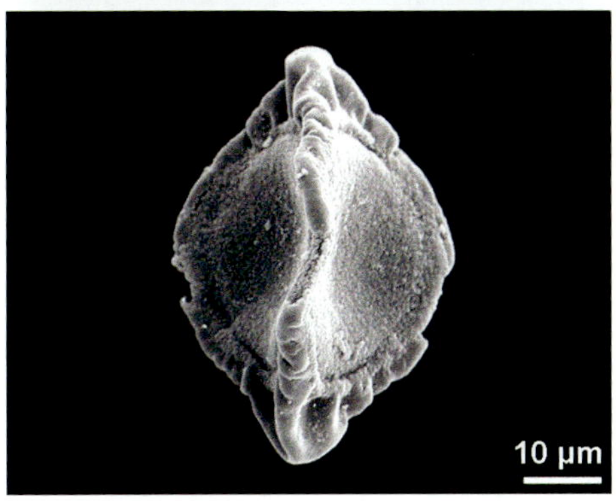

Trapa sp.
Trapaceae, fossil

equatorial view

Pollen Features can be Ambiguous

Case studies:

- — interpretation of specific aperture conditions
- — classification of infratectal structure characters
- — deviating pollen forms

Investigation of a **morphological series** within a genus can provide decisive nomenclature or at least support such a decision; studying only an isolated sample of a morphological series may easily lead to misinterpretations. *Passiflora* (Passifloraceae) pollen is an example of a morphological series concerning apertures. The apertures in *Passiflora* cf. *incarnata* may be interpreted as three ring-like apertures. An interpretation as porate-operculate is likewise possible, and probably more correct, if compared with pollen of other *Passiflora* species.

Not infrequently the apertures in angiosperms show indistinct margins, or appear as thin regions in the pollen wall. The Illustrated Glossary makes use of two terms, poroid and tenuitas, in describing superficially quite similar features.

A **poroid** is a circular or elliptic aperture with indistinct margin. A tenuitas is a general term for a pollen wall thinning, which has been applied to many different features (KREMP 1968, HARLEY 2004, PUNT et al. 2007).

A **tenuitas** (in angiosperms) is normally found additional to regular apertures (e.g., in some *Myosotis* species, see "Illustrated Glossary").

The harmomegathic effect may cause misinterpretations. A distinct infolding type suggests or pretends an erroneous aperture condition, while the correct aperture condition is inconspicuous or even hidden.

Pollen of *Sparganium erectum* (Sparganiaceae) is in dry stage infolded, boat-shaped, and would be considered as sulcate. In fact *Sparganium* pollen is ulcerate, the ulcus is seen clearly in the hydrated, spherical pollen stage.

Nymphaea alba (Nymphaeaceae) pollen has asymmetrical halves. The features

Ambiguous features

Passiflora cf. *incarnata*
Passifloraceae

left:
polar view

right:
equatorial view

left:
Passiflora citrina
Passifloraceae

stephanocolpate
operculate
polar view

right:
Passiflora suberosa
Passifloraceae

stephanocolpate
operculate, dry pollen

of the smaller distal half can be interpreted either as a large ulcus with a conspicuous operculum, or as a more-or-less equatorially situated ring-like aperture surrounding the polar area. Ultrastructural characters and germination experiments support the interpretation as a ring-like aperture (HESSE and ZETTER 2005).

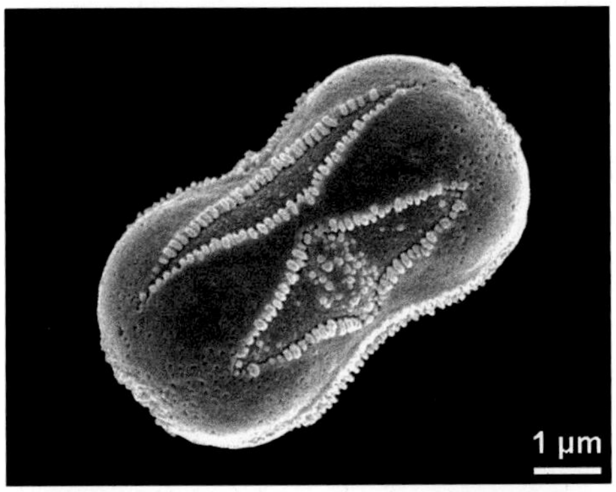

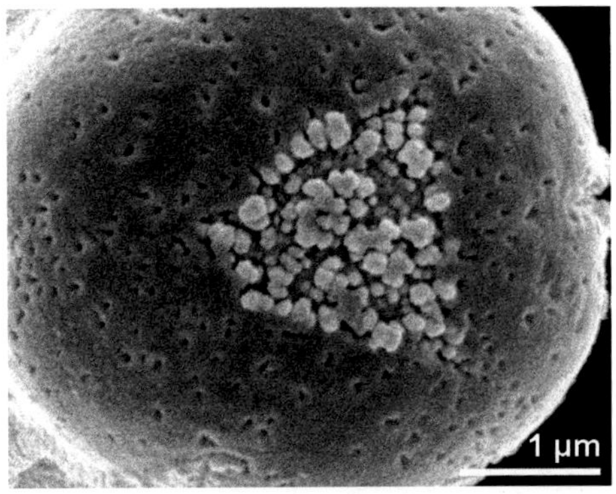

Ambiguous features

Myosotis palustris
Boraginaceae

left:
equatorial view
heteroaperturate, alternating colpori and colpi (pseudocolpi)

right:
polar view
polar area with triangular tenuitas

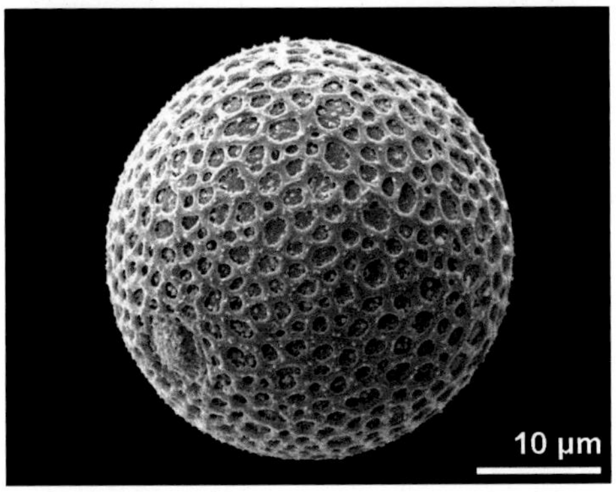

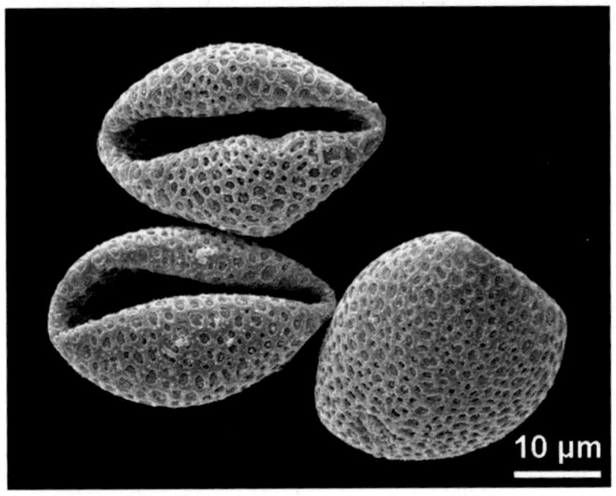

Ambiguous features

Sparganium erectum
Sparganiaceae
ulcerate

left:
equatorial view

right:
dry pollen
boat-shaped

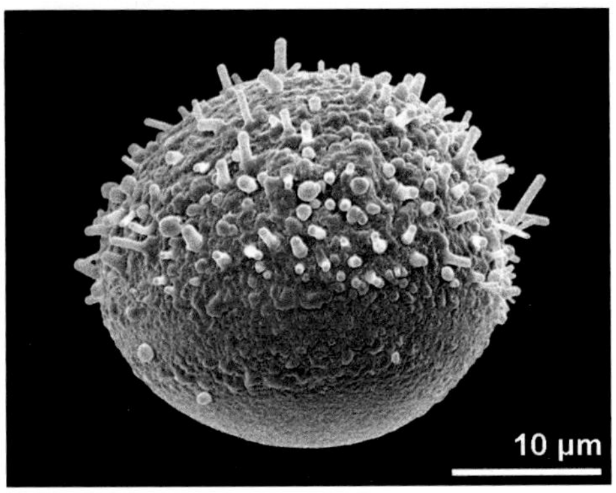

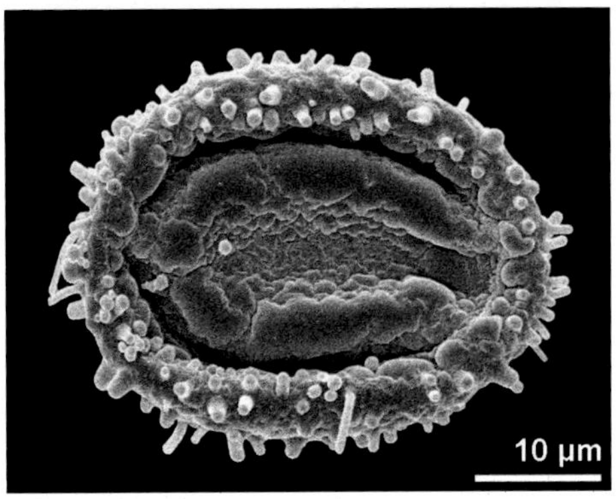

Ambiguous features

Nymphaea sp.
Nymphaeaceae
ring-like aperture

left:
equatorial view

right:
dry pollen
cup-shaped

Infratectum is a term which includes in fact a morphological series. The classical angiosperm character states simply comprise columellar and granular. However, as, e.g., DOYLE (2005) has pointed out, intermediate conditions are not uncommon. Even the areolate infratectum, usually restricted by definition to gymnosperms, is found in some angiosperms (see "Illustrated Glossary").

Deviating pollen forms – an underrated topic.

Abnormal pollen grains occur regularly in small percentages in nearly all anthers and may vary from one individual to another (heterogeneity of pollen forms). A much higher percentage of such deviating, more precisely, malformed pollen grains are found in many cultivated plants (ornamental plants, agricultural crops) or in plants with asexual reproduction (autogamic plants, apomicts). The effect also depends on the ploidy level, hybrids may produce a series of pollen types. Pollen variation is generally underreported, because most studies focus on normal rather than abnormal pollen morphology (BANKS et al. 2007).

Aberrant pollen may occur at a high percentage within a single pollen sac, anther or flower.

These aberrant, deformed pollen grains differ from normal pollen in shape and dimension, in number and form of apertures, in type and arrangement of ornamentation. Very often in hybrids, pollen ornamentation is intermediate between the typical forms of the parent species. Some species produce only malformed pollen, a feature typical for plants with asexual reproduction (e.g., some *Alchemilla* spp.).

The reasons for the production of deviating pollen forms are genetically, chemically or environmentally induced. While many reports dealing with the developmental causes, a tiny minority of papers refers predominantly to the habit of the deviating pollen forms (e.g., POZHIDAEV 2000a, b, focusing on the aperture patterning). As a consequence, these outliers should neither be ignored nor overrated, but we should take notice of this issue as an ordinary feature.

Controversial or Fuzzy Terms

Sometimes terms are ambiguous in definition or application, or may be simply misleading.

Acalymmate/Calymmate

In compound pollen grains or pollen units the individual grains are held together by different physical means of the pollen wall. Van CAMPO and GUINET (1961) recognized calymmate and acalymmate types. The term **calymmate** denotes a feature describing a dispersal unit of two or more monads enclosed by a continuous ektexine.

Calymmate

Chlorospatha kolbii
Araceae

tetrads

Acalymmate

Annona muricata
Annonaceae

tetrad, dry pollen

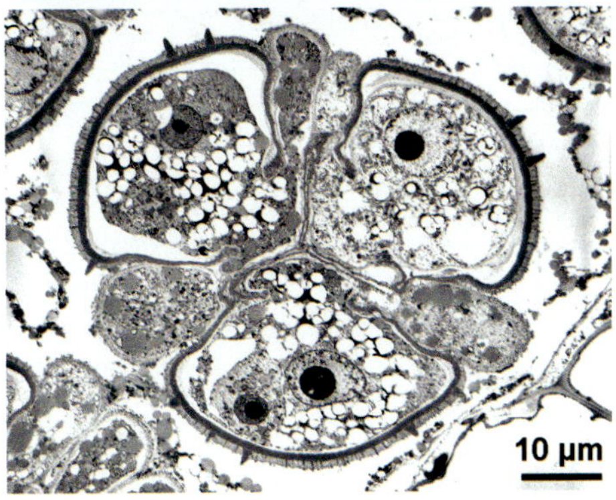

Drosera sp.
Droseraceae

tetrad

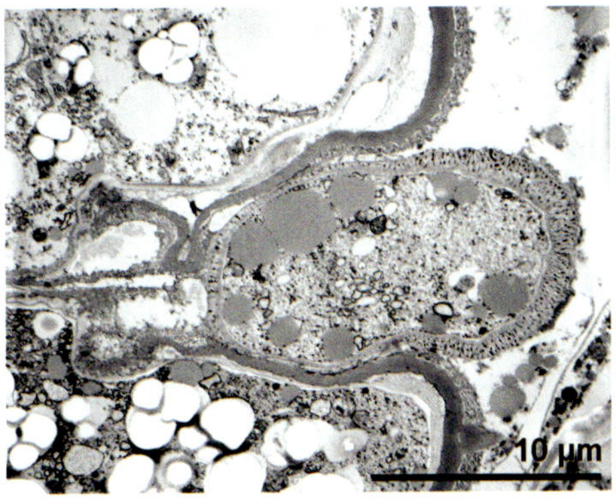

detail of pollen wall

Acalymmate denotes a feature describing a dispersal unit of two or more monads enclosed by an exine, which is discontinuous at the junctions between the monads, and is absent from the internal walls.

Since these basic types were described a lot of variations have been found, and the question was raised whether the basic distinction can remain (a survey and a detailed discussion was provided by KNOX and McCONCHIE 1986). The micrographs show examples for either a continuous (calymmate) or a discontinuous exine enclosure (acalymmate).

Areolae/Areolate

The term areola/areolate was originally applied to pollen grains ornamented with small, mostly convex exine areas separated by small grooves. Later on, the term was used describing pollen with large, variably ornamented exine areas separated by grooves; this type of ornamentation is better called **clypeate**, a pollen wall, in which the exine is subdivided into shields. Clypeate, however, also denotes a pollen class; for examples see "Illustrated Glossary".

Pseudocolpus

A pseudocolpus is a colpus in hetero-aperturate pollen grains and is presumed as non-functional. Pseudocolpi are always associated with colpori, never with colpi. Pseudocolpi mostly alternate with colpori (e.g., in Boraginaceae, Lythraceae) or are situated on both sides of a colporus (in Acanthaceae). Since germination experiments for pseudocolpi are absent, it cannot be assumed that pseudocolpi act as germination sites. Without doubt they play a role in harmomegathy, but their effects have been poorly studied; for examples see "Illustrated Glossary".

Retipilate

There are yet no known examples for retipilate, "a reticulum formed by rows of pila instead of muri". In contrast to earlier observations, investigations based on LM and SEM have revealed that the examples *Callitriche* (Plantaginaceae) in PUNT et al. (2007: "*describing a reticulum formed by rows of pila instead of muri*") and *Cuscuta lupuliformis* (Cuscutaceae) in ERDTMAN (1952, p. 128: „*with a reticuloid pattern with pila instead of muri*") do not fit their definition; in fact there are muri with prominent sculpture elements (a **reticulum cristatum**) but no isolated pila. Nota bene: while ERDTMAN (1952) provides a correct drawing, the diagram in PUNT et al. (2007) is erroneous.

Zona-, Zono- etc.

Terms combining the basic prefix zon- together with its linguistic derivatives are a source of endless confusion, misunderstanding and superfluous inflation of terms. The prefixes include **zon**- (in zonorate, for a ring-like endoaperture, the os, at the equator), the outdated, rarely used **zoni**- (however, with two quite different terminological applications), but especially **zona**- (indicating exclusively a ring-like feature situated anywhere) and **zono**- (indicating any feature located strictly equatorially).

A curious example deserves special attention. Terms for ring-like (aperture)

Retipilate

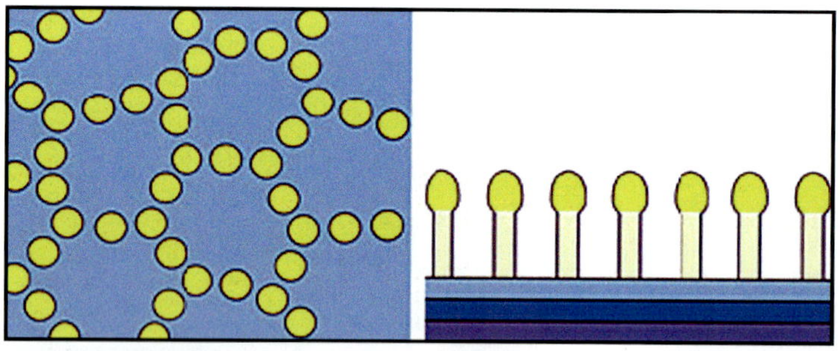

left: drawings from PUNT et al. (2007)

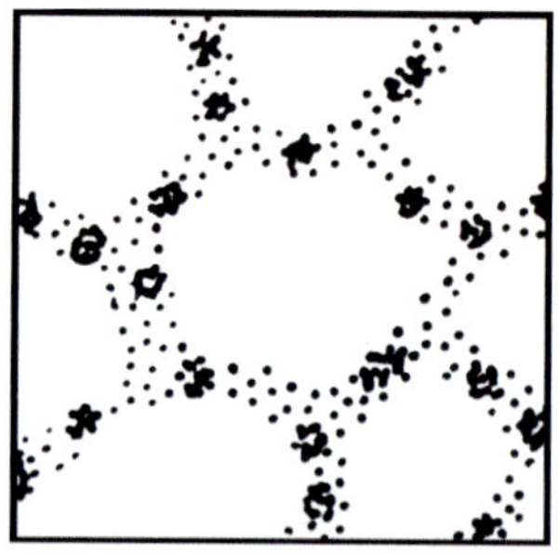

right: drawing from ERDTMAN (1952), p. 22, Fig. 5d

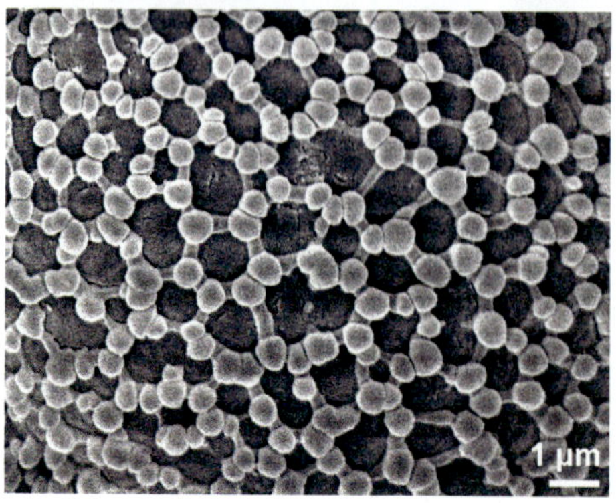

left: *Callitriche polymorpha* Plantaginaceae

reticulum cristatum with small gemmae on thin muri

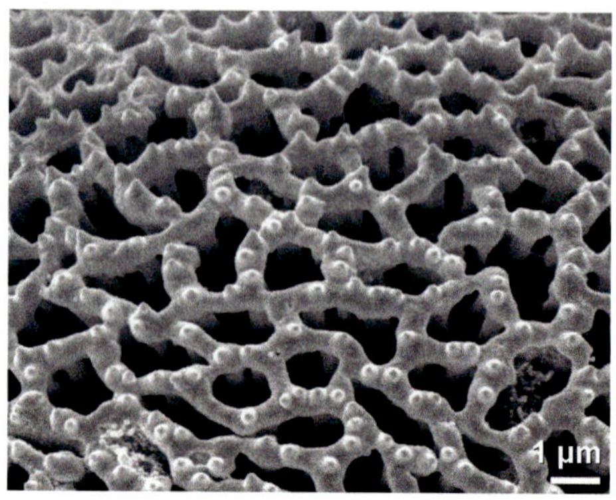

right: *Cuscuta lupuliformis* Cuscutaceae

reticulum cristatum with microechini

features include zona-aperturate, zona-sulculus (addressing the polarity by anazona-sulculus and catazona-sulculus), zona-sulcus, zonate, zono-aperturate, and also related names (e.g., "fully zonate condition" sensu GRAYUM 1992). Even the misleading and contradictory **zono**-sulcus (a sulcus cannot be situated equatorially) is used instead of the correct, but phonetically confusable, zona-sulcus. The trained palynologist may be overstrained. It is proposed that all these terms should be avoided.

To avoid any confusion, not more than two non-interchangeable terms are necessary, without combination of syllables, prefixes or suffixes. 1. any encircling aperture ("zona-aperturate"), irrespective of meridional or equatorial location, is simply called a **ring-like aperture**. The location or direction regarding the pollen grains polarity is not easy to address, since the orientation of the pollen grain in tetrad stage is relevant. 2. any case with more than three apertures at the equator ("zono-aperturate") is called **stephanoaperturate**.

A typical meeting debating controversial terms at any Thursday within the last three years

14:00 start of meeting, participants in time: MH, HH, SU, RB, AFR, MW

14:38 RZ comes delayed and wants to have "bisaccate" in the glossary

14:40 RZ falls into a deep sleep

14:50 MH wants to include "cryptoaperturate" to the glossary
all others: "noooooo..."

15:00 RZ wakes up and again calls for "bisaccate"
all others: „nooooooooooo...!"
RZ: "Wish to be a squirrel ... hide and find nuts ... no more pollen terminology!"

15:05 RB repeatedly yawns, because of low nicotin and coffeine levels

15:10 emotional discussion about "zona-" and "zono-aperturate";
SU resigns while writing the protocol...
MH to HH: "Heidi, think faster...!"
MW and SU suffer from paroxysm of laughter
AFR asks for more discipline

16:00 HH: "It's time for the four o'clock-coffee!"
MH: "Forget about it!"

16:15 RZ (enervated) again calls for "bisaccate"

16:30 MW asks the "tool for final decisions" about "bisaccate"

17:00 end of meeting

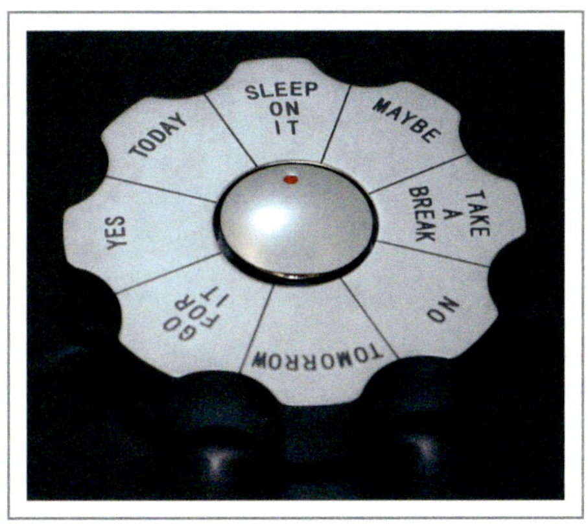

tool for final decisions

frustrate
Reinhard Zetter, Ralf Buchner
Heidemarie Halbritter, Andrea Frosch-Radivo
Martina Weber, Silvia Ulrich, Michael Hesse

Methods

Multiple techniques and methods should be used when investigating pollen grains in order to provide comprehensive and accurate information and help to avoid misinterpretations. Good examples of such an approach are the endexine (mentioned below) and the endoaperture, the latter visible only in the light microscope (LM) and hidden in the scanning electron microscope (SEM).

The selection of micrographs in "*Pollen Terminology. An illustrated Handbook*" in cludes both LM and EM pictures.

Scanning Electron Microscopy

SEM techniques cannot substitute LM but they can provide a great deal more information, especially about ornamentation. Methods of sample preparation for SEM should preserve the living condition of a pollen grain as far as possible. In addition, pollen coatings should be removed from the pollen surface in such a way that no details of the pollen grain are lost. For the SEM, acetolysis is not an optimal method for cleaning the pollen surface, as it will often destroy apertural details. Pollen with fragile exines may also be destroyed.

As a routine, all pollen grains should be observed in an air-dried condition, which gives the best information about the pollen grains at anthesis and their harmomegathic situation.

The best results have been obtained using 2,2-dimethoxypropane (DMP) (HALBRITTER 1998). This method can be used for fresh material (pollen grains should be collected when anthesis starts) and for herbarium samples after short rehydration in water. Unless stated otherwise, the pollen grains shown in "*Pollen Terminology. An illustrated Handbook*" represent the turgescent state.

Fresh anthers are put into a pouch made of filter paper and immediately transferred into acidified 2,2-dimethoxypropane (a drop of 0.2 M HCl added to 30 ml DMP). After 20–30 min in DMP (or up to 24 h) samples are transferred to pure acetone for a few minutes and critical-point dried in CO_2 using acetone as intermediate fluid. The dried pollen samples are then mounted on stubs using double-sided adhesive tape, sputter-coated with gold and observed with the SEM.

Acetolysis and Light Microscopy

Acetolysis is an indispensable method for illustrating pollen grains with the LM. Untreated or stained pollen grains will hide much of the important information for the description of a pollen grain.

Acetolysis is a combination of chlorination and acetylation:

For chlorination, the samples are transferred to a test tube and covered with a layer (1.5 cm) of glacial acetic acid and a layer (approx. 3 cm) of a freshly prepared solution of saturated sodium chlorate. After adding 3 or 4 drops of concentrated HCl, the mixture is stirred with a glass rod, heated in a bath of boiling water for 3 min, centrifuged, and the liquid fraction decanted. The residue is carefully washed to eliminate any remaining chemicals and then finally washed in concentrated acetic acid or acetic anhydride to remove the water.

For the acetylation step, the samples are put into a mixture of 9 parts acetic anhydride and 1 part concentrated sulfuric acid and heated to 100° C for approximately 4 min. After the mixture has been centrifuged and the liquid fraction decanted, the residue is washed in acetic acid and water. Glycerine is then added to the sample to form a suspension.

Single-Grain Technique

(ZETTER 1989, FERGUSON et al. 2007)

For fossil pollen grains a combined LM/SEM investigation should be used. After the LM investigation the same acetolyzed pollen grain is transferred to the SEM.

After extracting the pollen grains from the sediment, the samples undergo acetolysis (chlorination plus acetylation as described above). Glycerine is then added to the organic residue to form a suspension and a drop transferred to a glass slide. Using a dissecting needle to which a nasal hair has been affixed, grains of particular interest are brushed to the edge of the glycerine, where they can be picked up and transferred to another glass slide for photography under the LM. Because no cover slip is used, it is possible to photograph the same grain in various orientations. After this, the pollen is transferred to an SEM stub to which a drop of absolute ethanol has been added to remove all traces of the glycerine from the surface of the pollen grains, so that these can be examined in great detail under the SEM.

Transmission Electron Microscopy

For TEM studies of pollen from different plant species, more than one protocol for fixation and staining may be needed.

For fixation, anthers are placed in 2 % glutaraldehyde in 0.1 M phosphate buffer (pH 7.4) for 8–18 h at room temperature. After rinsing in buffer and distilled water, samples are postfixed in 2 % OsO_4 plus 0.8 % phosphate-buffered potassium ferrocyanide $K_4Fe(CN)_6$ 3 H_2O (2:1) for 8 h at 6° C. Samples are then washed in distilled water, dehydrated in 2,2-DMP followed by pure acetone, and embedded in Spurr's low-viscosity epoxy resin (SPURR 1969) or Agar low-viscosity resin.

Sections of pollen grains are routinely stained using the following methods:

U+Pb

URANYL ACETATE-LEAD CITRATE STAINING

In many species investigated uranyl acetate-lead citrate staining has not proved satisfactory.

Staining is carried out in uranyl acetate solution (Leica Ultrastain-1) for 45 min followed by lead citrate (Leica Ultrastain-2) for 1 min (all steps at room temperature).

TCH+SP, PA+TCH+SP

THIOCARBOHYDRAZIDE-SILVER PROTEINATE STAINING (TCH+SP) AND PERIODIC ACID-THIOCARBOHYDRAZIDE-SILVER PROTEINATE STAINING (PA+TCH+SP)

The endexine is a frequently misinterpreted layer of the pollen wall. Using standard TEM staining techniques (uranyl acetate and lead citrate), ektexine and endexine may differ in their electron opaqueness in that the endexine is higher in electron density than the ektexine, or the situation may be reversed [1]. But in many species, especially when the endexine is thin and less compact or discontinuous, the differentiation of the two layers is insufficient.

The endexine can be differentiated from the ektexine and the intine by thiocarbohydrazide-silver proteinate (TCH+SP) staining in osmium-fixed material and periodic acid-thiocarbohydrazide-silver proteinate (PA+TCH+SP) staining in osmium-free material. The endexine stains electron dense after TCH+SP staining [2], indicating lipidic compounds, and electron translucent after PA+TCH+SP staining [3], excluding this layer as part of the intine, as it is well known that intine reacts positively for polysaccharides.

A morphological characteristic of the endexine is its increasing thickness close to the aperture.

PA+TCH+SP staining (localization of neutral polysaccharides):
Sections (80–100 nm) from osmium-free material are placed on gold grids and treated with 1 % PA for 45 min, 0.2 % TCH for 8–15 h, and 1 % SP for 30 min (THIÉRY 1967).

TCH+SP-staining (detection of unsaturated lipids):
Sections (80–100 nm) from osmified material are placed on gold grids and treated with 0.2 % TCH for 8–15 h and 1 % SP for 30 min (ROWLEY and DAHL 1977, WEBER 1992).

PA+TCH+SP (short)

MODIFIED THIÉRY TEST

The modified (short) Thiéry test (WEBER and FROSCH 1995) is especially effective after fixation of specimens with osmium and potassium ferrocyanide and is a good method for general enhancement of contrast in the cytoplasm and the pollen wall.

Sections 80–100 nm thick are collected on gold grids and stained with 1 % periodic acid (PA) for 10 min, 0.2 % thiocarbohydrazide (TCH) for 15 min, and 1 % silver proteinate (SP) for 10 min (all steps at room temperature). After the staining steps with PA and SP, sections are intensively washed in distilled water; after staining with TCH, washes are in 3 % acetic acid followed by distilled water.

Acetocarmine Staining for Light Microscopy

For the detection of the generative nucleus and the sperm nuclei, fresh pollen grains are put into a drop of acetocarmine, warmed for a few seconds and observed with the LM (GERLACH 1969).

Staining methods

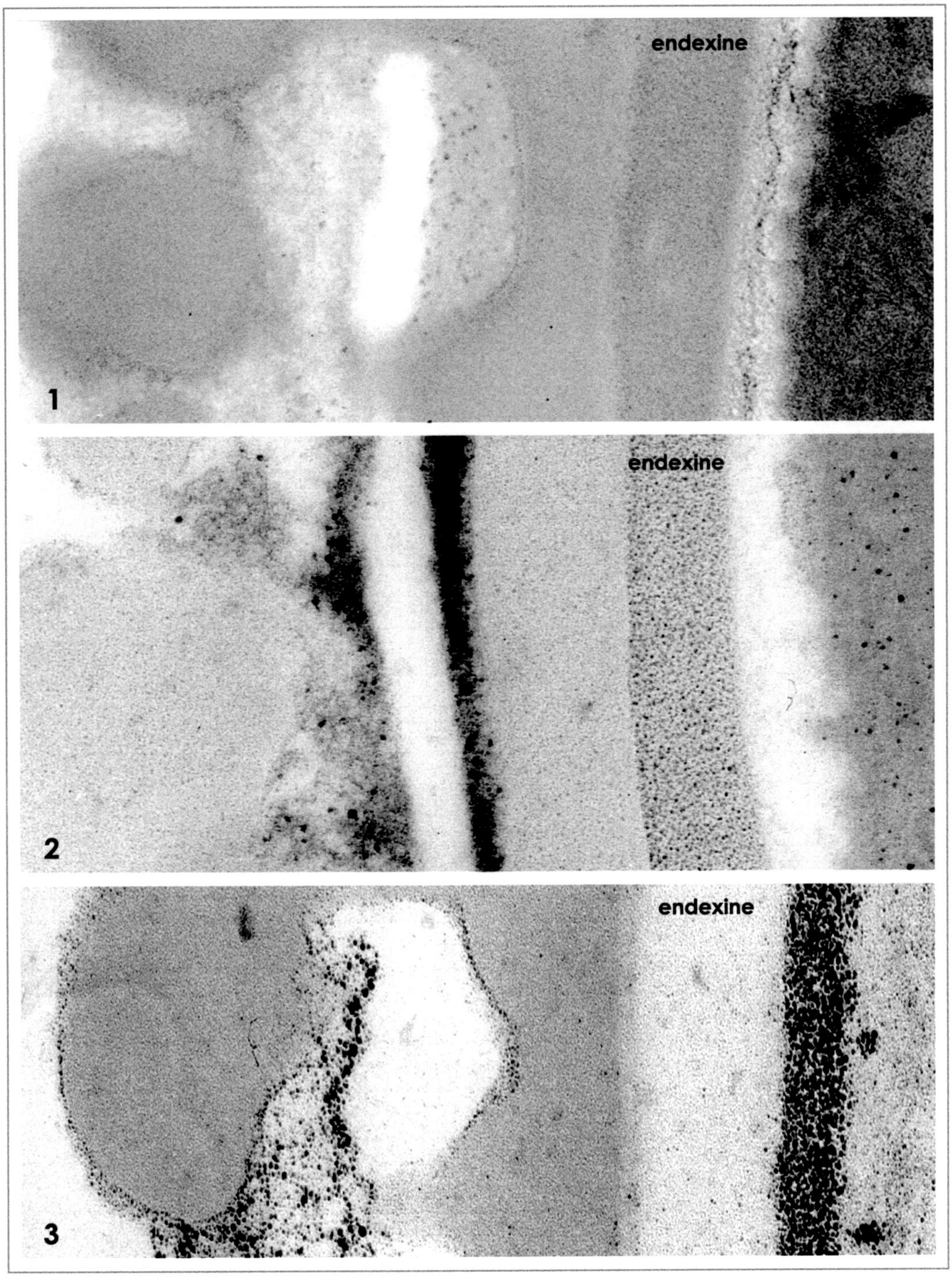

U+Pb

TCH+SP

PA+TCH+SP

How to Describe a Pollen Grain

The minimal features for describing a pollen grain are:

size, shape, aperture condition, ornamentation, and the method, how the pollen grain was prepared for LM and EM, respectively.

LM- and SEM-diagnosis may be different from each other, due to the methods and techniques used.

LM-diagnosis

plant name: *Centaurea jacea* (Asteraceae)
method: pollen grains acetolyzed
size: medium (longest axis about 40 µm)
shape (determined by equatorial and polar view)**:** prolate
aperture: tricolporate
ornamentation: echinate
peculiarities: remarkable endoaperture

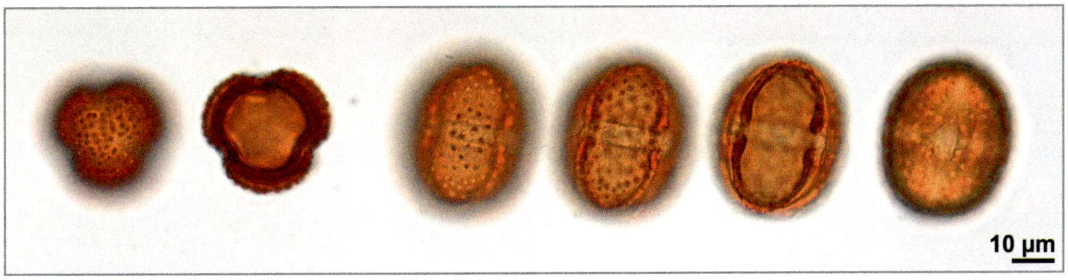

polar view equatorial view

SEM-diagnosis

method: pollen grains critical-point-dried (hydrated condition) and air-dried, respectively
size: medium (longest axis about 40 µm)
shape: spheroidal (in hydrated condition); prolate, lobate (in dry condition)
aperture: tricolporate
ornamentation: echinate, perforate
peculiarities: —

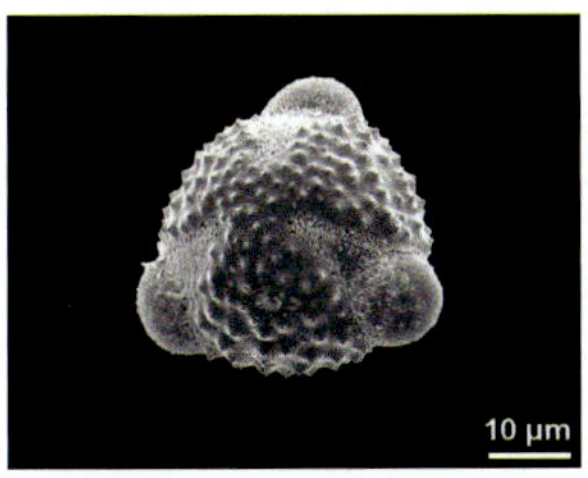

polar view

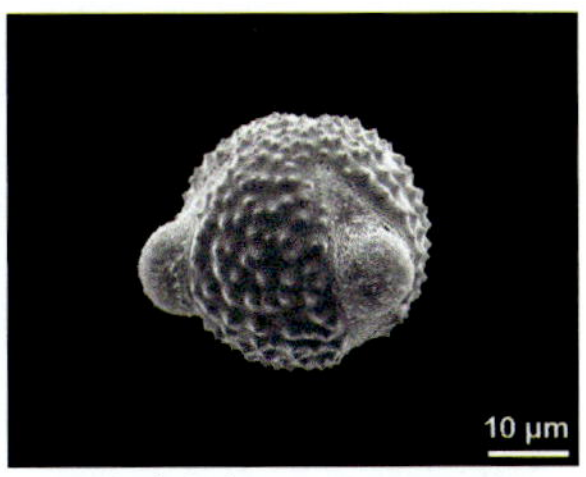

equatorial view

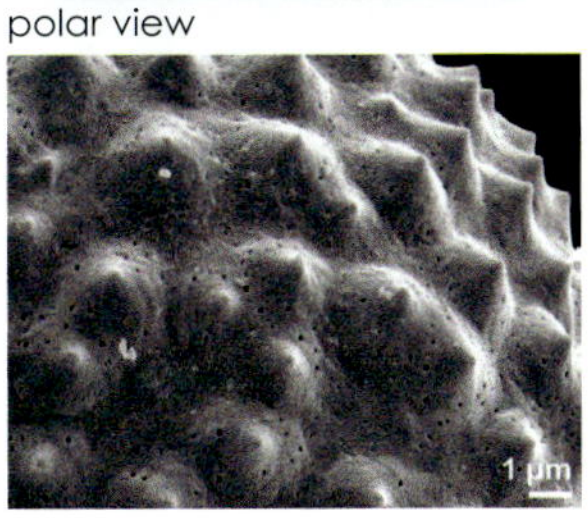

detail of exine ornamentation

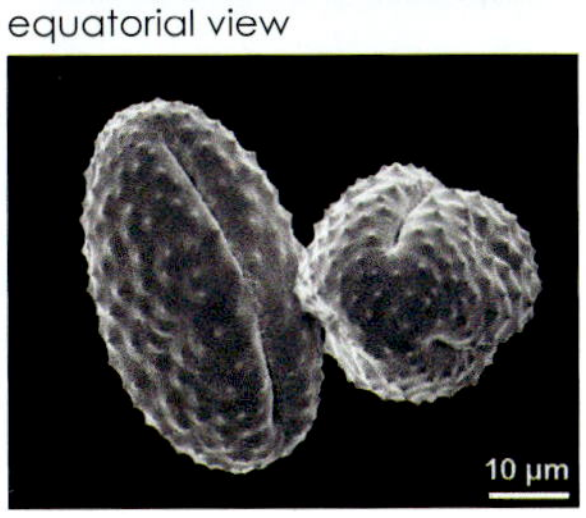

dry pollen

ILLUSTRATED GLOSSARY

Fritzsche über den Pollen.

CONTENT

LM SEM TEM mo ana fnc

monad: dispersal unit consisting of a single pollen grain.

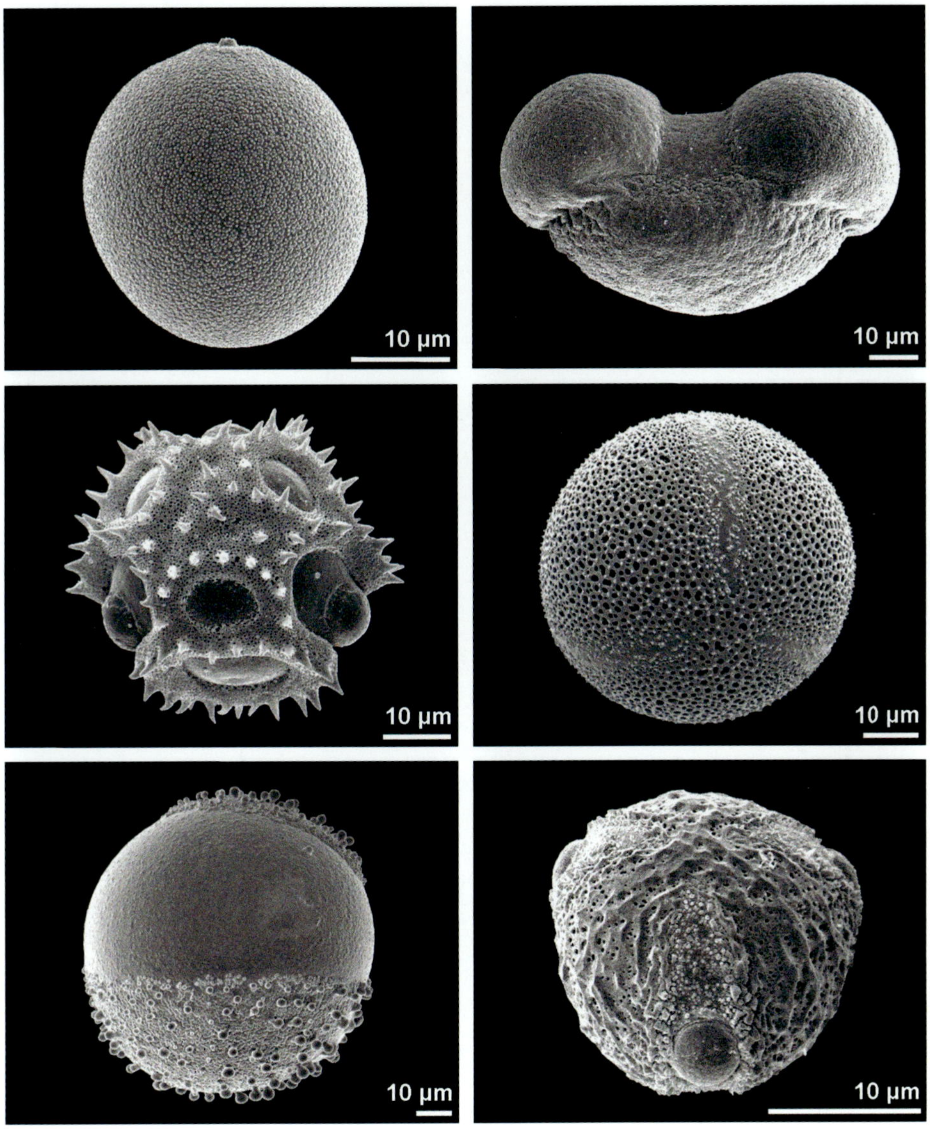

Dactylis glomerata
Poaceae
ulcerate, equatorial view

Pinus strobus
Pinaceae
bisaccate, equatorial view

Tragopogon orientalis
Asteraceae
tricolporate, oblique polar view

Leuchtenbergia principis
Cactaceae
tricolpate, oblique polar view

Iris pumila
Iridaceae
sulcate, oblique distal polar view

Exacum affine
Gentianaceae
tricolporate, oblique polar view

LM SEM TEM mo ana fnc

dyad: dispersal unit of two pollen grains.

Polypleurum stylosum
Podostemaceae

Polypleurum stylosum
Podostemaceae
acetolyzed, pollen collapsed

Zeylanidium olivaceum
Podostemaceae

Zeylanidium subulatum
Podostemaceae

Thelethylax minutiflora
Podostemaceae

Thelethylax minutiflora
Podostemaceae
acetolyzed, pollen collapsed

LM SEM TEM mo ana fnc

pseudomonad: dispersal unit of a permanent tetrad with three rudimentary pollen grains.

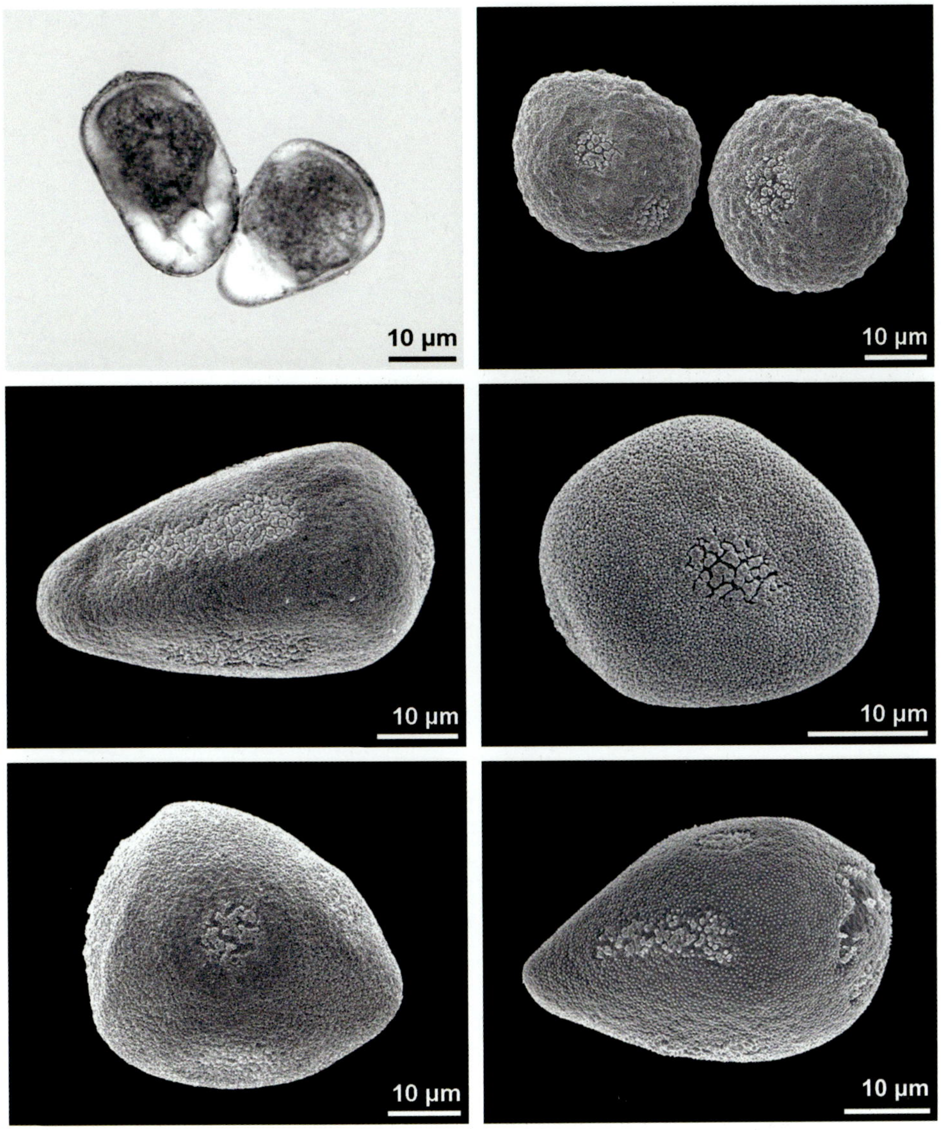

Carex sp.
Cyperaceae

Carex atrata
Cyperaceae

Schoenoplectus lacustris
Cyperaceae

Scirpus sylvaticus
Cyperaceae

Carex distans
Cyperaceae

Cyperus longus
Cyperaceae

LM SEM TEM mo ana fnc

tetrad tetrahedral: dispersal unit of four pollen grains in which the centers of the grains define a tetrahedron.

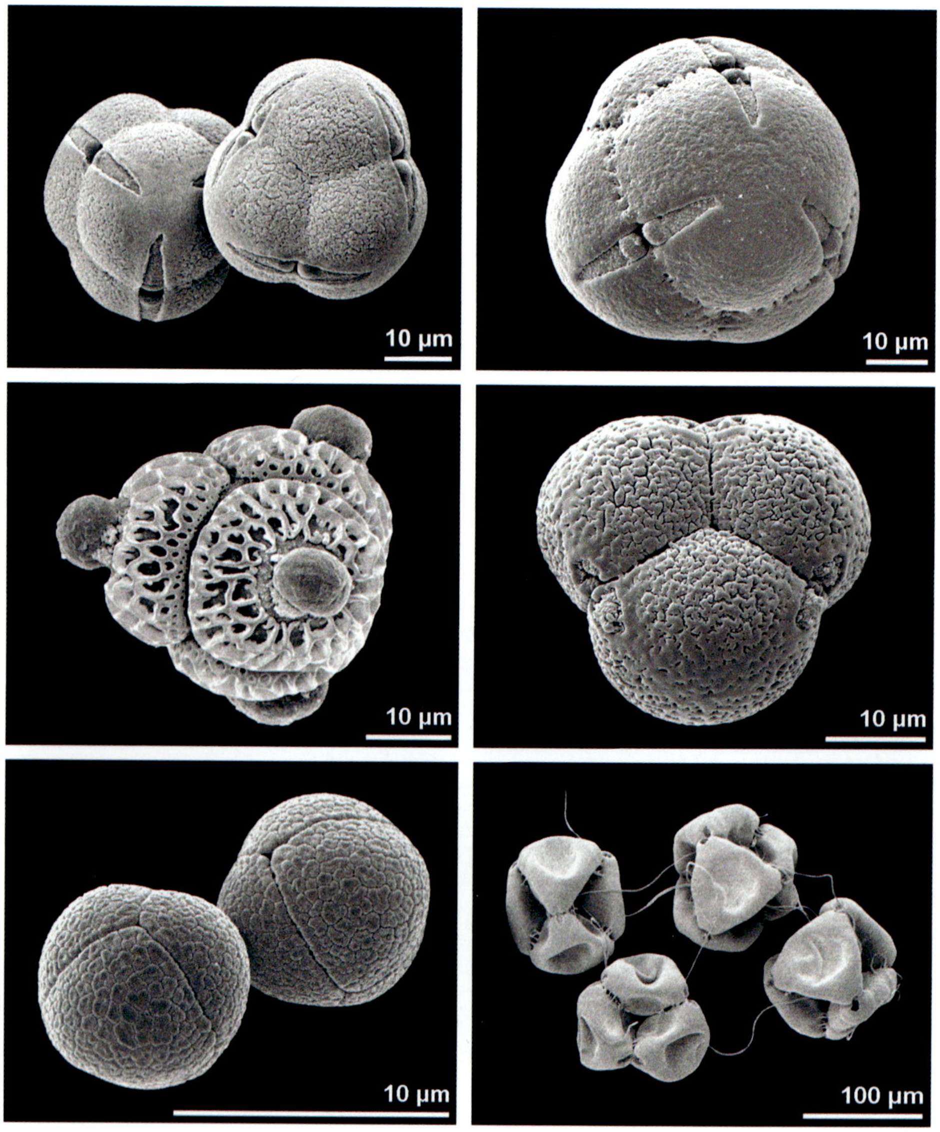

Erica herbacea
Ericaceae

Drimys granatensis
Winteraceae

Mimosa pudica
Mimosaceae

Arbutus unedo
Ericaceae

Moneses uniflora
Ericaceae

Epilobium parviflorum
Onagraceae
dry pollen, viscin threads

tetrad tetrahedral: dispersal unit of four pollen grains in which the centers of the grains define a tetrahedron.

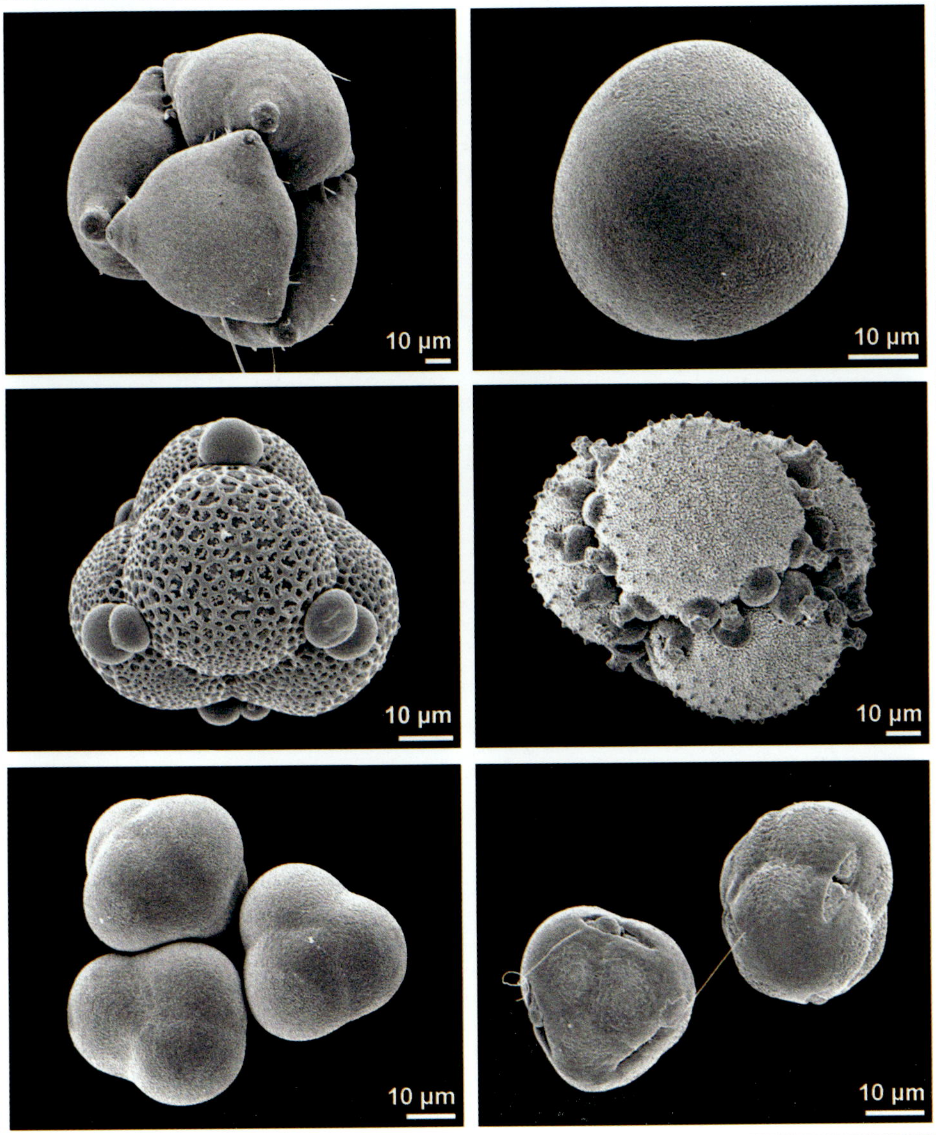

Epilobium montanum
Onagraceae
viscin threads

Oxyanthus subpunctatus
Rubiaceae

Juncus effusus
Juncaceae

Luzula campestris
Juncaceae

Dionaea muscipula
Droseraceae

Rhododendron hirsutum
Ericaceae
viscin threads

LM SEM TEM mo ana fnc

tetrad: dispersal unit of four pollen grains (spores).

Chlorospatha kolbii
Araceae
tetrads planar (tetrahedral and T-shaped)

Chlorospatha ceronii
Araceae
tetrad planar (left) and decussate (right)

Chlorospatha dodsonii
Araceae
tetrad planar

Chlorospatha dodsonii
Araceae
tetrad decussate

Catalpa bungei
Bignoniaceae
tetrad decussate

Epipactis helleborine
Orchidaceae
tetrads decussate

LM SEM TEM mo ana fnc

tetrad planar: dispersal unit of four pollen grains arranged in one plane; can be: tetragonal, T-shaped, linear.

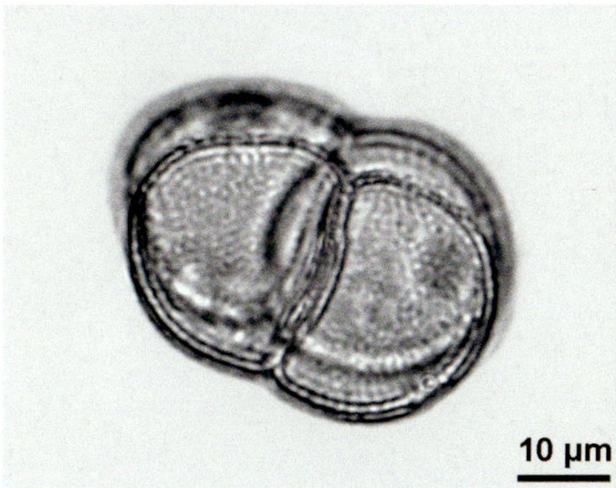

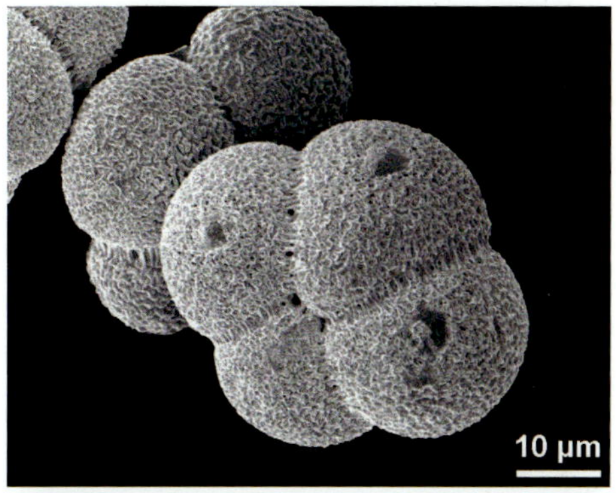

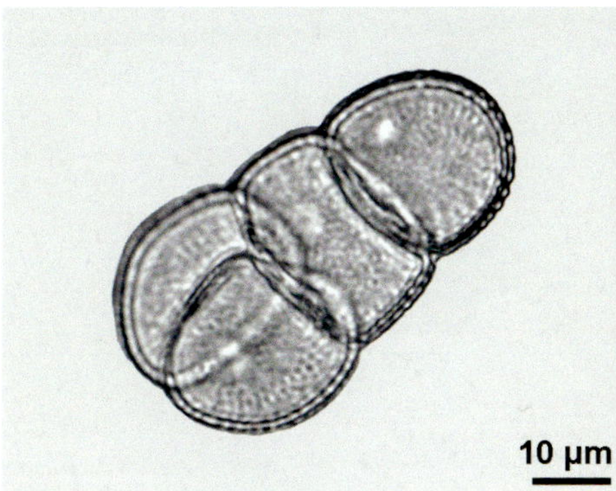

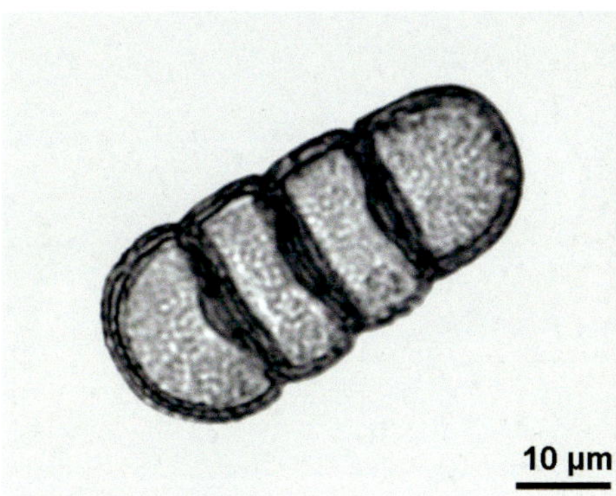

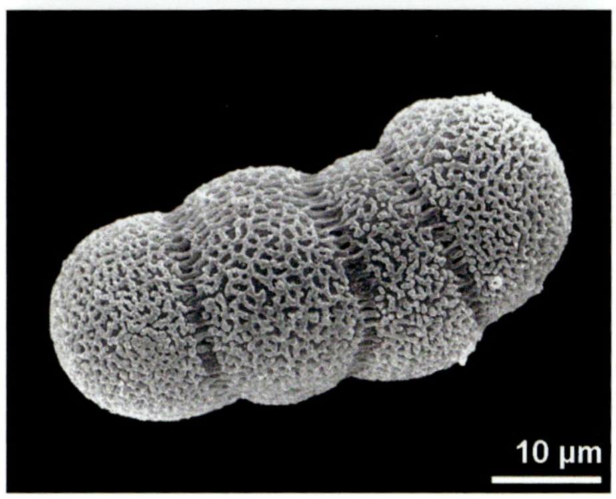

Typha latifolia
Typhaceae
tetrad tetragonal

Typha latifolia
Typhaceae
tetrad T-shaped

Typha latifolia
Typhaceae
tetrad linear

Typha latifolia
Typhaceae
tetrad tetragonal

Typha latifolia
Typhaceae
tetrad T-shaped

Typha latifolia
Typhaceae
tetrad linear

LM SEM TEM mo ana fnc

massula: dispersal unit of more than four pollen grains and fewer than the locular content.

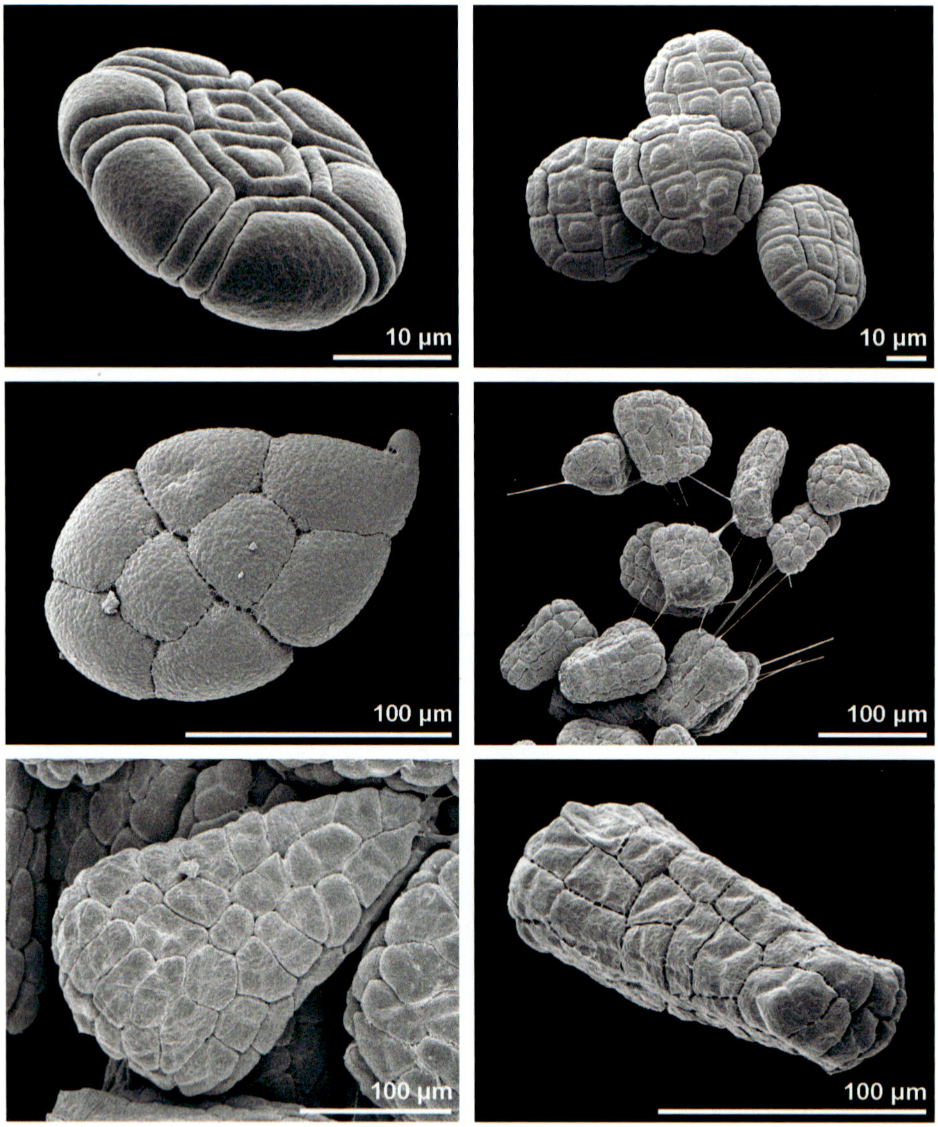

Acacia myrtifolia
Mimosaceae

Acacia binerva
Mimosaceae

Calliandra emarginata
Mimosaceae

Nigritella rhellicani
Orchidaceae
massulae connected by elastoviscin threads

Anteriorchis coriophora
Orchidaceae

Platanthera bifolia
Orchidaceae

LM SEM TEM mo ana fnc

pollinium: dispersal unit of a more or less interconnected loculiform pollen mass.

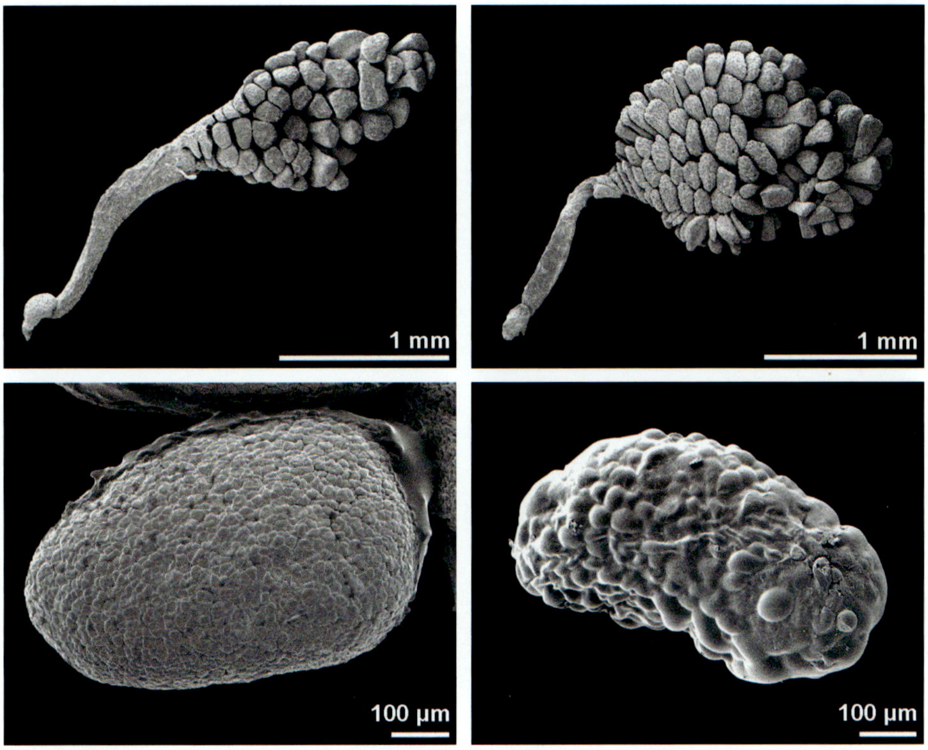

pollinarium: dispersal unit of pollinium (or pollinia) and a single interconnecting sterile appendage.

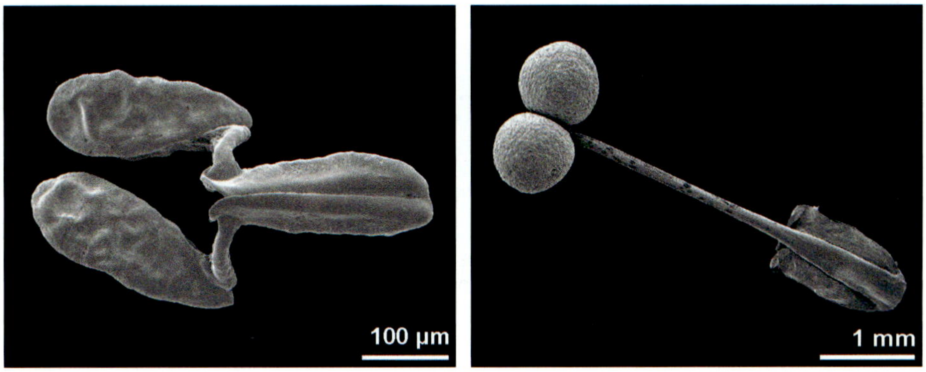

Ophrys sphegodes
Orchidaceae

Anteriorchis coriophora
Orchidaceae

Polystachya sp.
Orchidaceae

Stephanotis floribunda
Asclepiadaceae

Vincetoxicum hirundinaria
Asclepiadaceae

Aerides multiflora
Orchidaceae

LM SEM TEM mo ana fnc

pollen class: artificial grouping of pollen grains that share a single distinctive character.
pollen class: clypeate

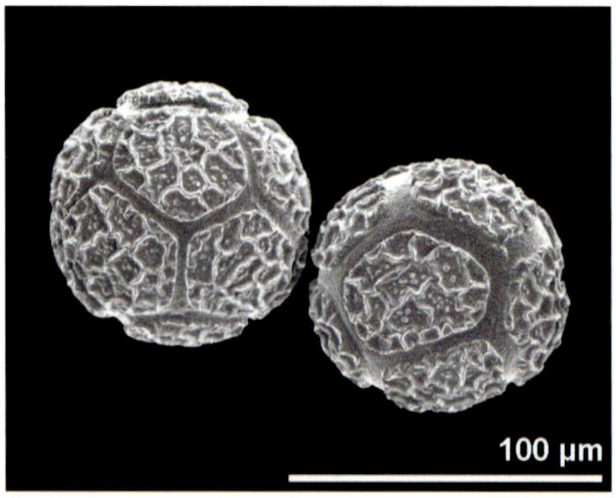

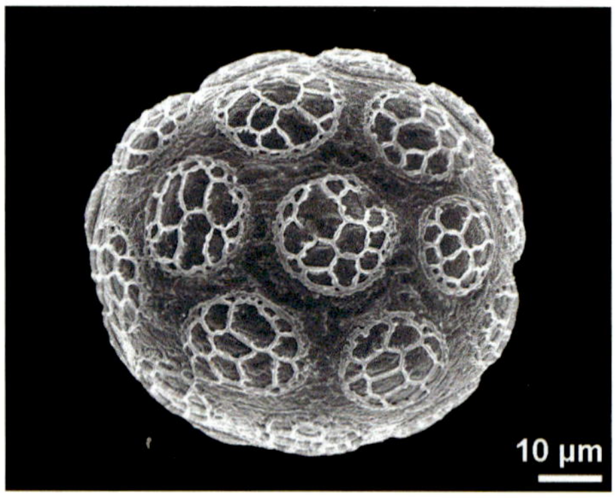

pollen class: colpate

pollen class: colporate

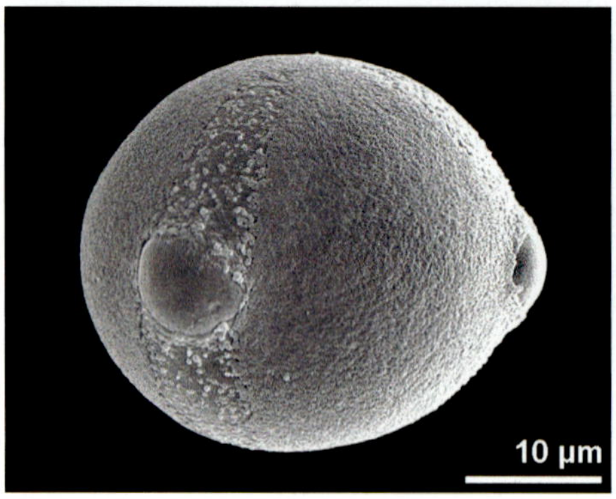

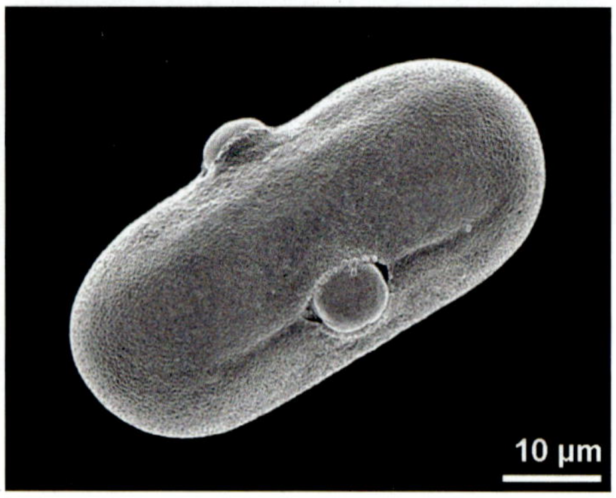

Iris bucharica Iridaceae	*Ibicella lutea* Martyniaceae
Bunias orientalis Brassicaceae	*Corylopsis glabrescens* Hamamelidaceae
Viola alba Violaceae	*Orlaya grandiflora* Apiaceae

pollen class: dyad

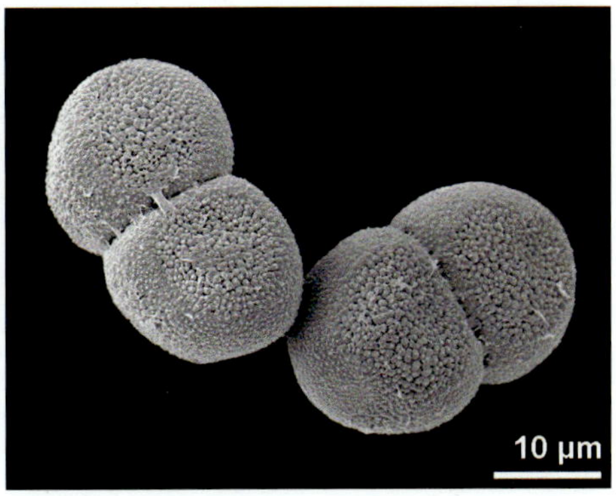

pollen class: inaperturate

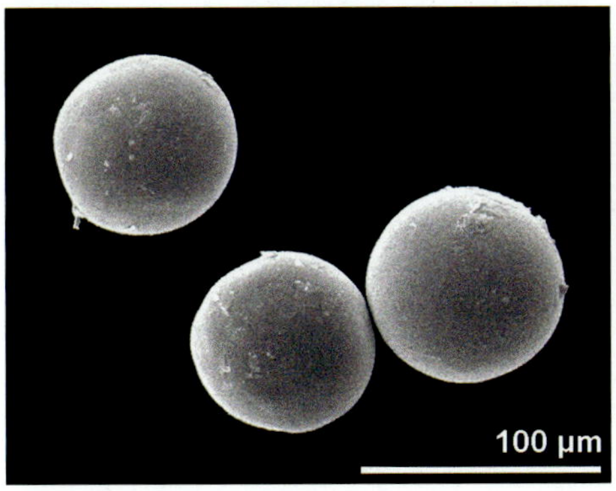

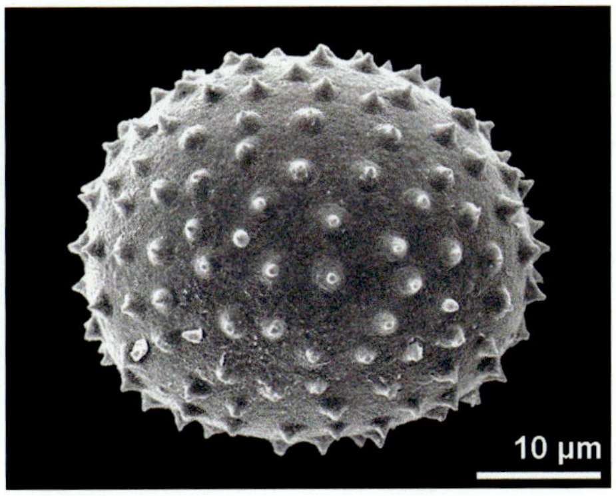

pollen class: lophate

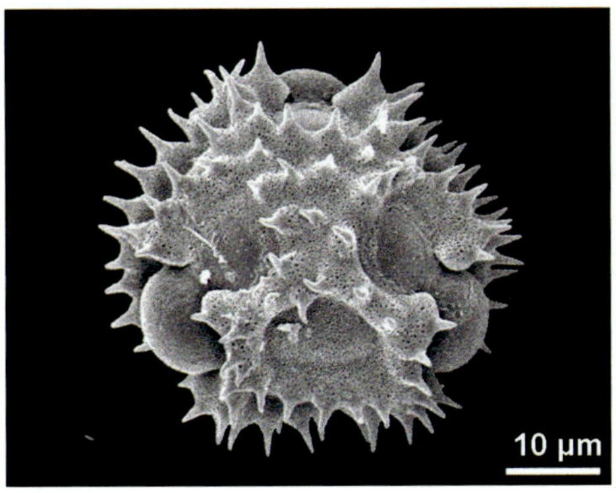

Zeylanidium subulatum
Podostemaceae

Polypleurum munnarense
Podostemaceae

Hedychium gardnerianum
Zingiberaceae

Sauromatum venosum
Araceae

Prenanthes purpurea
Asteraceae

Gazania sp.
Asteraceae

LM SEM TEM mo ana fnc

pollen class: plicate

pollen class: polyad

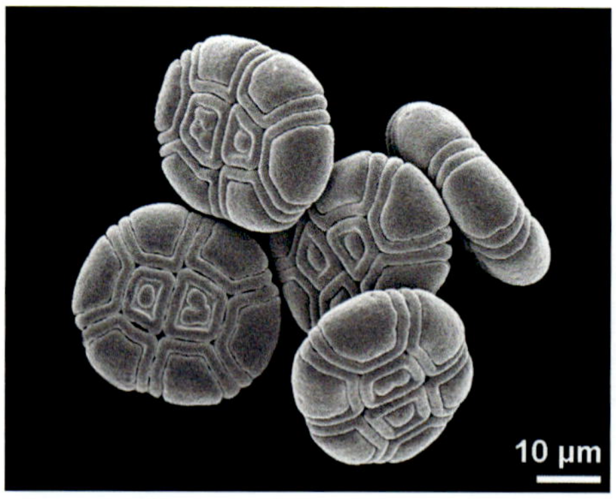

pollen class: porate

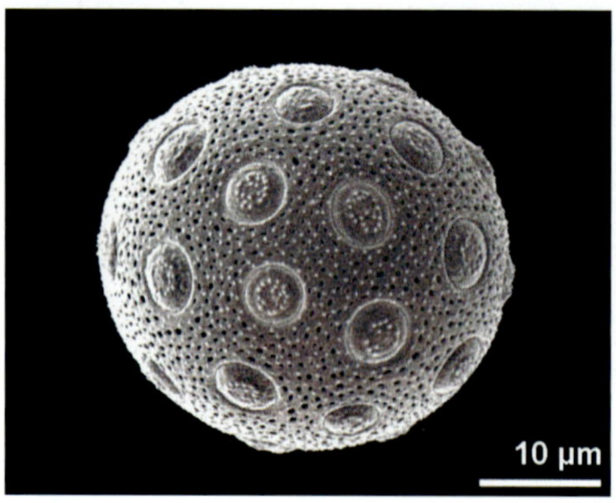

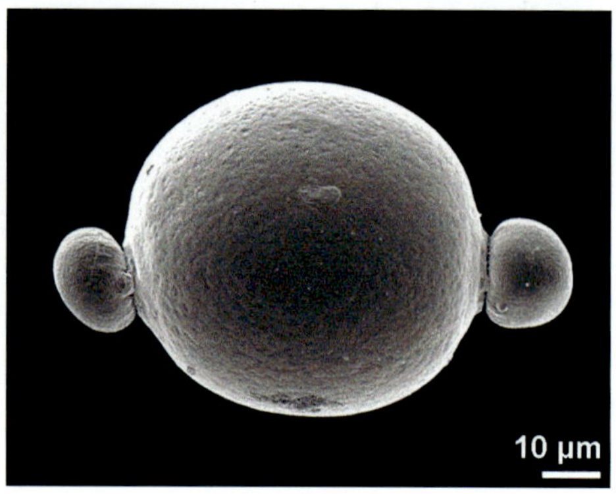

Ephedra distachya Ephedraceae	*Welwitschia mirabilis* Welwitschiaceae
Acacia myrtifolia Mimosaceae	*Himantoglossum adriaticum* Orchidaceae
Lychnis flos-cuculi Caryophyllaceae	*Pachypodium saundersii* Apocynaceae

LM SEM TEM mo ana fnc

pollen class: saccate

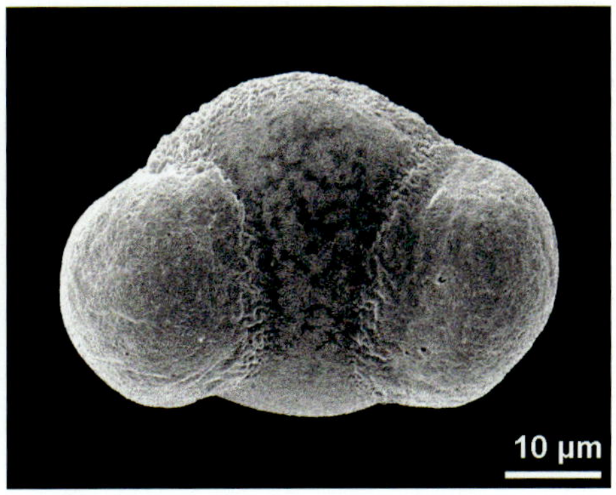

pollen class: spiraperturate

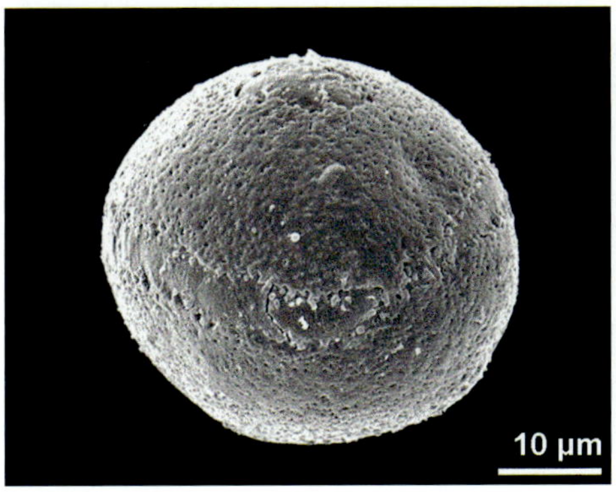

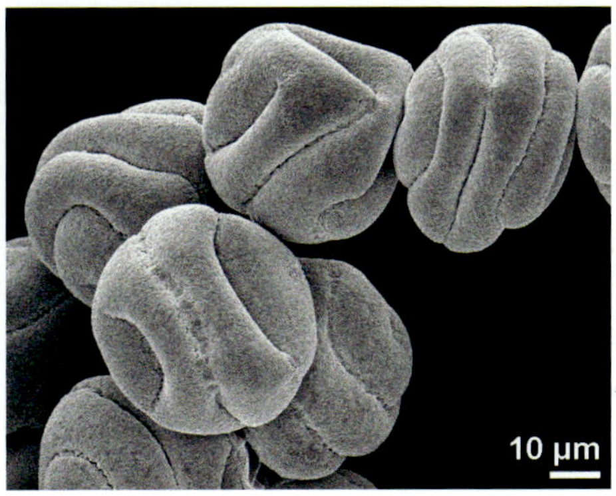

pollen class: sulcate

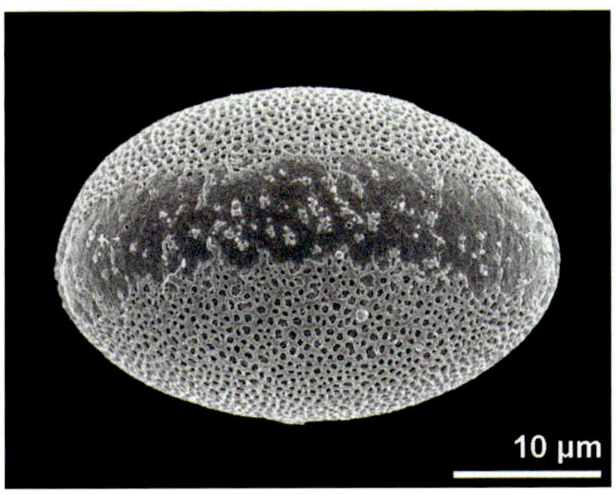

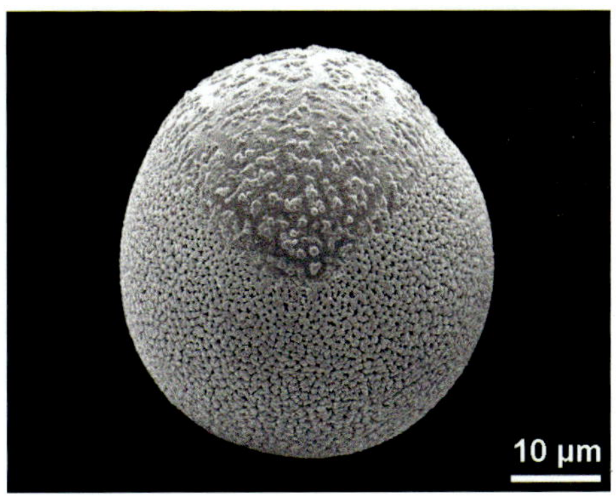

Abies cephalonica
Pinaceae
dry pollen

Pinus mugo
Pinaceae

Berberis thunbergii
Berberidaceae

Thunbergia alata
Acanthaceae
dry pollen

Sandersonia aurantiaca
Colchicaceae

Gagea villosa
Liliaceae

LM SEM TEM mo ana fnc

pollen class: synaperturate

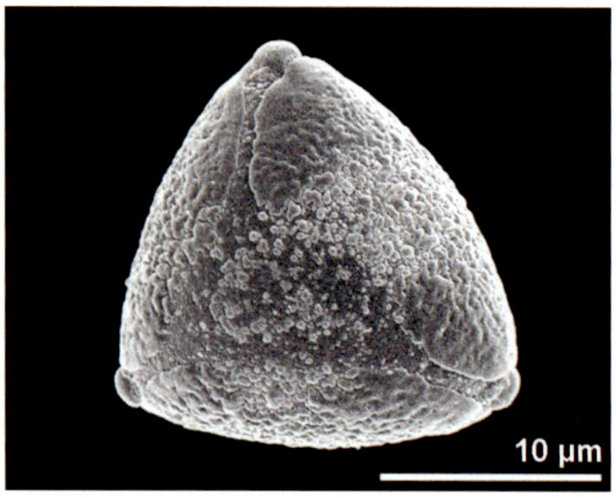

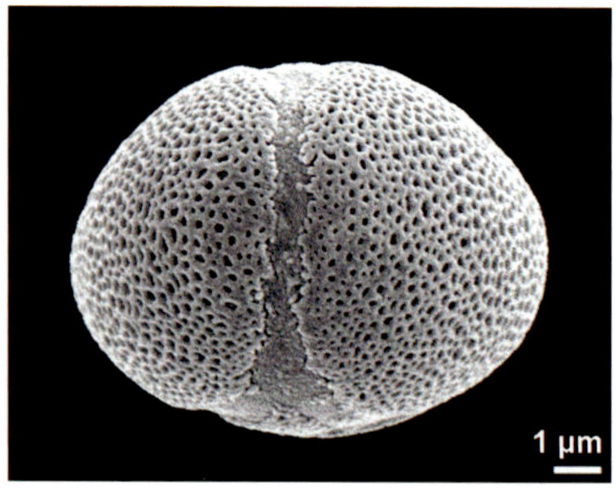

pollen class: tetrad

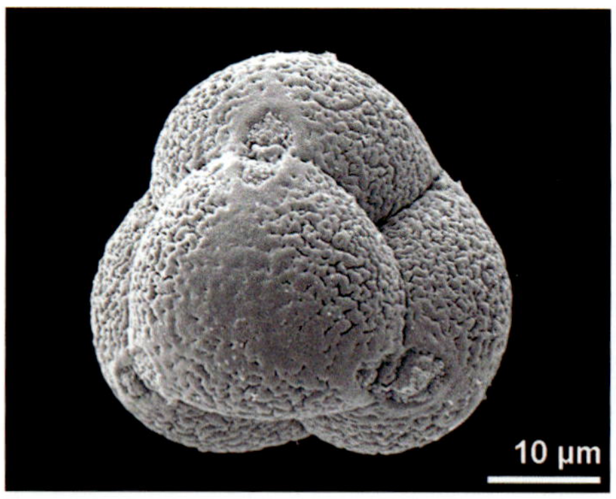

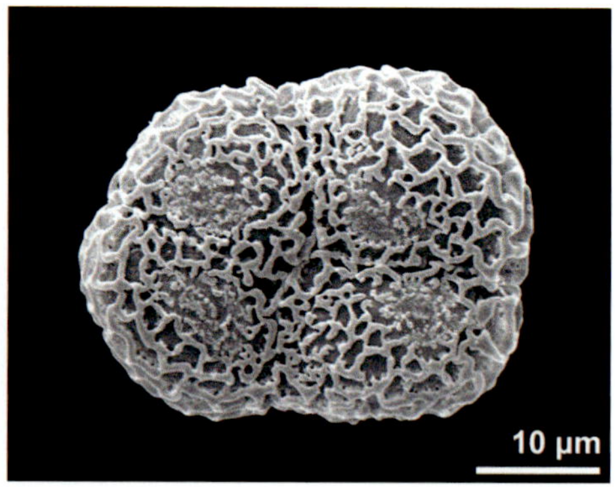

pollen class: ulcerate

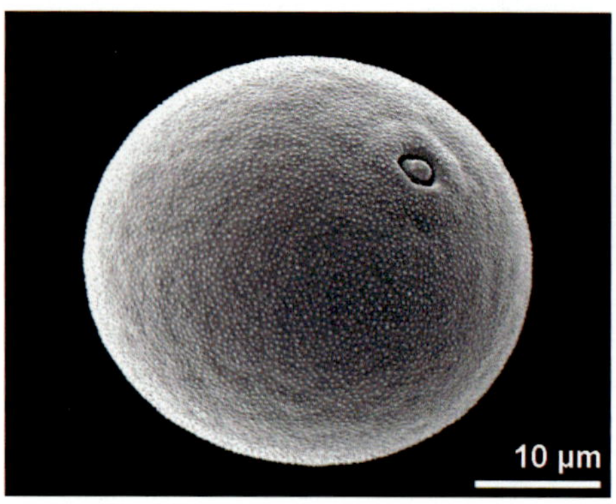

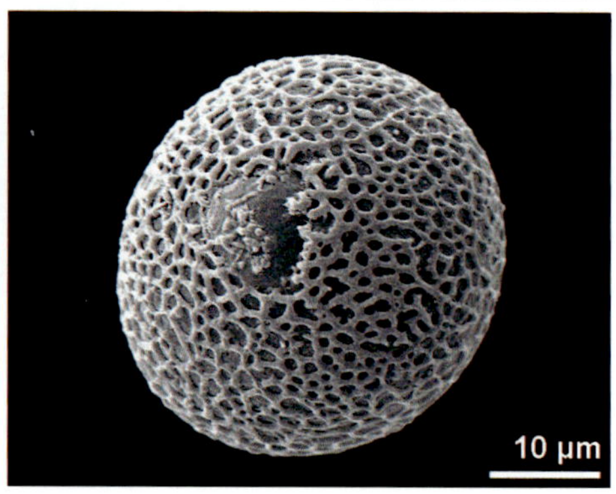

Acca sellowiana
Myrtaceae
polar view

Primula farinosa
Primulaceae
equatorial view

Moneses uniflora
Ericaceae

Listera ovata
Orchidaceae

Sesleria uliginosa
Poaceae

Typha minima
Typhaceae

LM SEM TEM mo ana fnc

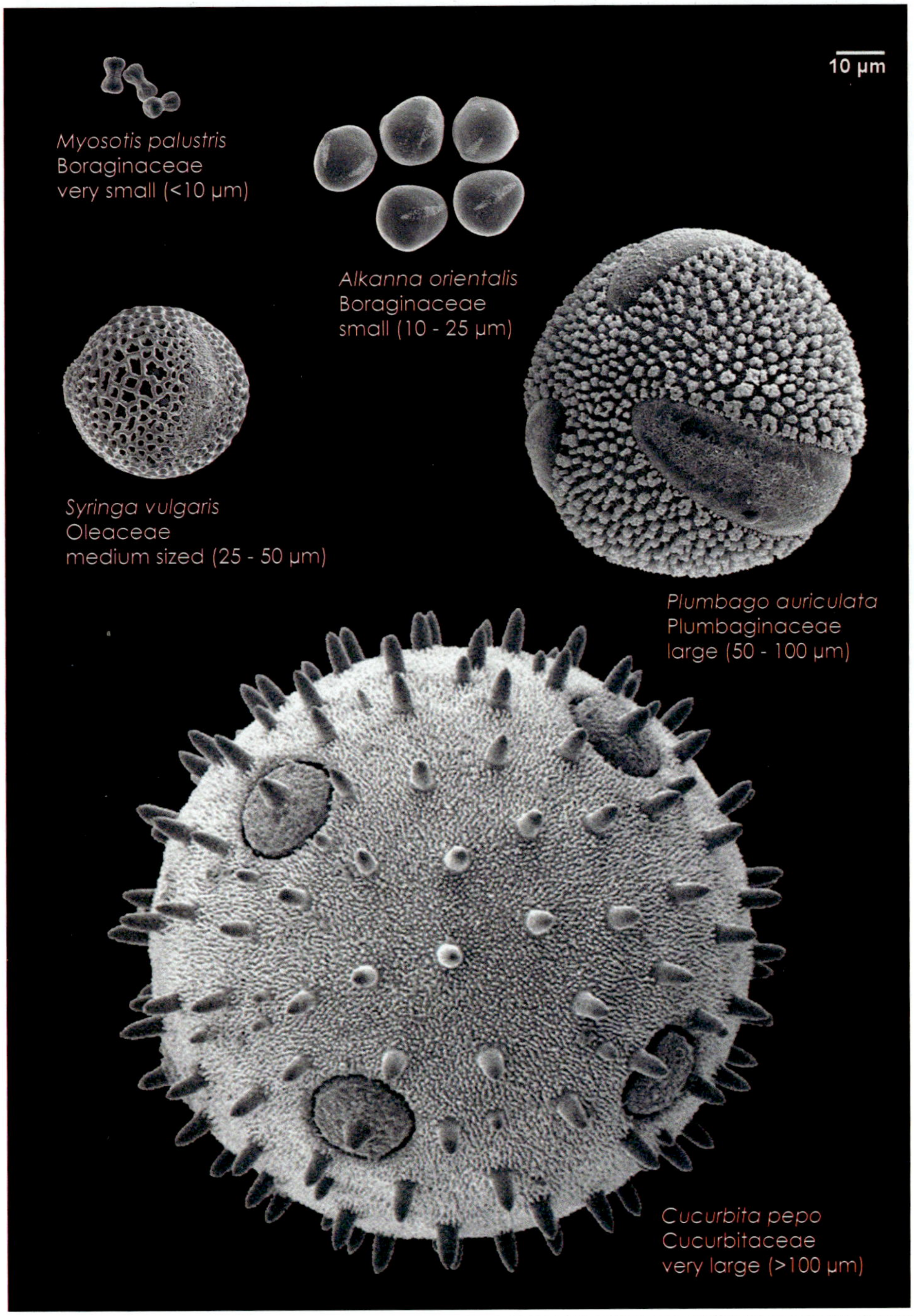

LM SEM TEM mo ana fnc

spheroidal

Lavatera thuringiaca
Malvaceae
pantoporate

Silene alba
Caryophyllaceae
pantoporate

Cirsium oleraceum
Asteraceae
tricolporate, equatorial view

Phlox paniculata
Polemoniaceae
pantoporate

Pinellia ternata
Araceae
inaperturate

Luffa cylindrica
Cucurbitaceae
tricolporate, equatorial view

LM SEM TEM mo ana fnc

spheroidal

Oxalis acetosella
Oxalidaceae
tricolpate, equatorial view

Adansonia gregorii
Bombacaceae
triporate

Stellaria holostea
Caryophyllaceae
pantoporate

Cistus creticus
Cistaceae
tricolporate, polar view

Ruellia macrantha
Acanthaceae
tricolporate, polar view

Sagittaria sagittifolia
Alismataceae
pantoaperturate

LM SEM TEM mo ana fnc

prolate: pollen grain with the polar axis longer than the equatorial diameter.

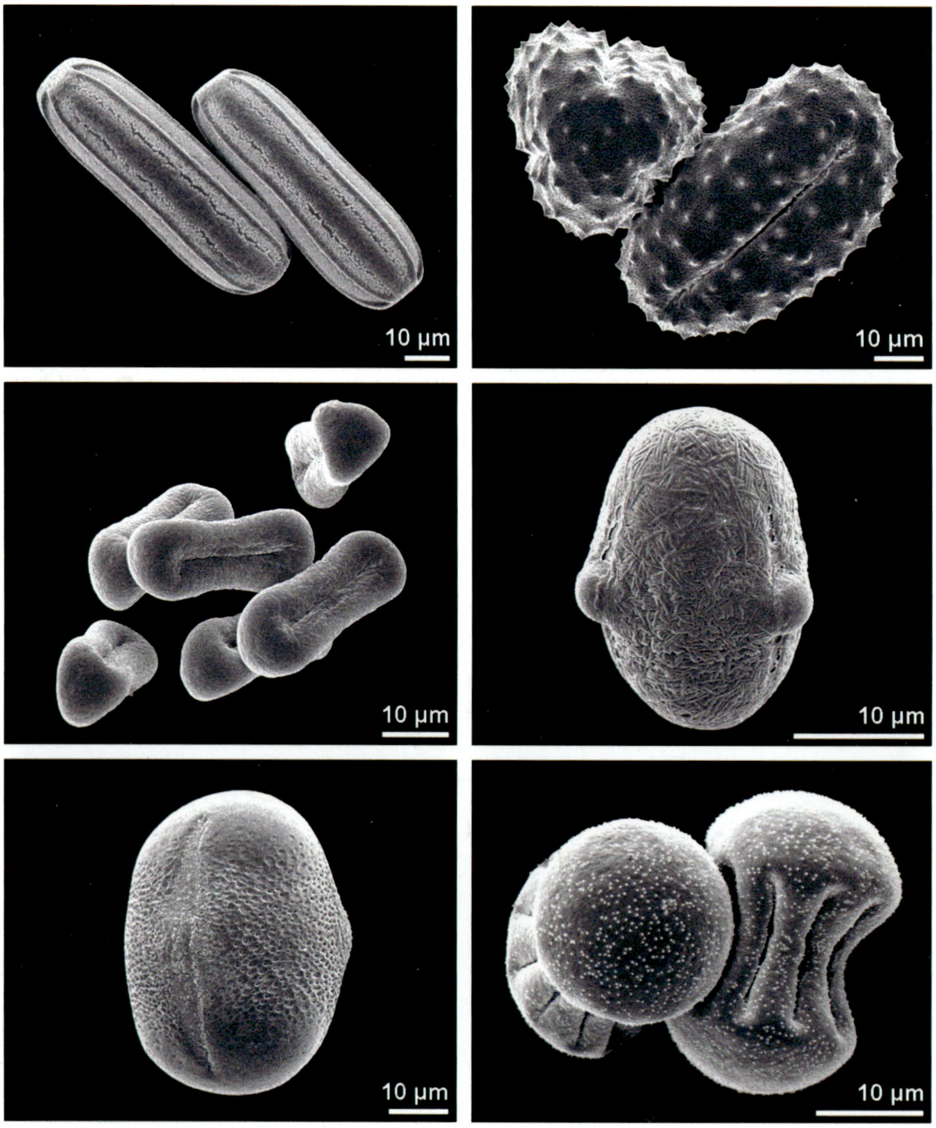

Crossandra flava
Acanthaceae
equatorial view

Jurinea mollis
Asteraceae
dry pollen

Torilis arvensis
Apiaceae
dry pollen

Peucedanum cervaria
Apiaceae
equatorial view

Astragalus onobrychis
Fabaceae
equatorial view

Symphytum officinale
Boraginaceae
dry pollen

prolate: pollen grain with the polar axis longer than the equatorial diameter.

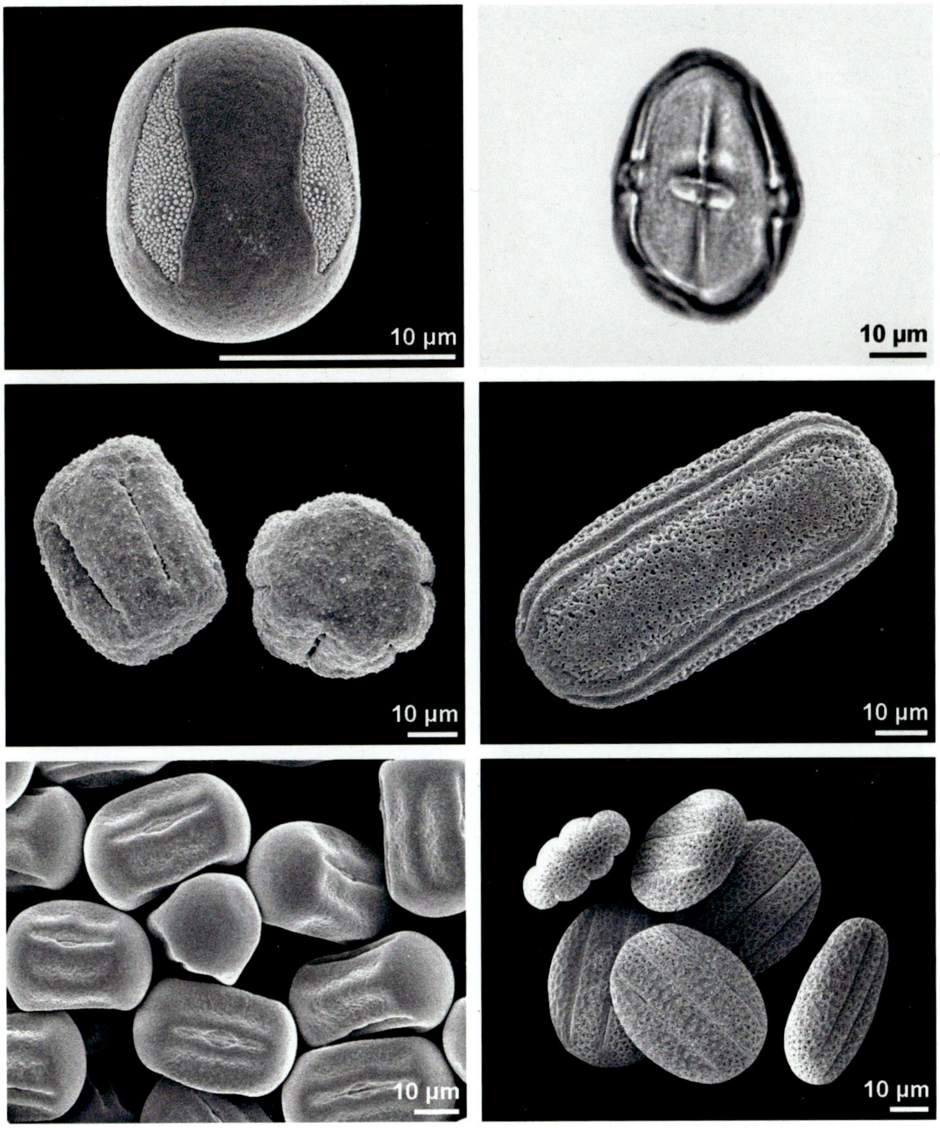

Buglossoides purpurocaerulea
Boraginaceae
equatorial view

Platycodon grandiflorum
Campanulaceae
dry pollen

Lathyrus tuberosus
Fabaceae
dry pollen

indet.
Sapotaceae, fossil
equatorial view

Stenandriopsis guineensis
Acanthaceae
equatorial view

Salvia sclarea
Lamiaceae
dry pollen

LM SEM TEM mo ana fnc

oblate: pollen grain with the polar axis shorter than the equatorial diameter.

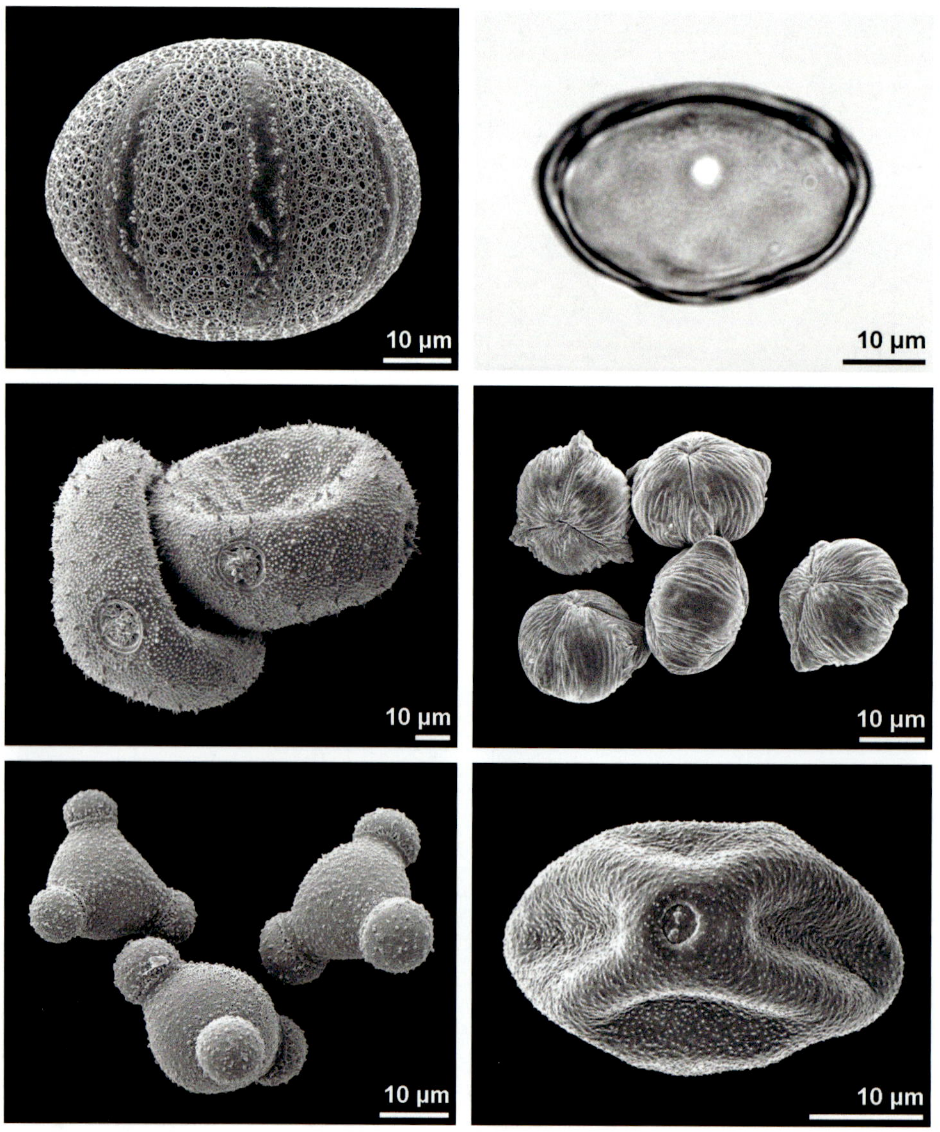

Salvia argentea
Lamiaceae
equatorial view

Carya sp.
Juglandaceae, fossil
equatorial view

Knautia drymeia
Dipsacaceae
dry pollen

Cuphea purpurea
Lythraceae
dry pollen

Hakea kippistiana
Proteaceae

Corylus avellana
Betulaceae
dry pollen, equatorial view

oblate: pollen grain with the polar axis shorter than the equatorial diameter.

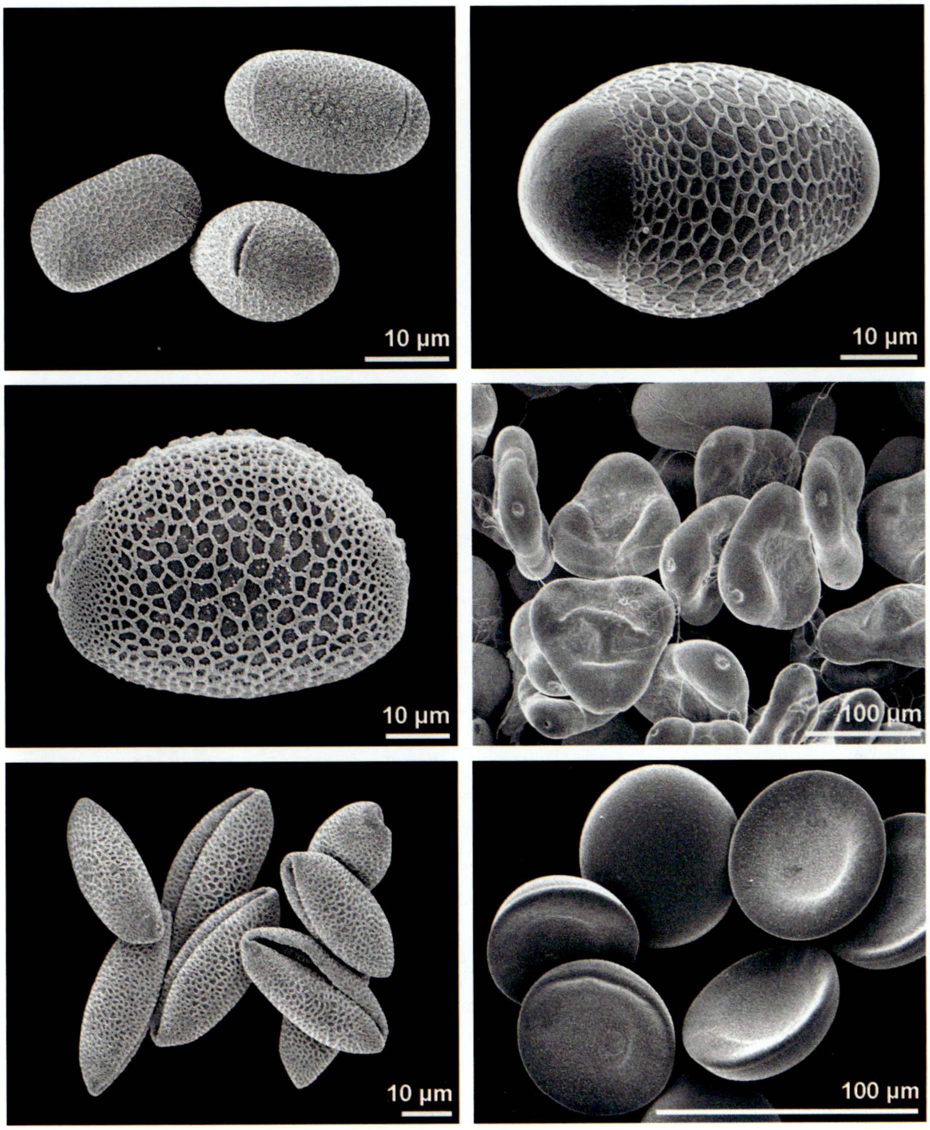

Impatiens glandulifera
Balsaminaceae

Vriesea neoglutinosa
Bromeliaceae
equatorial view

Veratrum album
Melanthiaceae
dry pollen

Aechmea caudata
Bromeliaceae

Godetia purpurea
Onagraceae
dry pollen

Heliconia sp.
Heliconiaceae
dry pollen

LM SEM TEM mo ana fnc

outline circular

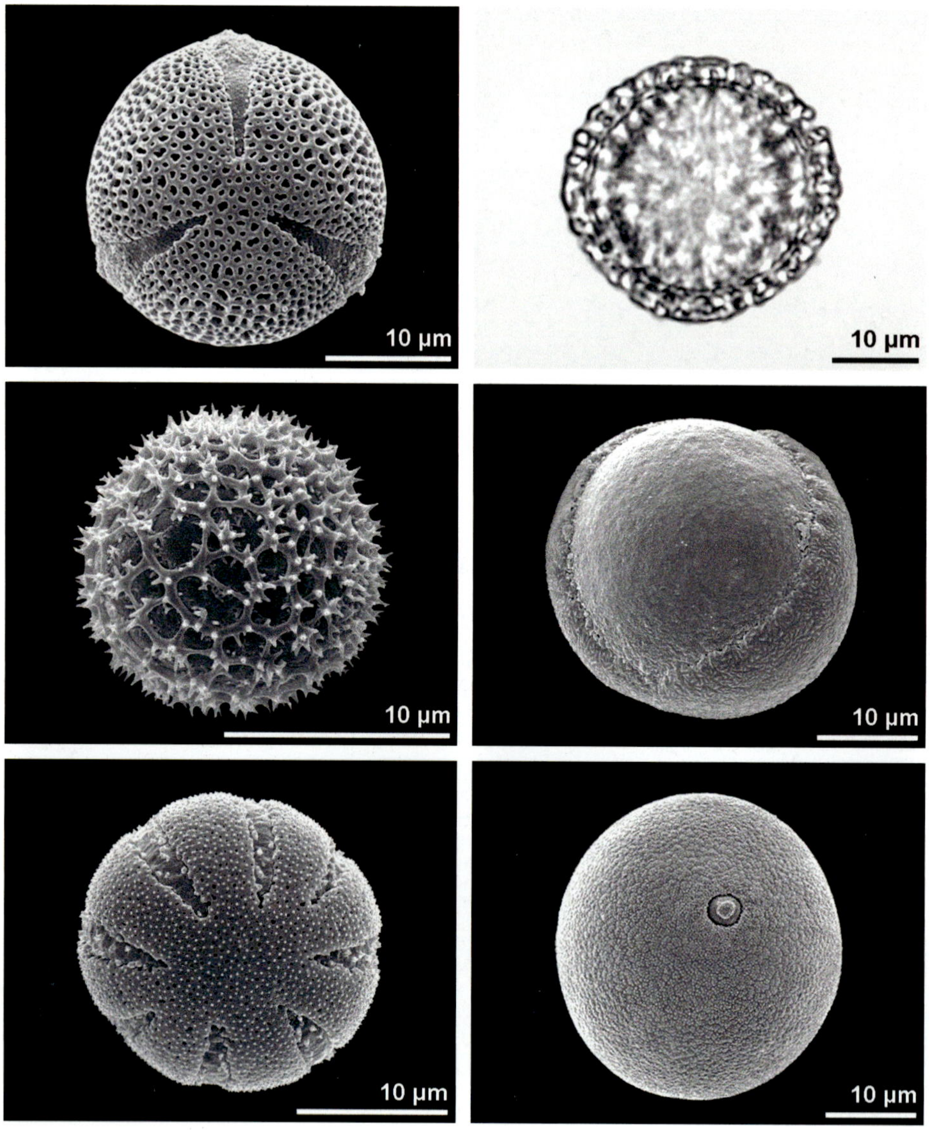

Fraxinus ornus
Oleaceae
polar view

Ligustrum sp.
Oleaceae, fossil
equatorial view

Anthurium trisulcatum
Araceae

Ginkgo biloba
Ginkgoaceae
oblique distal polar view

Galium lucidum
Rubiaceae
polar view

Phleum pratense
Poaceae
distal polar view

LM SEM TEM mo ana fnc

outline elliptic

Allium oleraceum
Alliaceae
distal polar view

Aechmea dealbata
Bromeliaceae

Impatiens parviflora
Balsaminaceae
polar view

Salvia coccinea
Lamiaceae
polar view

Galeopsis tetrahit
Lamiaceae
dry pollen

Physostegia virginiana
Lamiaceae
dry pollen

LM SEM TEM mo ana fnc

outline quadrangular

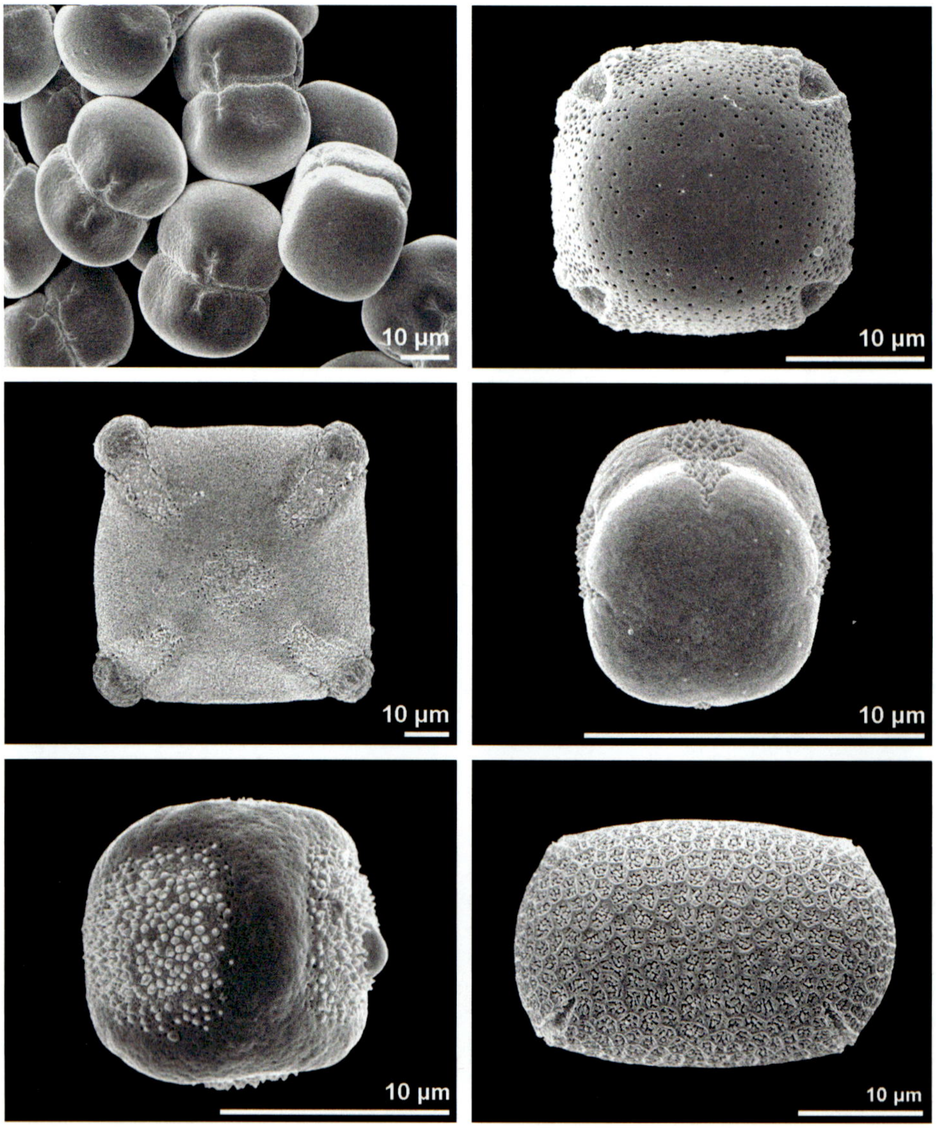

Anchusa officinalis
Boraginaceae
dry pollen

Viola tricolor
Violaceae
polar view

Herniaria glabra
Caryophyllaceae

Nonea pulla
Boraginaceae
polar view

Lithospermum officinale
Boraginaceae
polar view

Impatiens glandulifera
Balsaminaceae
polar view

LM SEM TEM mo ana fnc

outline polygonal

Stellaria holostea
Caryophyllaceae
dry pollen

Arbutus unedo
Ericaceae
tetrads, dry pollen

Talinum paniculatum
Portulacaceae
dry pollen

Viola arvensis
Violaceae
polar view

Opuntia basilaris
Cactaceae
dry pollen

Silene nutans
Caryophyllaceae
dry pollen

LM SEM TEM mo ana fnc

outline triangular

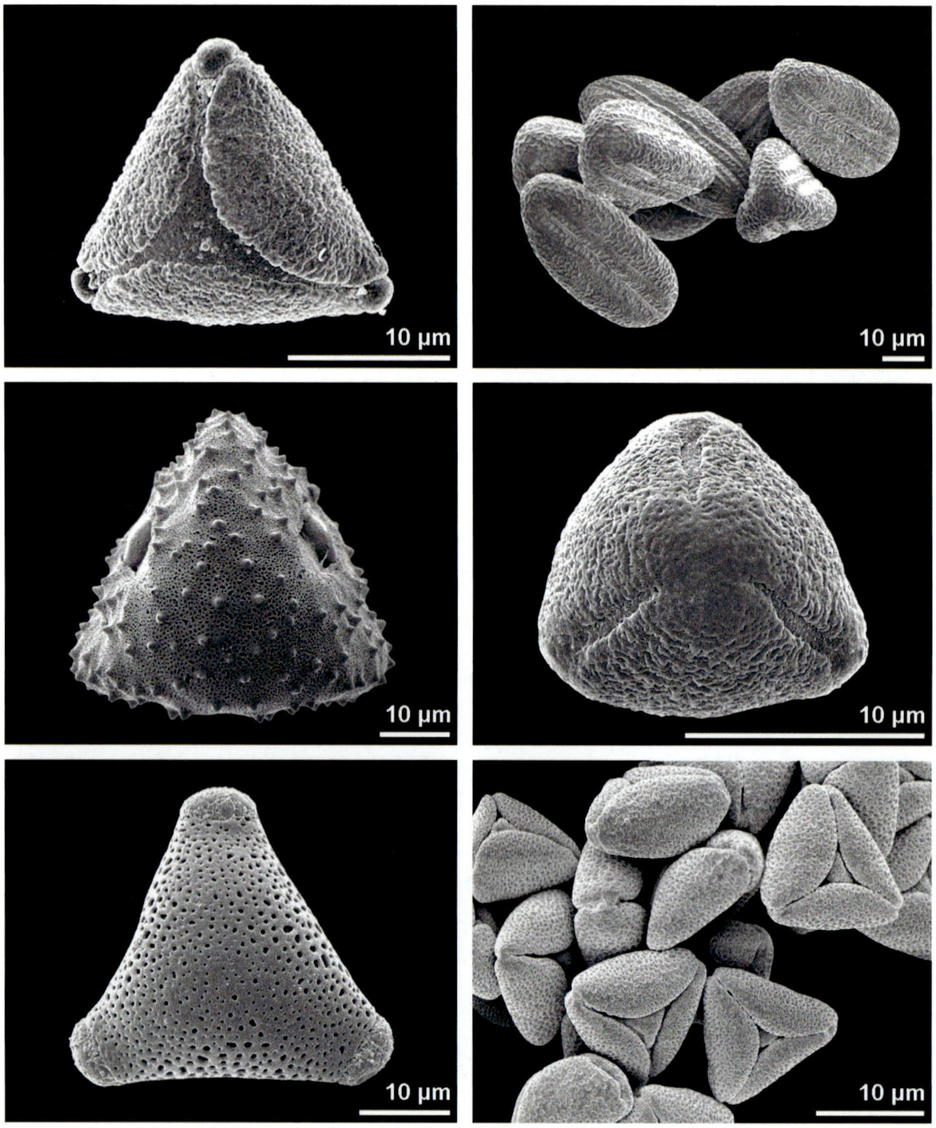

Callistemon coccineus
Myrtaceae
polar view

Hypoestes phyllostachya
Acanthaceae
dry pollen

Echinops ritro
Asteraceae
polar view

Bupleurum rotundifolium
Apiaceae
polar view

Paullinia tomentosa
Sapindaceae
polar view

Primula denticulata
Primulaceae
dry pollen

outline triangular

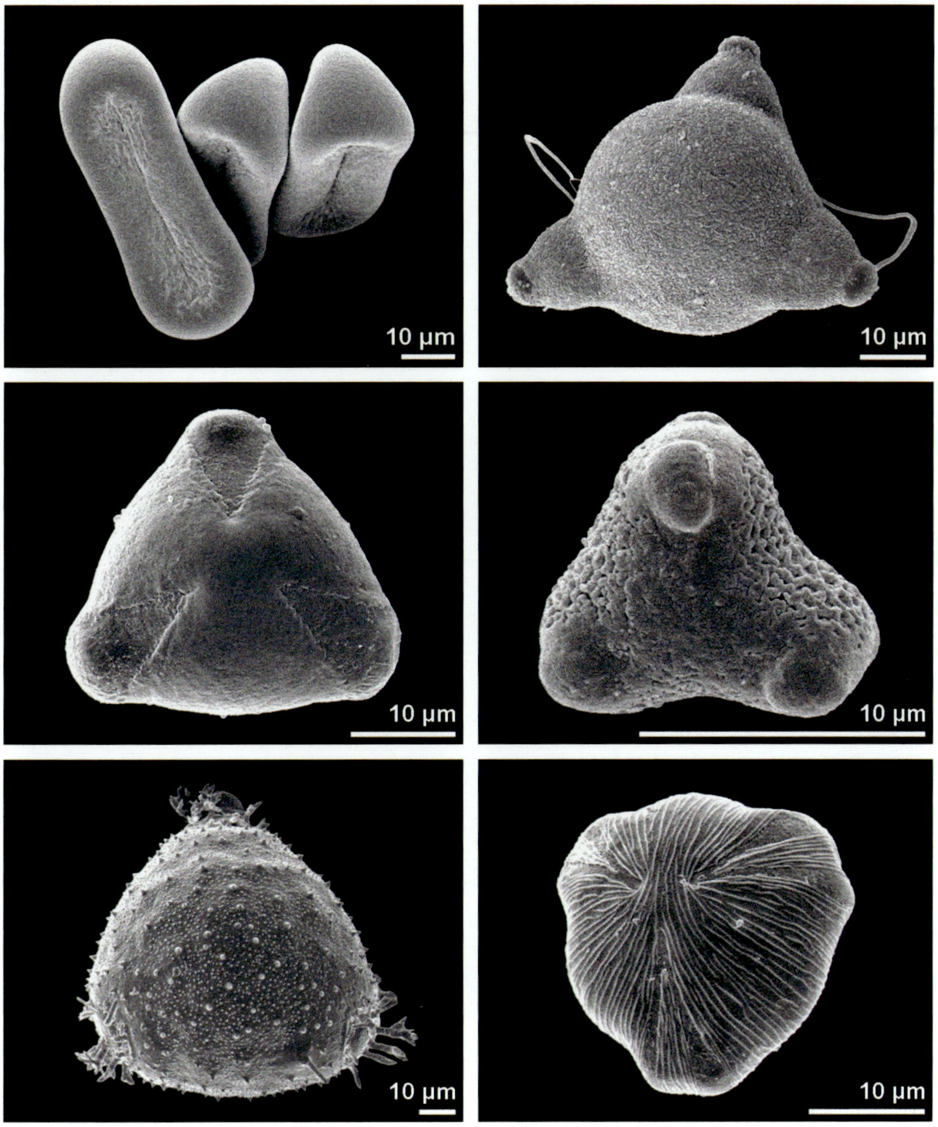

Orlaya grandiflora
Apiaceae
dry pollen

Jovibarba hirta
Crassulaceae
polar view

Dipsacus fullonum
Dipsacaceae
polar view

Circaea lutetiana
Onagraceae
polar view, viscin threads

Cunonia capensis
Cunoniaceae
polar view

Potentilla inclinata
Rosaceae
polar view

LM SEM TEM mo ana fnc

outline lobate

Acer pseudoplatanus
Sapindaceae
dry pollen

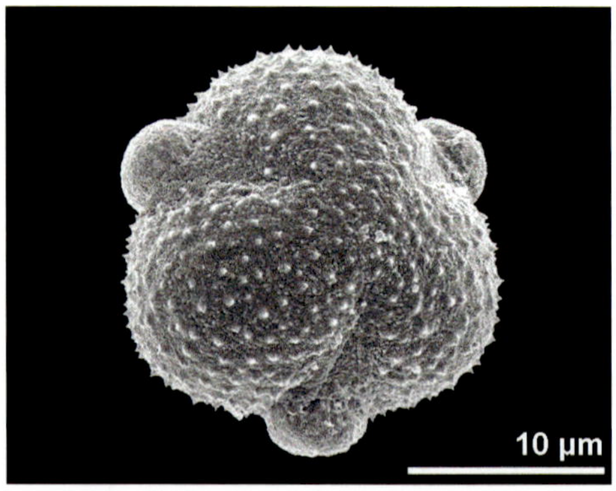

Artemisia pontica
Asteraceae
polar view

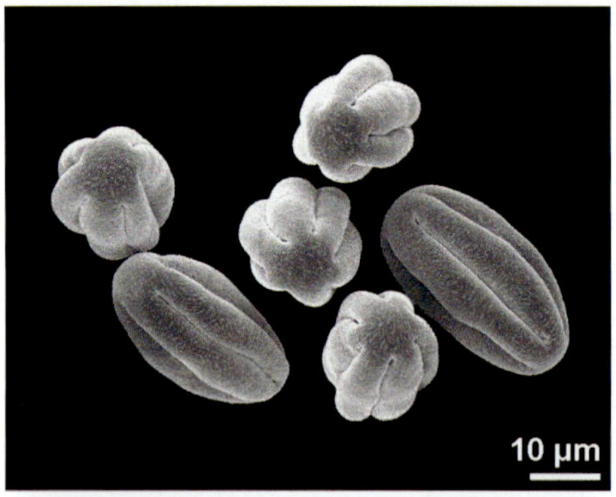

Sanguisorba officinalis
Rosaceae
dry pollen

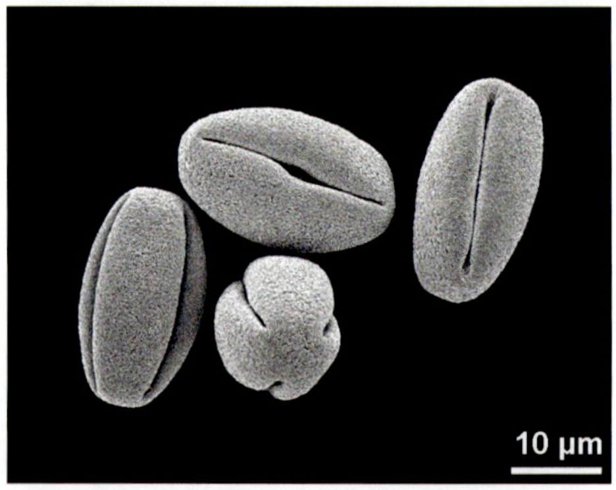

Orthilia secunda
Ericaceae
dry pollen

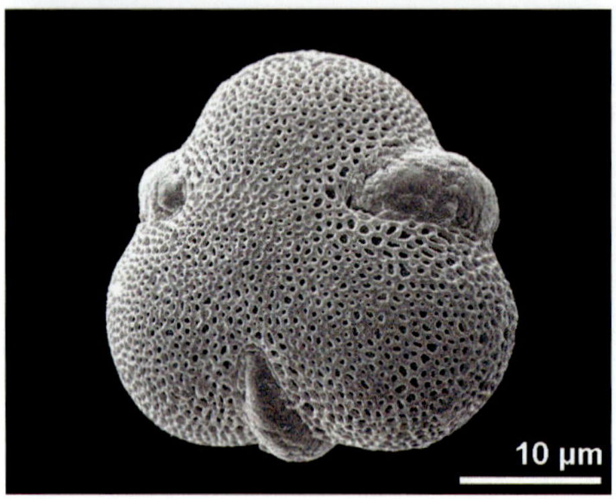

Gunnera chilensis
Gunneraceae
polar view

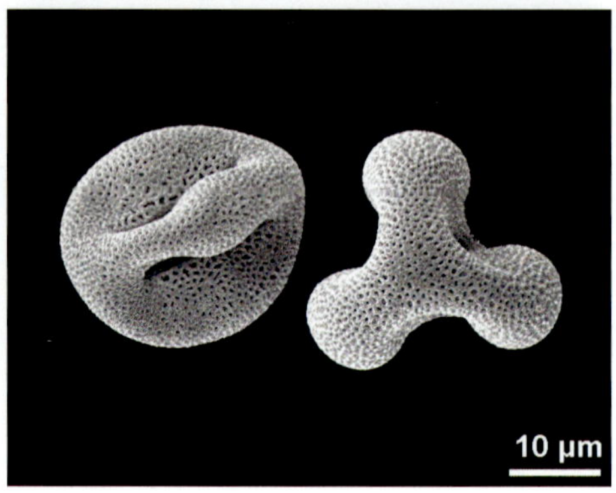

Gunnera chilensis
Gunneraceae
dry pollen, equatorial (left) and polar view (right)

LM SEM TEM mo ana fnc

outline lobate

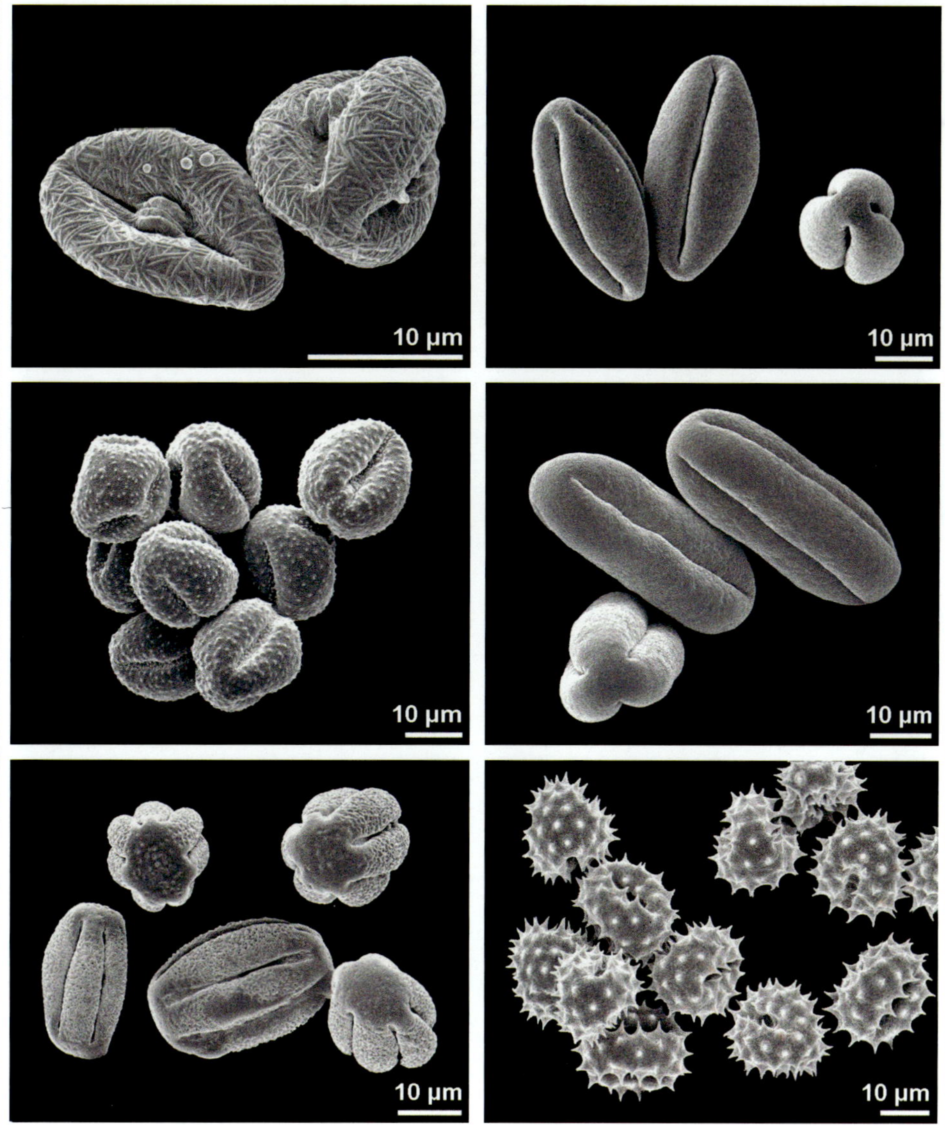

Sedum rupestre
Crassulaceae
dry pollen

Viola alba
Violaceae
dry pollen

Clematis heracleifolia
Ranunculaceae
dry pollen

Sanicula europaea
Apiaceae
dry pollen

Pinguicula ehlersiae
Lentibulariaceae
dry pollen

Bellis perennis
Asteraceae
dry pollen

LM SEM TEM mo ana fnc

outline, pollen infoldings, irregular

Urtica dioica
Urticaceae
dry pollen

Sesleria albicans
Poaceae
dry pollen

Coriaria sinica
Coriariaceae
dry pollen

Populus alba
Salicaceae
dry pollen

Anthurium radicans
Araceae
dry pollen

Orobanche hederae
Orobanchaceae
dry pollen

LM SEM TEM mo ana fnc

outline, pollen infoldings, apertures sunken

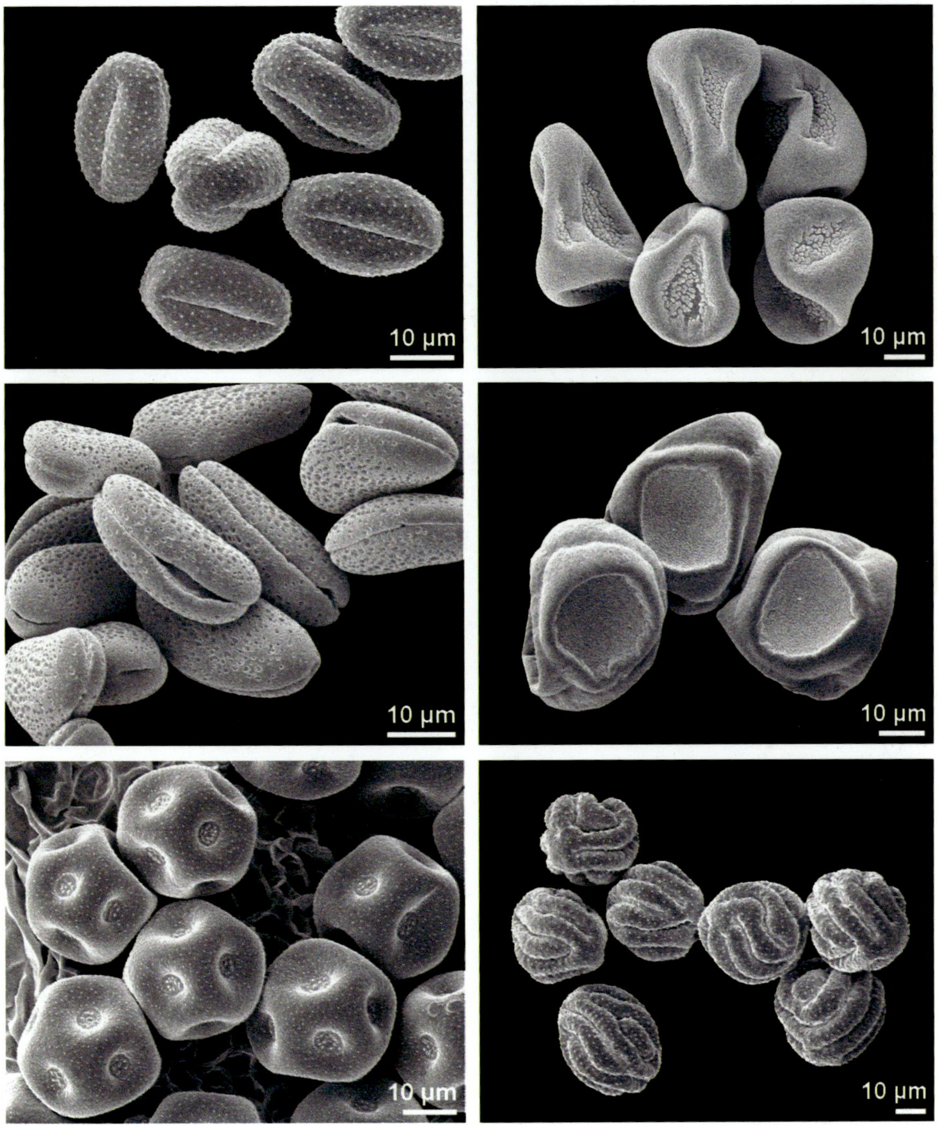

Artemisia pontica
Asteraceae
tricolporate, dry pollen

Lachenalia aloides
Liliaceae
sulcate, dry pollen

Moehringia muscosa
Caryophyllaceae
pantoporate, dry pollen

Carex alba
Cyperaceae
pseudomonads with poroids, dry pollen

Luzula sylvestris
Juncaceae
tetrads ulcerate, dry pollen

Anemone hortensis
Ranunculaceae
stephanocolpate to spiraperturate, dry pollen

LM SEM TEM mo ana fnc

outline, pollen infoldings, interapertural area sunken

Alnus glutinosa
Betulaceae
dry pollen

Bupleurum rotundifolium
Apiaceae
dry pollen

Leucadendron discolor
Proteaceae
dry pollen

Erica arborea
Ericaceae
dry pollen

Melampyrum arvense
Scrophulariaceae
dry pollen

Verbena officinalis
Verbenaceae
dry pollen

outline, pollen infoldings, cup-shaped: characteristic shape of pollen grains caused by infoldings as a consequence of harmomegathy; see "Pollen Morphology"

Bougainvillea sp.
Nyctaginaceae
dry pollen

Tilia euchlora
Tiliaceae
dry pollen

Luzula campestris
Juncaceae
tetrads, dry pollen

Heliconia sp.
Heliconiaceae
dry pollen

Elaeagnus angustifolia
Elaeagnaceae
dry pollen

Tsuga canadensis
Pinaceae
dry pollen

LM SEM TEM mo ana fnc

outline, pollen infoldings, boat-shaped: characteristic shape of sulcate pollen grains caused by an infolding as a consequence of harmomegathy; see "Pollen Morphology".

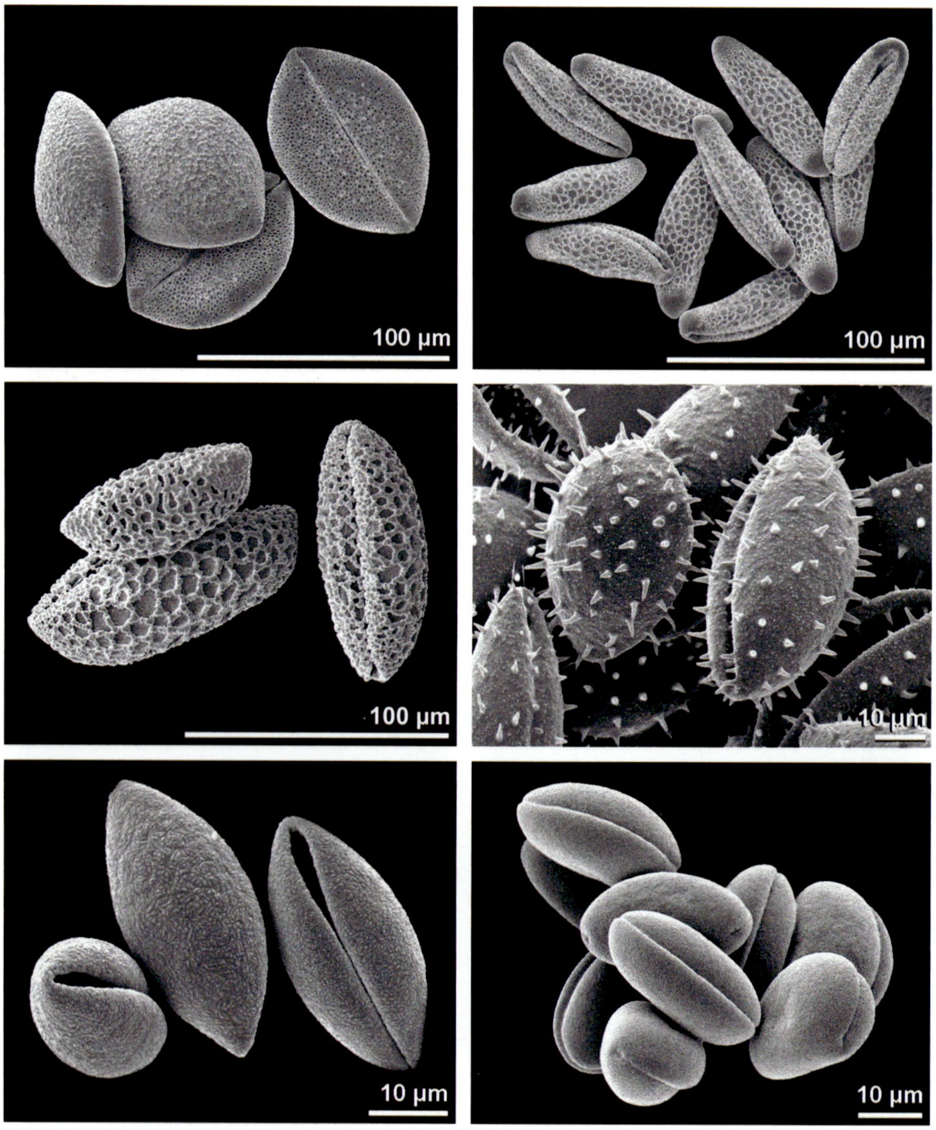

Asphodeline lutea
Asphodelaceae
sulcate, dry pollen

Lilium candidum
Liliaceae
sulcate, dry pollen

Ginkgo biloba
Ginkgoaceae
sulcate, dry pollen

Billbergia seidelii
Bromeliaceae
sulcate, dry pollen

Nuphar luteum
Nymphaeaceae
sulcate, dry pollen

Galanthus nivalis
Amaryllidaceae
sulcate, dry pollen

LM SEM TEM mo ana fnc

outline, pollen infoldings, boat-shaped: characteristic shape of sulcate pollen grains caused by an infolding as a consequence of harmomegathy; see "Pollen Morphology".

Lysichiton americanus
Araceae
sulcate, dry pollen

Gagea lutea
Liliaceae
sulcate, dry pollen

Dioon edule
Zamiaceae
sulcate, dry pollen

Piper nigrum
Piperaceae
sulcate, dry pollen

Sparganium erectum
Sparganiaceae
ulcerate (exception), dry pollen

Symplocarpus foetidus
Araceae
sulcate, dry pollen

LM SEM TEM mo ana fnc

isopolar: pollen grain with identical proximal and distal faces.

Cerinthe minor
Boraginaceae
dry pollen

Viburnum tinus
Viburnaceae
equatorial view

Aesculus flava
Hippocastanaceae
equatorial view

Asperula tinctoria
Rubiaceae
dry pollen

Myosotis palustris
Boraginaceae
dry pollen

Monotropa hypopitys
Ericaceae
equatorial view

LM SEM TEM mo ana fnc

heteropolar: pollen grain with different proximal and distal faces.

Onosma visianii
Boraginaceae

Microrrhinum minus
Scrophulariaceae
dry pollen

Sesleria albicans
Poaceae
oblique view

Billbergia seidelii
Bromeliaceae
equatorial view

Limnanthes douglasii
Limnanthaceae
dry pollen

Elaeagnus angustifolia
Elaeagnaceae
oblique polar view

LM SEM TEM mo ana fnc

heteropolar: pollen grain with different proximal and distal faces.

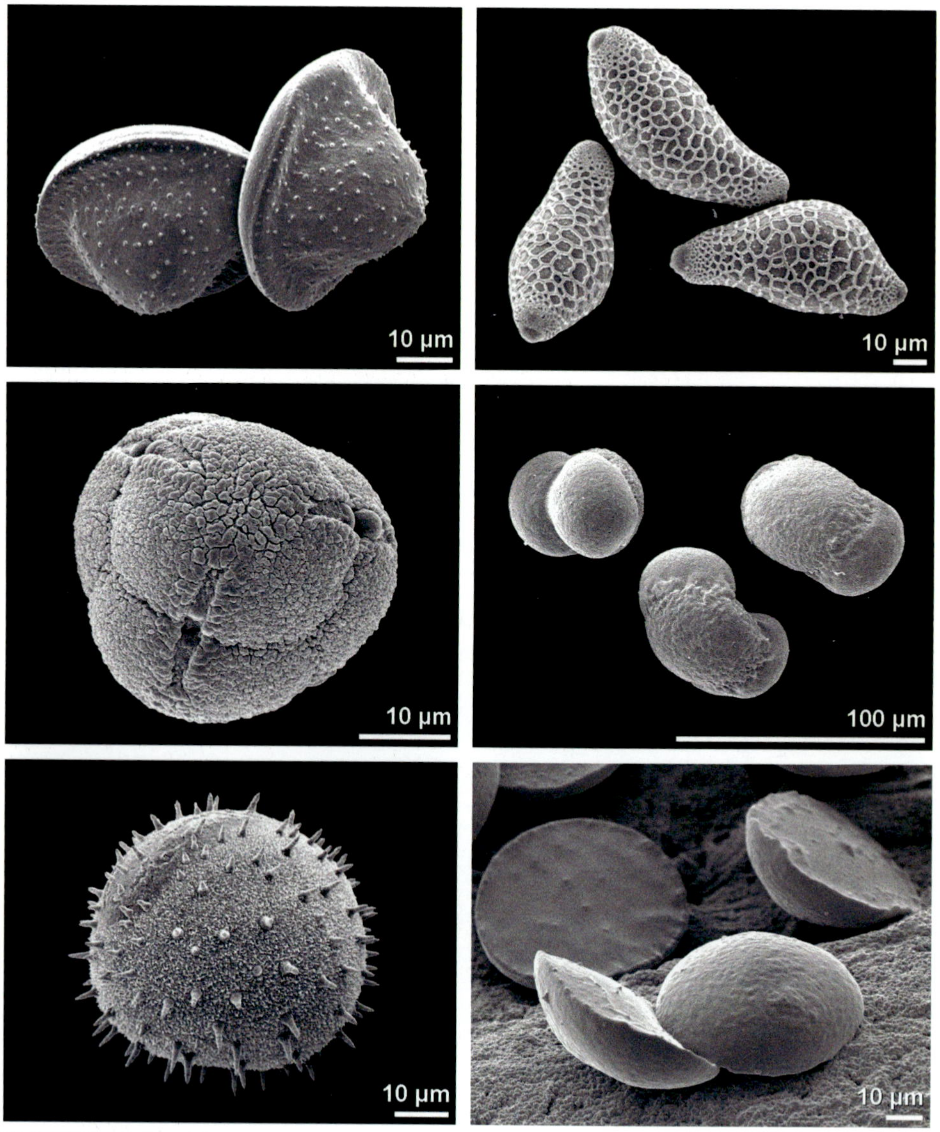

Heliconia sp.
Heliconiaceae
dry pollen

Erica arborea
Ericaceae
tetrad

Nuphar luteum
Nymphaeaceae
equatorial view

Quesnelia augusto-coburgii
Bromeliaceae

Pinus strobus
Pinaceae

Sansevieria dooneri
Dracaenaceae
dry pollen

LM SEM TEM mo ana fnc

saccus: exinous expansion forming an air sac.

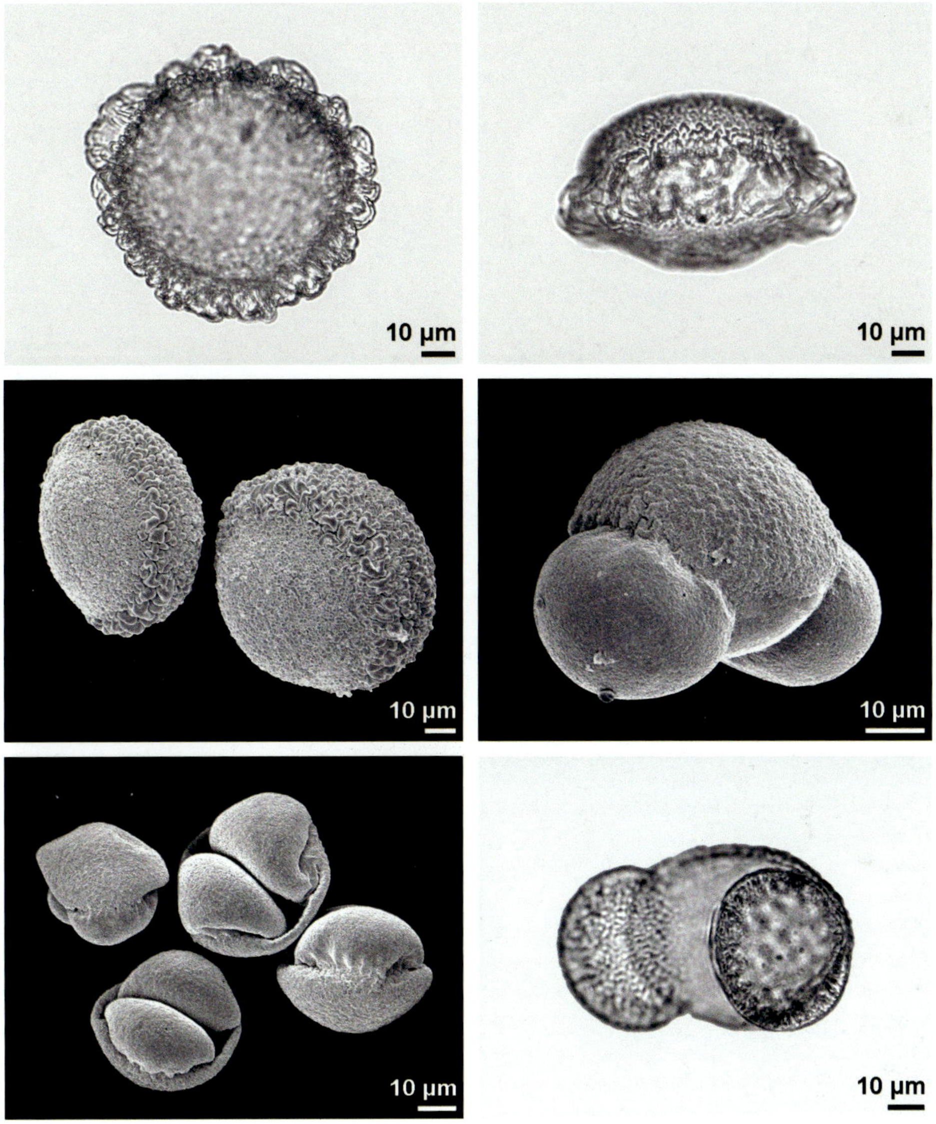

Tsuga sp.
Pinaceae, fossil
monosaccate, polar view

Tsuga sp.
Pinaceae, fossil
monosaccate, equatorial view

Tsuga canadensis
Pinaceae
monosaccate

Pinus nigra
Pinaceae
bisaccate, equatorial view

Pinus heldreichii
Pinaceae
bisaccate, dry pollen

Pinus sp.
Pinaceae, fossil
bisaccate, oblique distal polar view

LM SEM TEM mo ana fnc

saccus: exinous expansion forming an air sac.

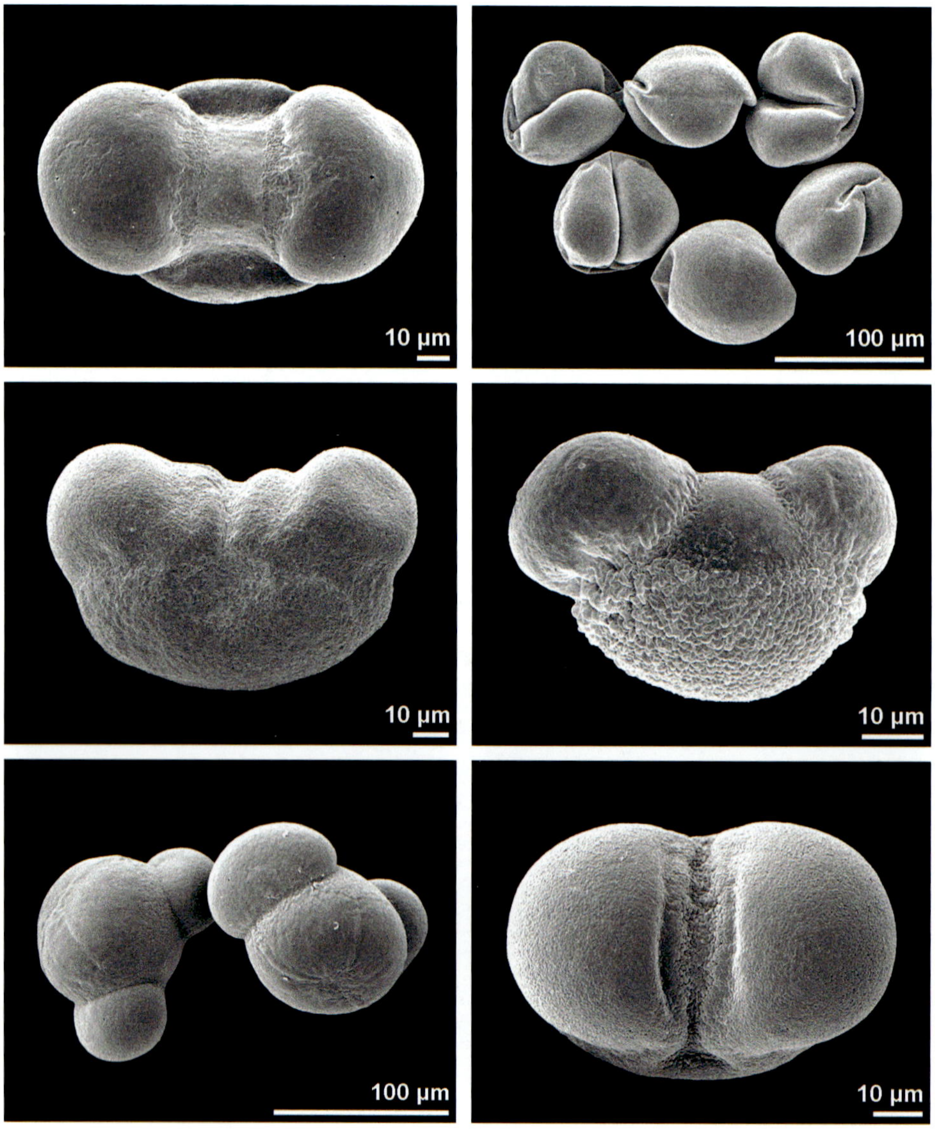

Abies cephalonica
Pinaceae
bisaccate, distal polar view

Picea abies
Pinaceae
bisaccate, equatorial view

Abies nordmanniana
Pinaceae
bisaccate

Picea abies
Pinaceae
bisaccate, dry pollen

Pinus mugo
Pinaceae
bisaccate, equatorial view

Picea pungens
Pinaceae
bisaccate, distal polar view

saccus: exinous expansion forming an air sac.

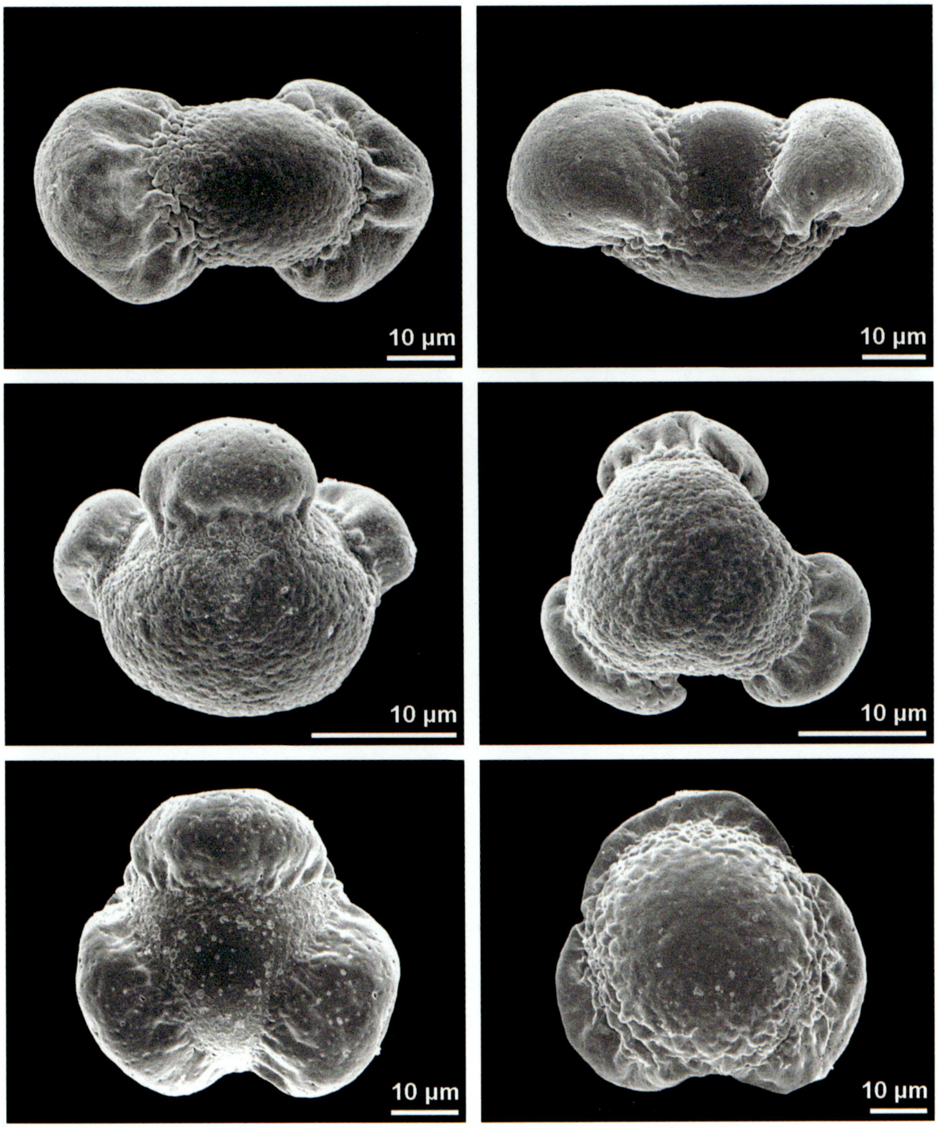

Podocarpus sp.
Podocarpaceae
bisaccate, proximal polar view

Podocarpus sp.
Podocarpaceae
bisaccate, oblique equatorial view

Microstrobus niphophilus
Podocarpaceae
trisaccate, equatorial view

Microstrobus niphophilus
Podocarpaceae
trisaccate, proximal polar view

Dacrycarpos dacrydioides
Podocarpaceae
trisaccate, distal polar view

Dacrycarpos dacrydioides
Podocarpaceae
trisaccate, proximal polar view

LM SEM TEM mo ana fnc

arcus: a curved wall thickening interconnecting apertures.

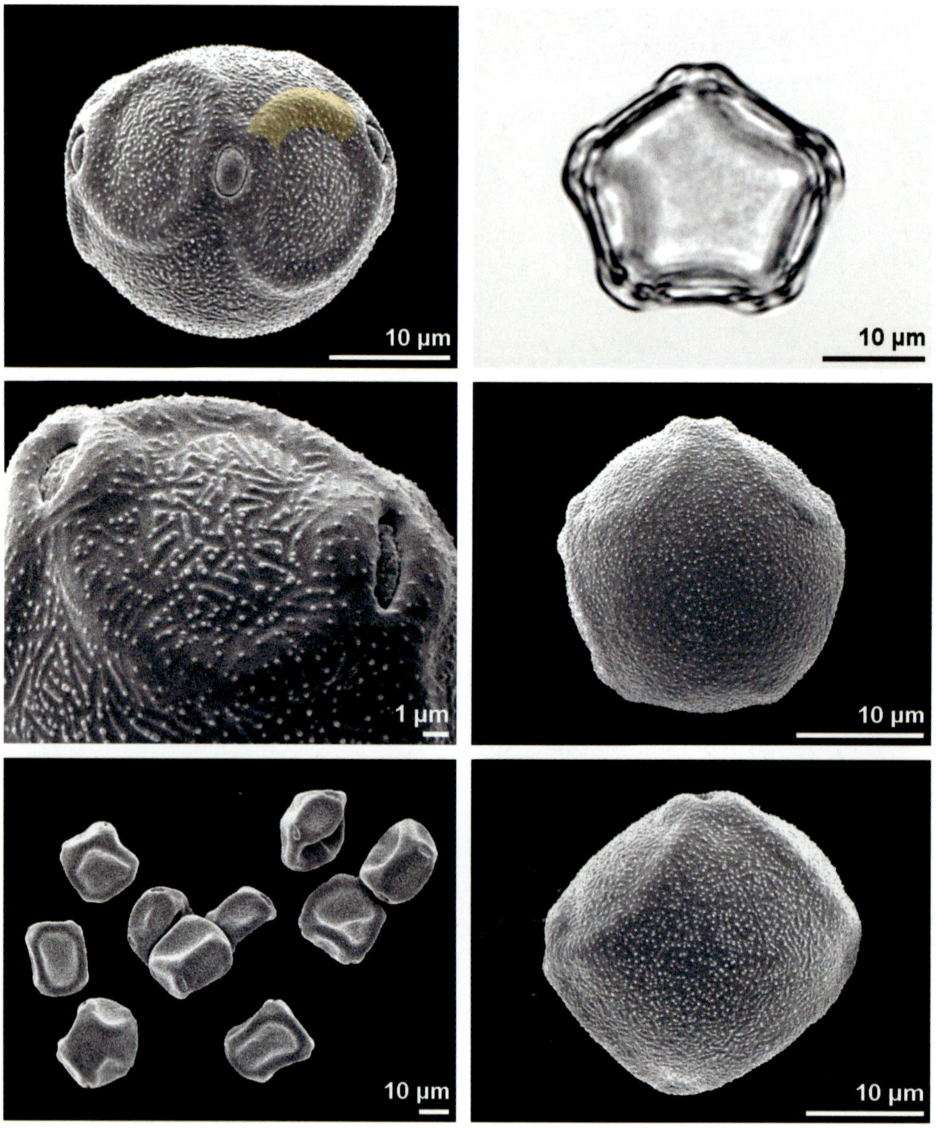

Alnus glutinosa
Betulaceae
equatorial view

Alnus glutinosa
Betulaceae

Alnus incana
Betulaceae
dry pollen

Alnus sp.
Betulaceae, fossil
polar view

Alnus viridis
Betulaceae
polar view

Alnus incana
Betulaceae
polar view

LM SEM TEM mo ana fnc

aperture: region of the pollen wall which differs significantly morphologically and/or anatomically from the rest of the pollen wall, presumed to function usually as germination site and to play a role in harmomegathy.

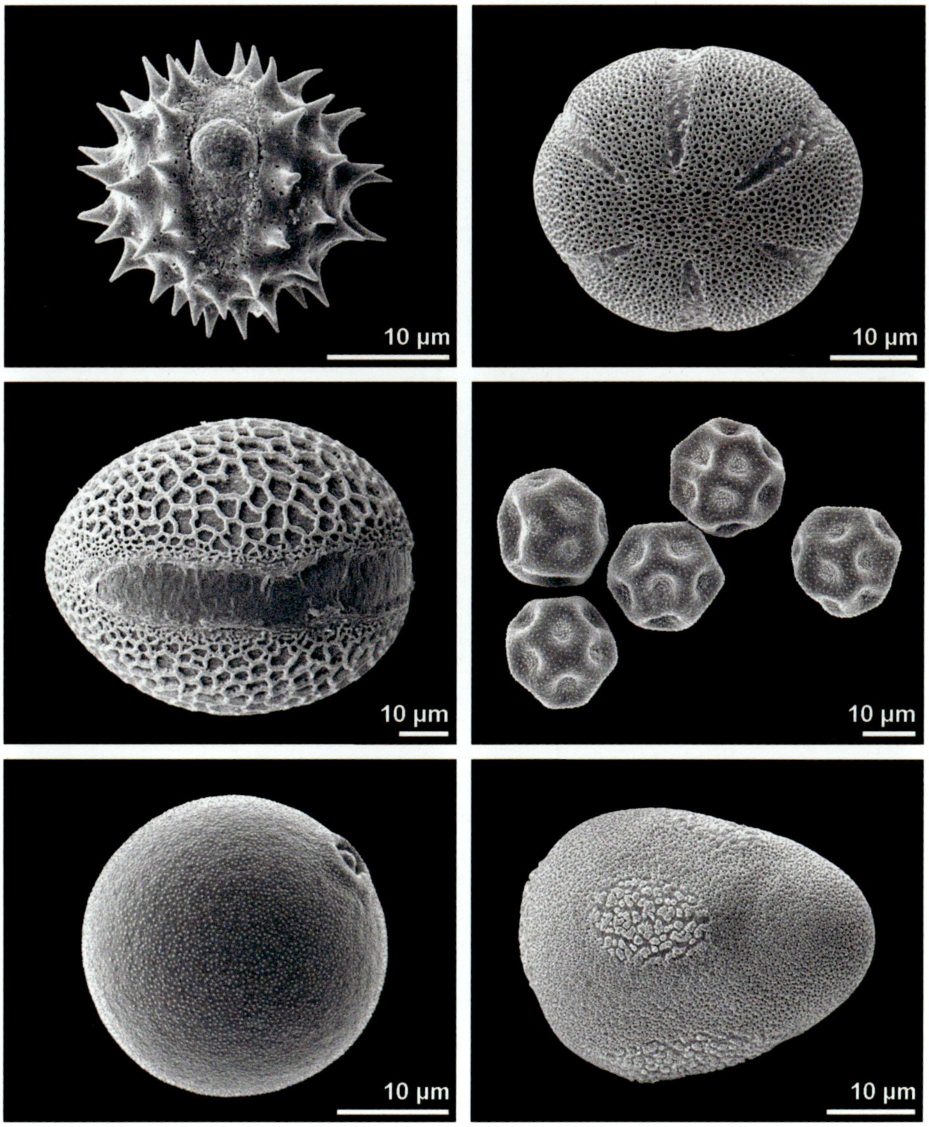

Eupatorium cannabinum
Asteraceae
tricolporate, equatorial view

Eucharis grandiflora
Amaryllidaceae
sulcate, distal polar view

Sesleria sadleriana
Poaceae
ulcerate, equatorial view

Clinopodium vulgare
Lamiaceae
hexacolpate, polar view

Alisma plantago-aquatica
Alismataceae
pantoporate, dry pollen

Carex alba
Cyperaceae
poroids, pseudomonad

LM SEM TEM mo ana fnc

aperture: region of the pollen wall which differs significantly morphologically and/ or anatomically from the rest of the pollen wall, presumed to function usually as germination site and to play a role in harmomegathy.

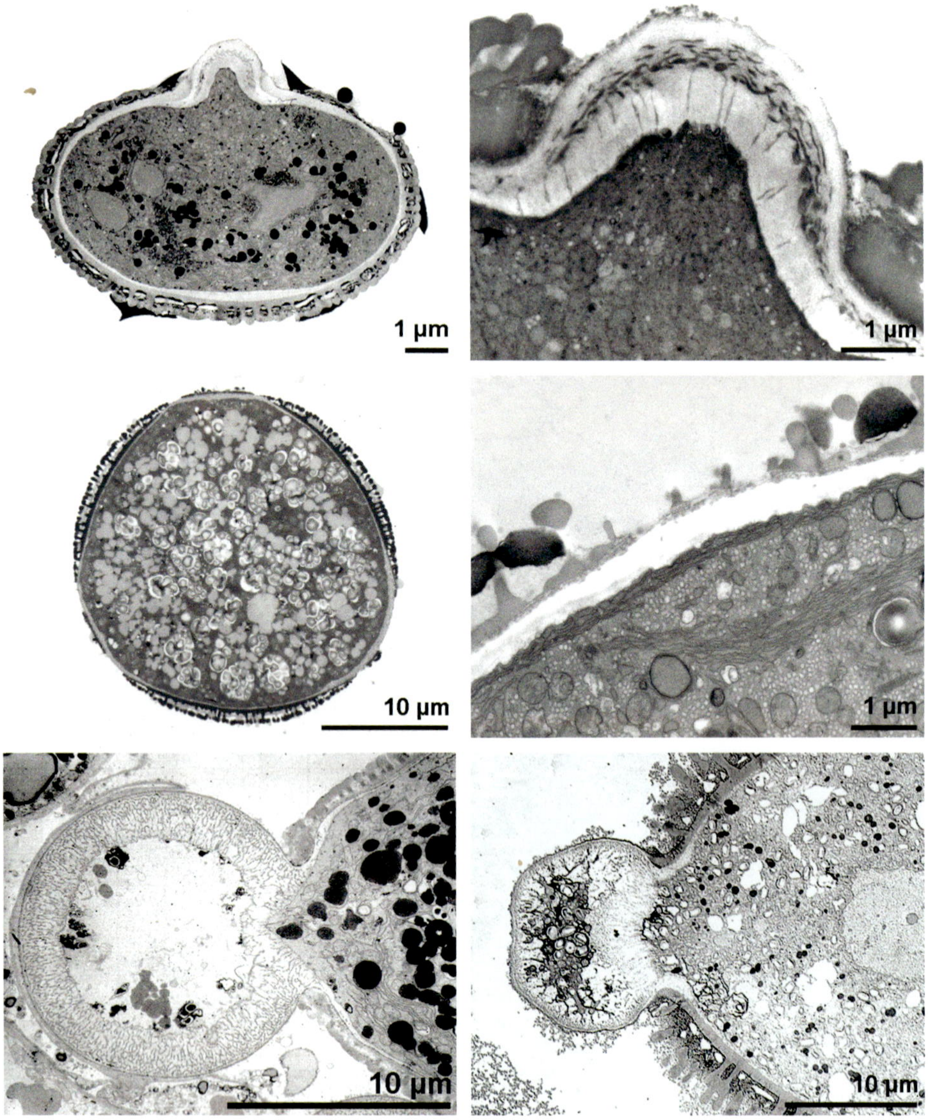

Apium nodiflorum
Apiaceae
PA+TCH+SP

Pseudolysimachion barrelieri
Scrophulariaceae
$KMnO_4$

Ophiorrhiza sp.
Rubiaceae
PA+TCH+SP
apertural intine protrusion; becomes eliminated during pollen maturation

Apium nodiflorum
Apiaceae
U+Pb

Mentha aquatica
Lamiaceae
PA+TCH+SP (short)

Geranium robertianum
Geraniaceae
PA+TCH+SP
apertural intine protrusion

inaperturate: pollen grain without distinct aperture(s).

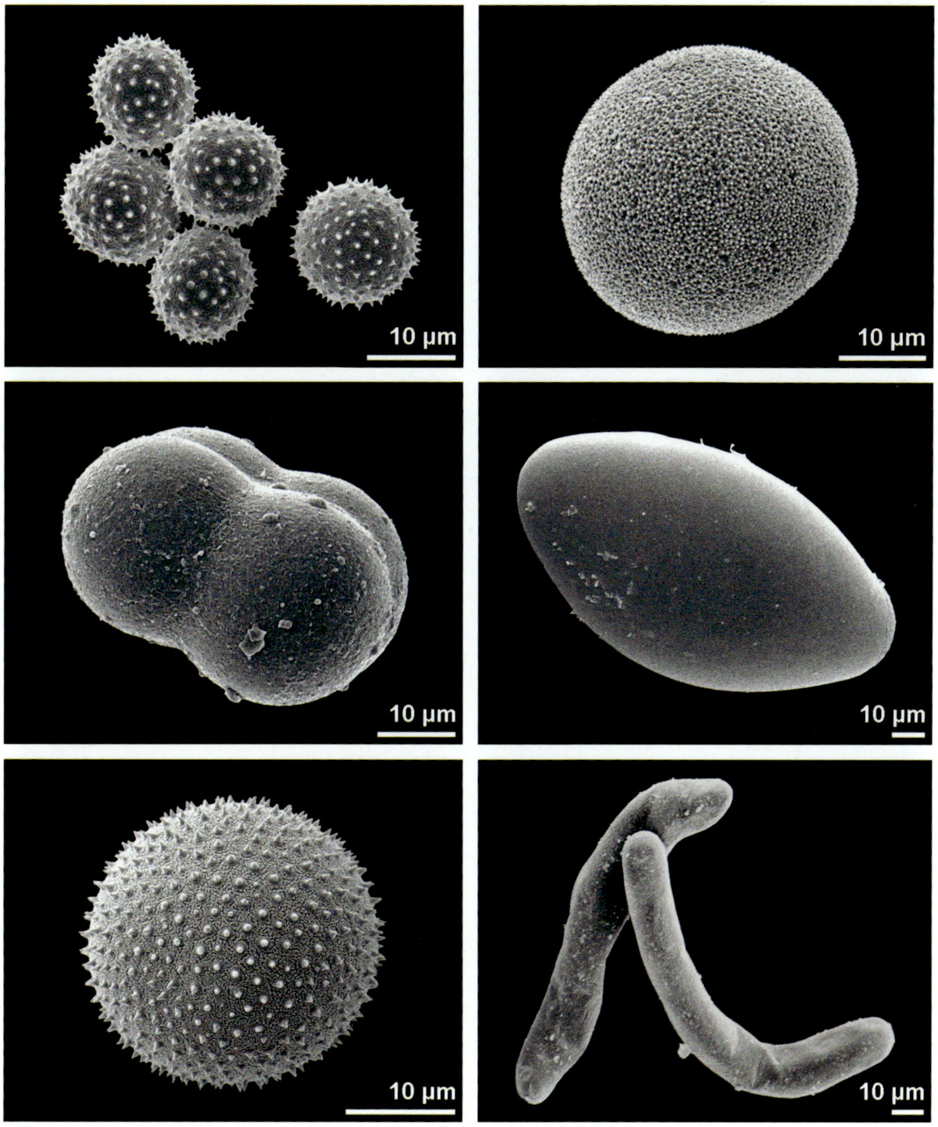

Pinellia ternata
Araceae

Populus alba
Salicaceae

Chlorospatha dodsonii
Araceae
tetrad

Aglaodorum griffithii
Araceae

Phoebe sheareri
Lauraceae

Posidonia sp.
Posidoniaceae

LM SEM TEM mo ana fnc

angulaperturate: pollen grain with an angular outline, where the apertures are situated at the angles.

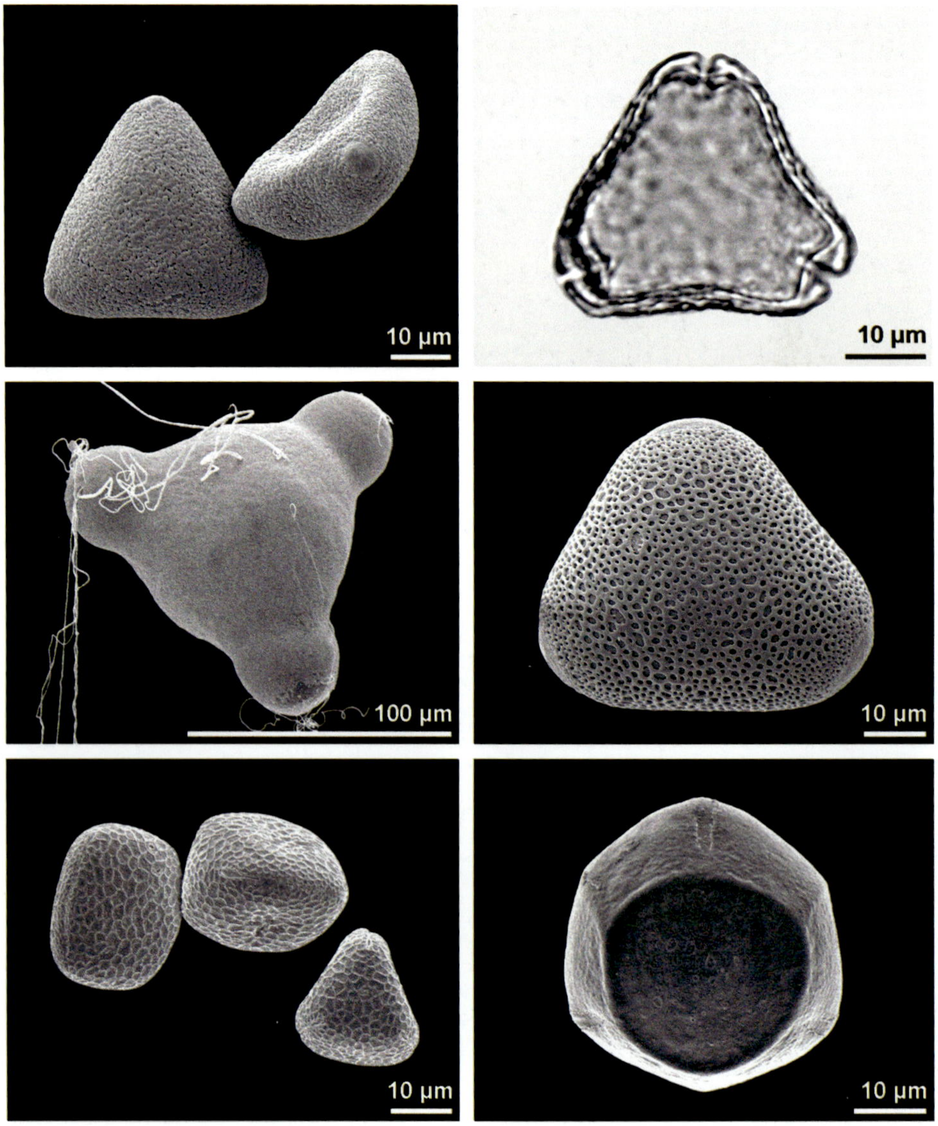

Leucadendron brunoides
Proteaceae
dry pollen

Oenothera biennis
Onagraceae
polar view, viscin threads

Tropaeolum majus
Tropaeolaceae
dry pollen

Symplocos sp.
Symplocaceae, fossil
polar view

Isopogon dawsonii
Proteaceae
polar view

Anthyllis vulneraria
Fabaceae
polar view

LM SEM TEM mo ana fnc

planaperturate: pollen grain with an angular outline, where the apertures are situated in the middle of the sides.

Centaurea cyanus
Asteraceae
polar view

Pachira aquatica
Bombacaceae
tricolpate, polar view

Beloperone guttata
Acanthaceae
dry pollen

Persicaria bistorta
Polygonaceae
dry pollen

Pachira aquatica
Bombacaceae

Pedilanthus smallii
Euphorbiaceae
polar view

LM SEM TEM mo ana fnc

stephanoaperturate: apertures situated at the equator (term usually used for more than three apertures).

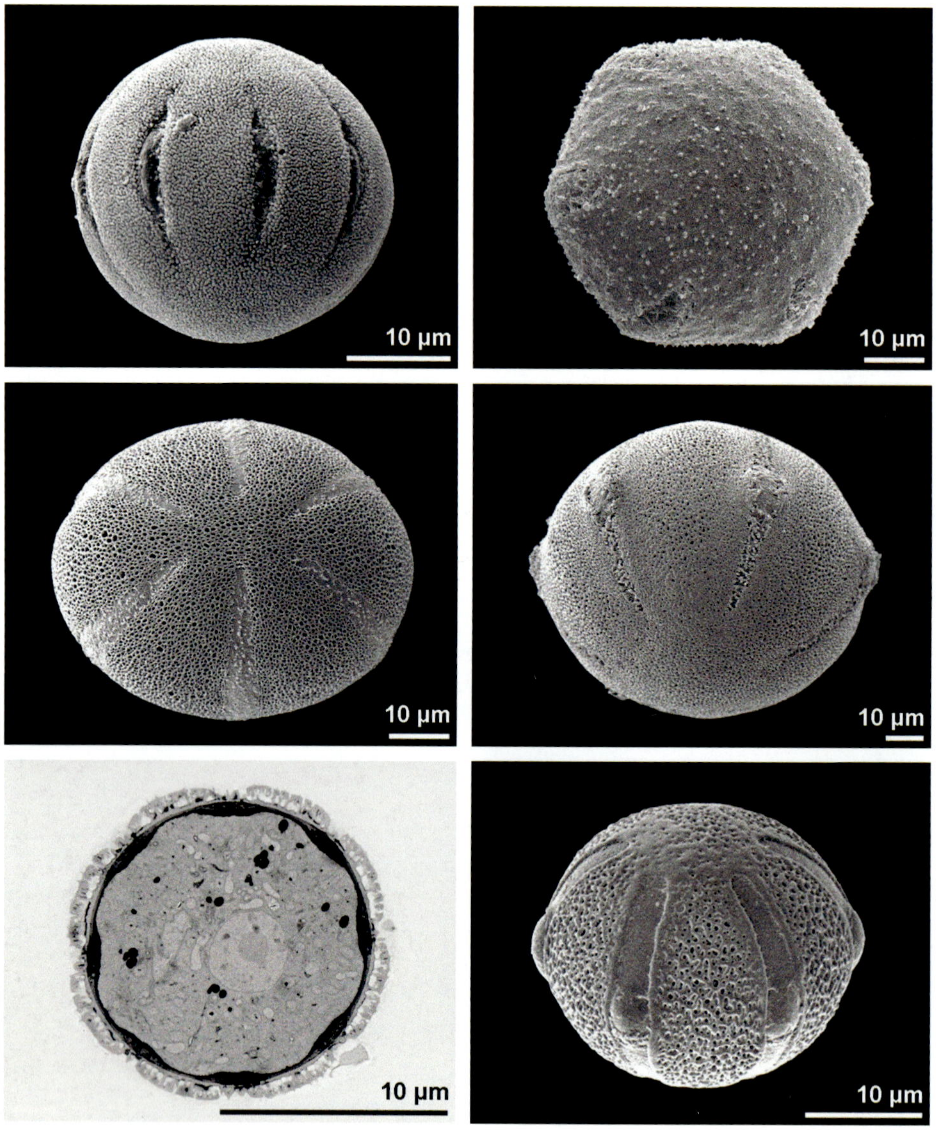

Borago officinalis
Boraginaceae
equatorial view

Dracocephalum austriacum
Lamiaceae
polar view

Galium mollugo
Rubiaceae
U+Pb
cross section

Platycodon grandiflorum
Campanulaceae
polar view

Cyclanthera pedata
Cucurbitaceae
oblique equatorial view

Pinguicula ehlersiae
Lentibulariaceae
oblique equatorial view

LM SEM TEM mo ana fnc

stephanoaperturate: apertures situated at the equator (term usually used for more than three apertures).

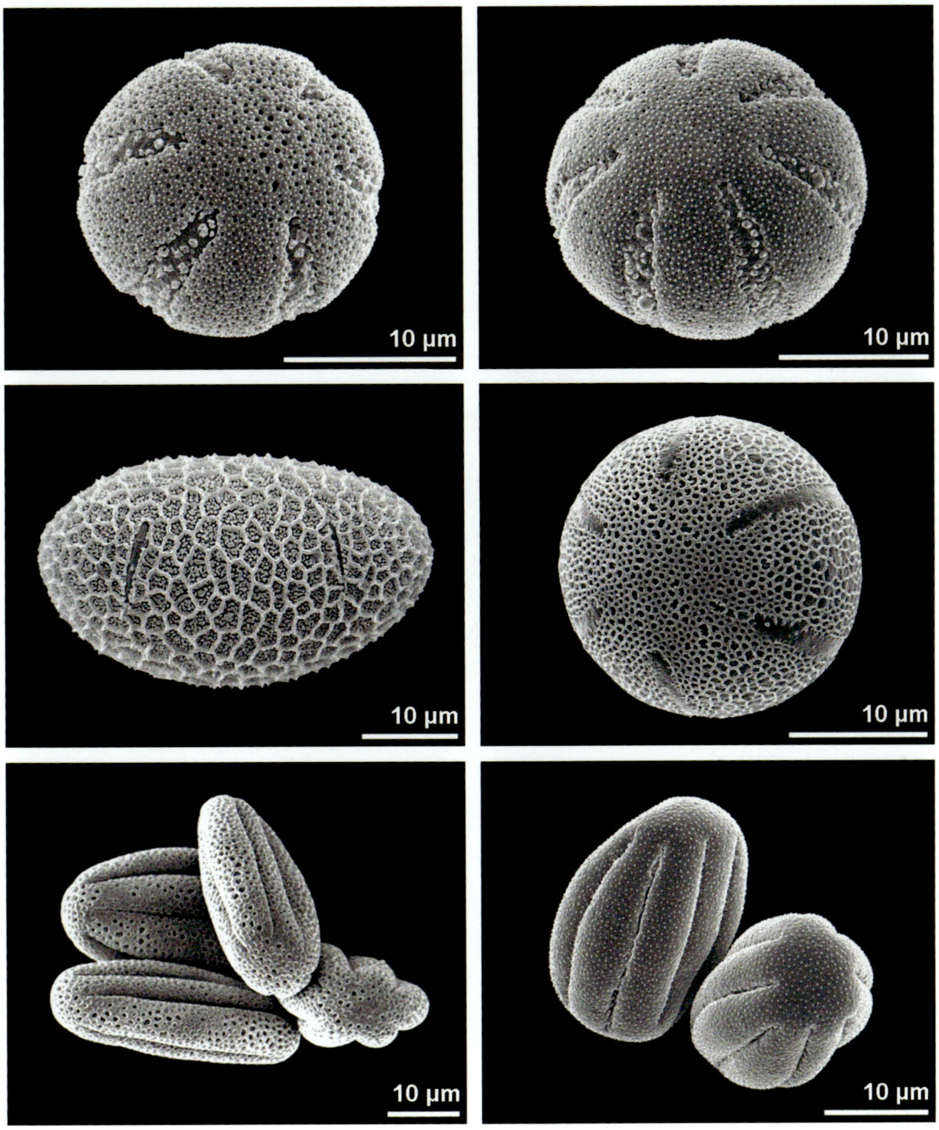

Asperula tinctoria
Rubiaceae
hexacolpate, polar view

Galium glaucum
Rubiaceae
polar view

Impatiens parviflora
Balsaminaceae
tetracolpate, equatorial view

Primula veris
Primulaceae
hexacolpate, polar view

Lycopus europaeus
Lamiaceae
hexacolpate, dry pollen

Galium lucidum
Rubiaceae
dry pollen

LM SEM TEM mo ana fnc

stephanoaperturate: apertures situated at the equator (term usually used for more than three apertures).

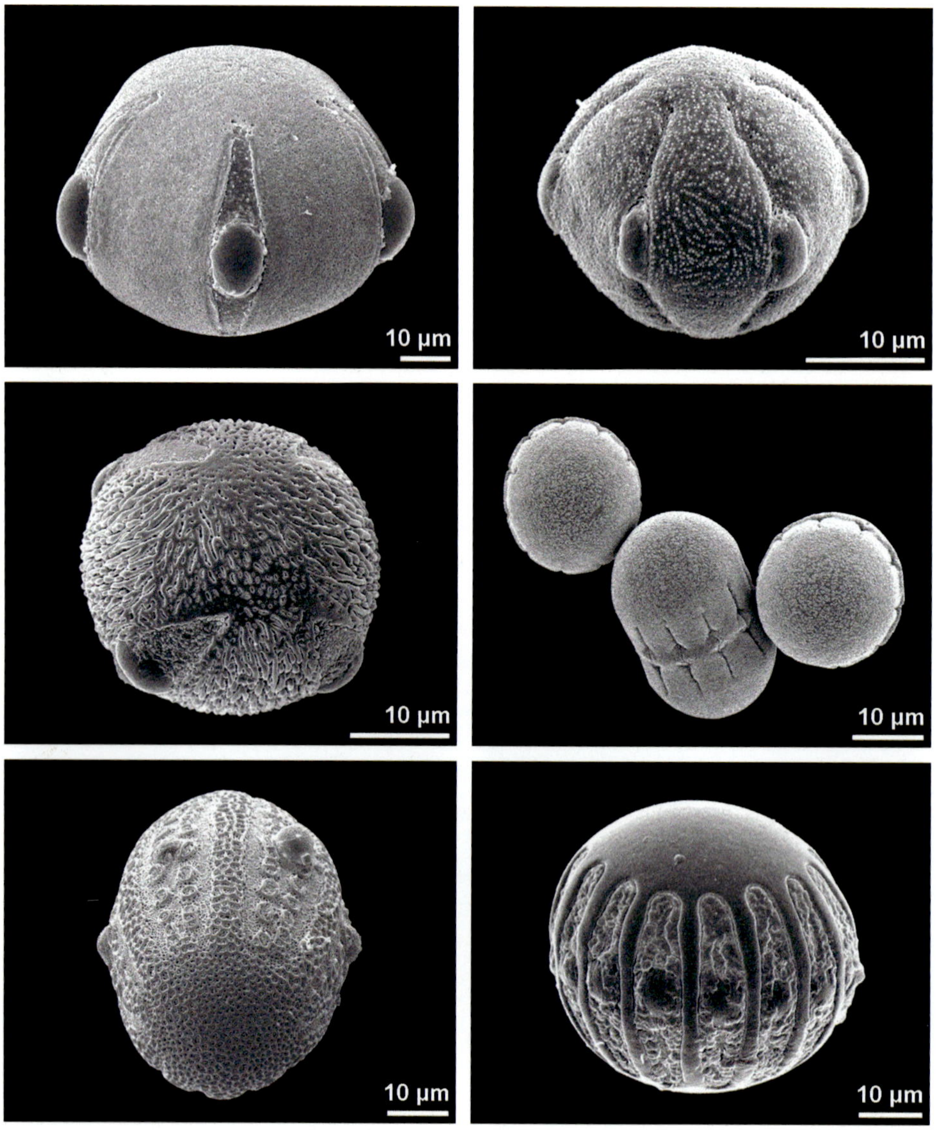

Viola arvensis
Violaceae
pentacolporate, equatorial view

Nicotiana tabacum
Solanaceae
tetracolporate, polar view

Justicia menesii
Acanthaceae
polar view

Sanguisorba officinalis
Rosaceae
equatorial view

Symphytum caucasicum
Boraginaceae
dry pollen

Polygala chamaebuxus
Polygalaceae
equatorial view

stephanoaperturate: apertures situated at the equator (term usually used for more than three apertures).

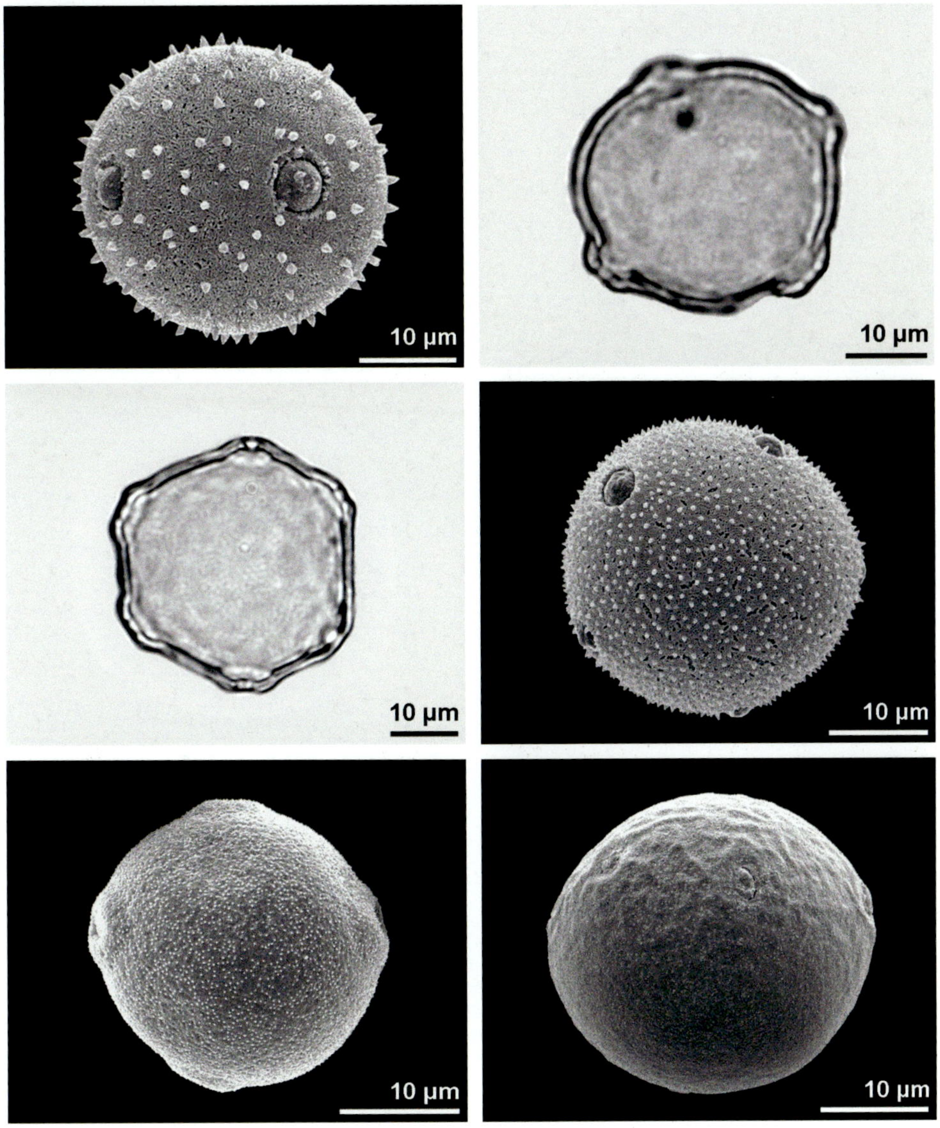

Campanula alpina
Campanulaceae
equatorial view

Pterocarya sp.
Juglandaceae, fossil
polar view

Myriophyllum spicatum
Haloragaceae
polar view

Carpinus sp.
Betulaceae, fossil
polar view

Legousia speculum-veneris
Campanulaceae
polar view

Ulmus minor
Ulmaceae
equatorial view

LM SEM TEM mo ana fnc

pantoaperturate: pollen grain with apertures distributed more or less regularly over the whole surface.

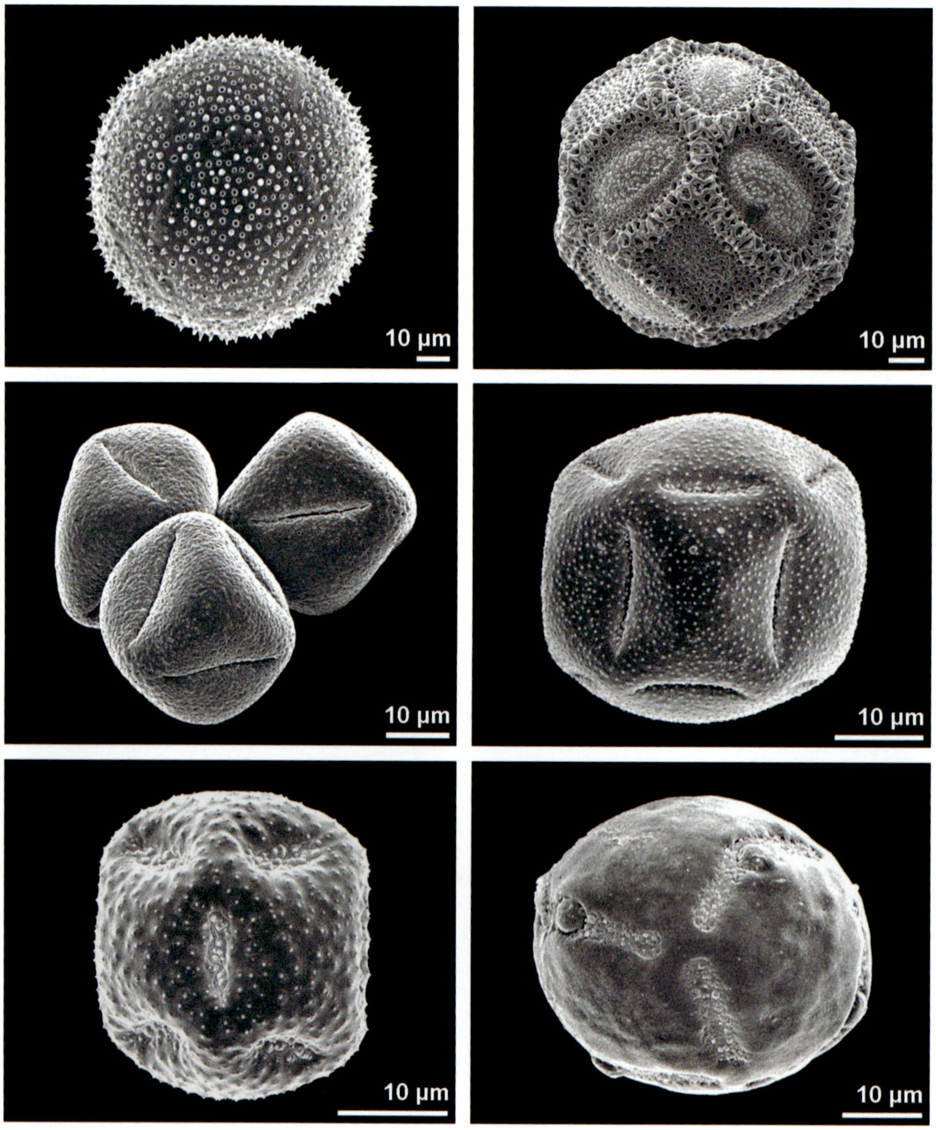

Portulaca grandiflora
Portulacaceae
pantocolpate

Sideritis syriaca
Lamiaceae
pantocolpate, hexacolpate, dry pollen

Ranunculus lanuginosus
Ranunculaceae
pantocolpate, dry pollen

Opuntia basilaris
Cactaceae
pantocolpate

Talinum paniculatum
Portulacaceae
pantocolpate, dry pollen

Banisteria argentea
Malpighiaceae
pantocolporate

LM SEM TEM mo ana fnc

pantoaperturate: pollen grain with apertures distributed more or less regularly over the whole surface.

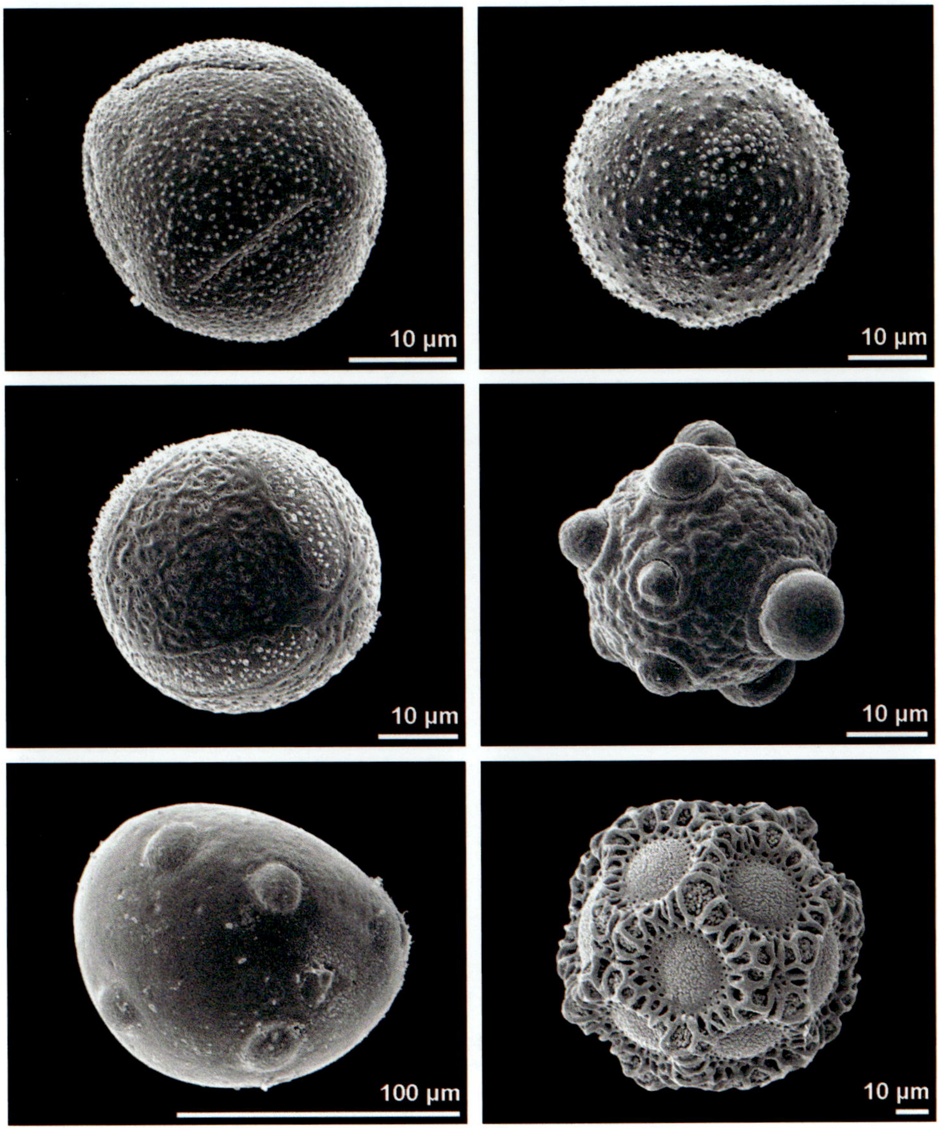

Trichostigma peruvianum
Phytolaccaceae
pantocolpate, dry pollen

Corydalis cava
Fumariaceae
pantocolpate, hexacolpate

Costus barbatus
Zingiberaceae
pantoporate

Hepatica transsylvanica
Ranunculaceae
pantocolpate

Fumaria vaillantii
Fumariaceae
pantoporate

Opuntia sp.
Cactaceae
pantoporate

LM SEM TEM mo ana fnc

pantoaperturate: pollen grain with apertures distributed more or less regularly over the whole surface.

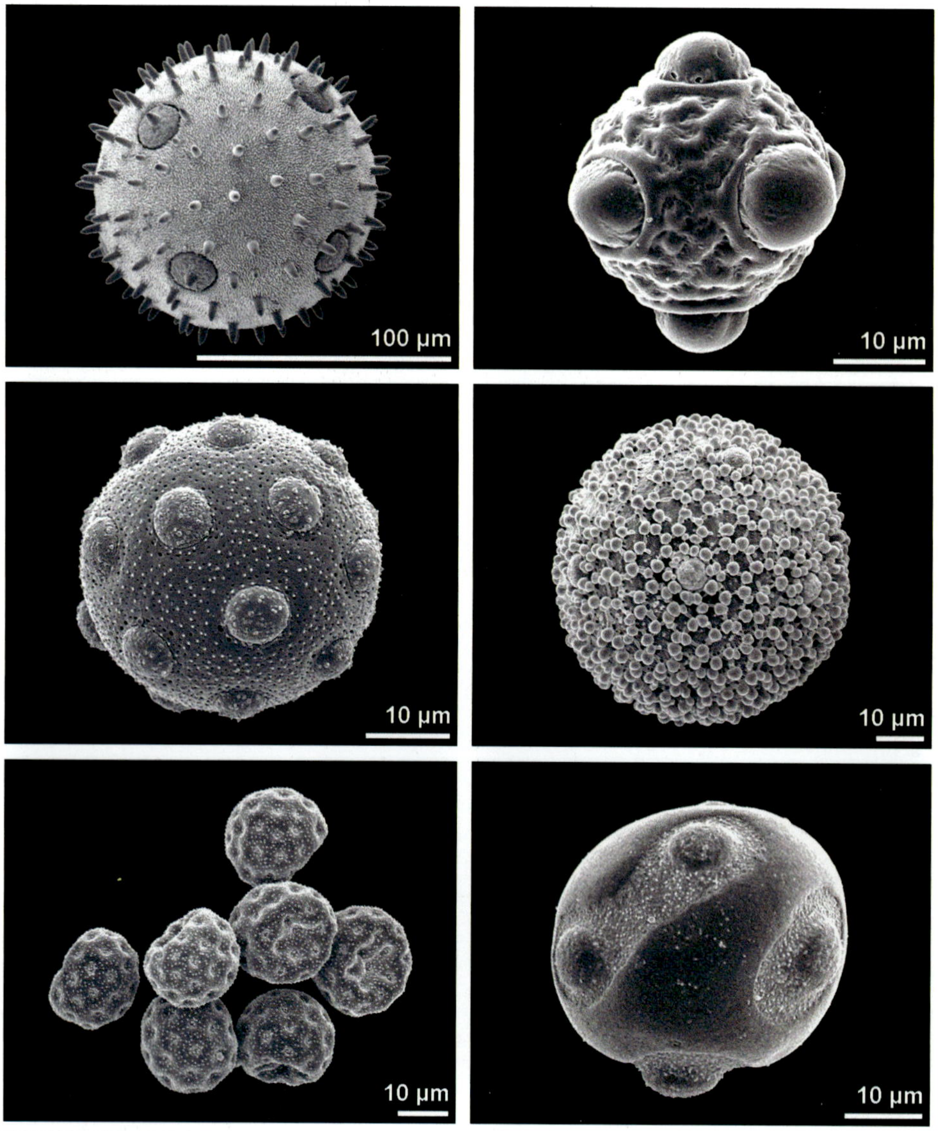

Cucurbita pepo
Cucurbitaceae
pantoporate

Stellaria graminea
Caryophyllaceae
pantoporate

Chenopodium hybridum
Chenopodiaceae
pantoporate, dry pollen

Fumaria officinalis
Fumariaceae
pantoporate

Phaleria capitata
Thymelaeaceae
pantoporate

Ribes aureum
Grossulariaceae
pantoporate

LM SEM TEM mo ana fnc

brevicolpus: short colpus
brevicolporus: short colpus in a compound aperture.

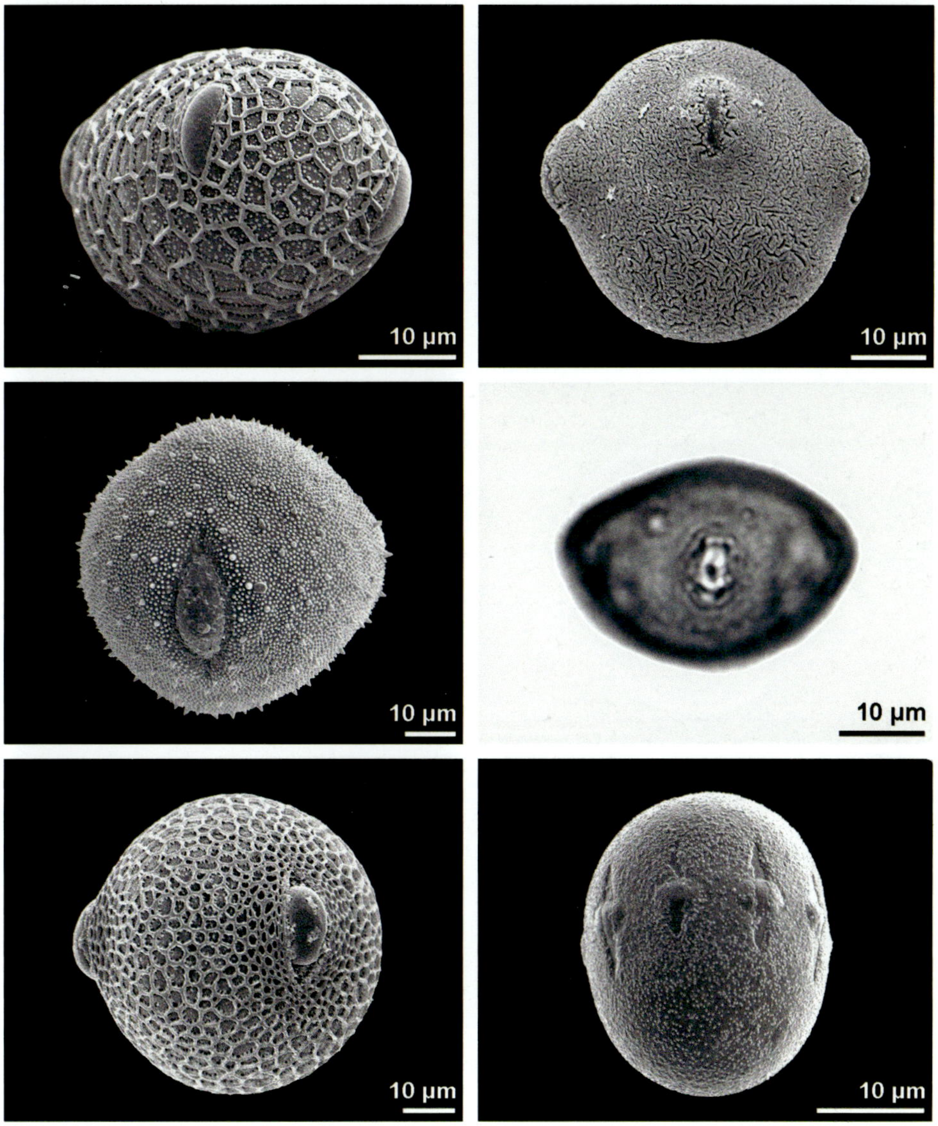

Impatiens columbaria
Balsaminaceae
brevicolpate, equatorial view

Scabiosa ochroleuca
Dipsacaceae
brevicolpate, equatorial view

Dalechampia roezliana
Euphorbiaceae
brevicolporate, equatorial view

Mendoncia albida
Acanthaceae
brevicolpate, equatorial view

Tilia americana
Tiliaceae
brevicolporate, equatorial view, acetolyzed

Symphytum orientale
Boraginaceae
brevicolporate, equatorial view

LM SEM TEM mo ana fnc

synaperturate: pollen grain with anastomosing apertures.

Pedicularis verticillata
Scrophulariaceae
syncolpate, dry pollen

Cuphea purpurea
Lythraceae
syncolporate, polar view

Cassia pulcherrima
Caesalpiniaceae
syncolporate

Nymphoides peltata
Menyanthaceae
syncolpate, polar view

Onosma visianii
Boraginaceae
syncolporate

Callistemon coccineus
Myrtaceae
syncolporate, equatorial view

spiraperturate: pollen grain with one or more spiral aperture(s).

Mimulus guttatus
Scrophulariaceae

Mimulus guttatus
Scrophulariaceae
dry pollen

Thunbergia alata
Acanthaceae

Crocus speciosus
Iridaceae

Claytonia perfoliata
Portulacaceae
dry pollen

Berberis vulgaris
Berberidaceae

LM SEM TEM mo ana fnc

heteroaperturate: pollen grain with two different types of apertures; only one type presumed to function as germination site.

Comment:
the term "heterocolpate" is commonly used for pollen grains with alternating colpi and colpori but "heterocolpate" means two different types of colpi; therefore we prefer the more general term "heteroaperturate".

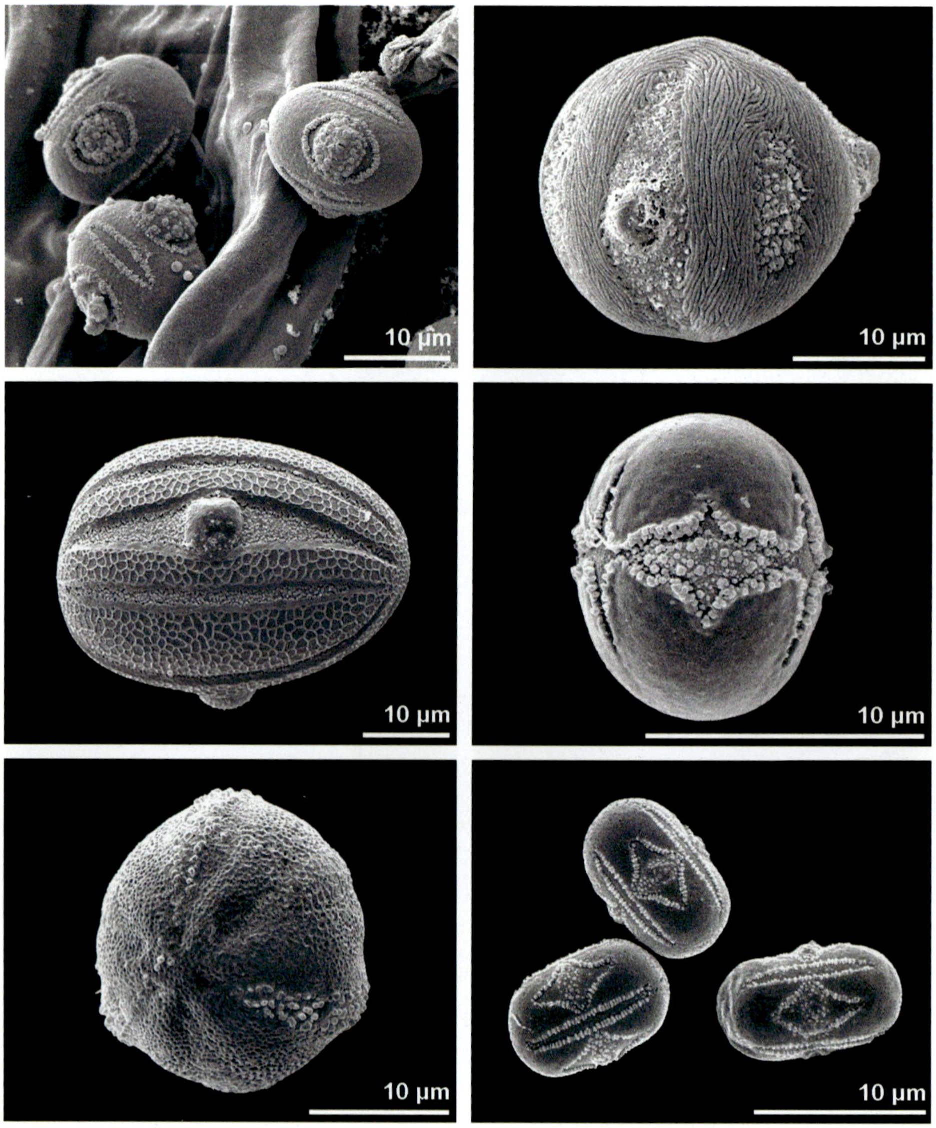

Pardoglossum sp.
Boraginaceae

Tetramerium nervosum
Acanthaceae
equatorial view

Phacelia tanacetifolia
Hydrophyllaceae
polar view

Lythrum hyssopifolia
Lythraceae
equatorial view

Cynoglossum officinale
Boraginaceae
equatorial view

Myosotis ramosissima
Boraginaceae

LM SEM TEM mo ana Inc

pseudocolpus: colpus in heteroaperturate pollen grains, presumably non-functional.

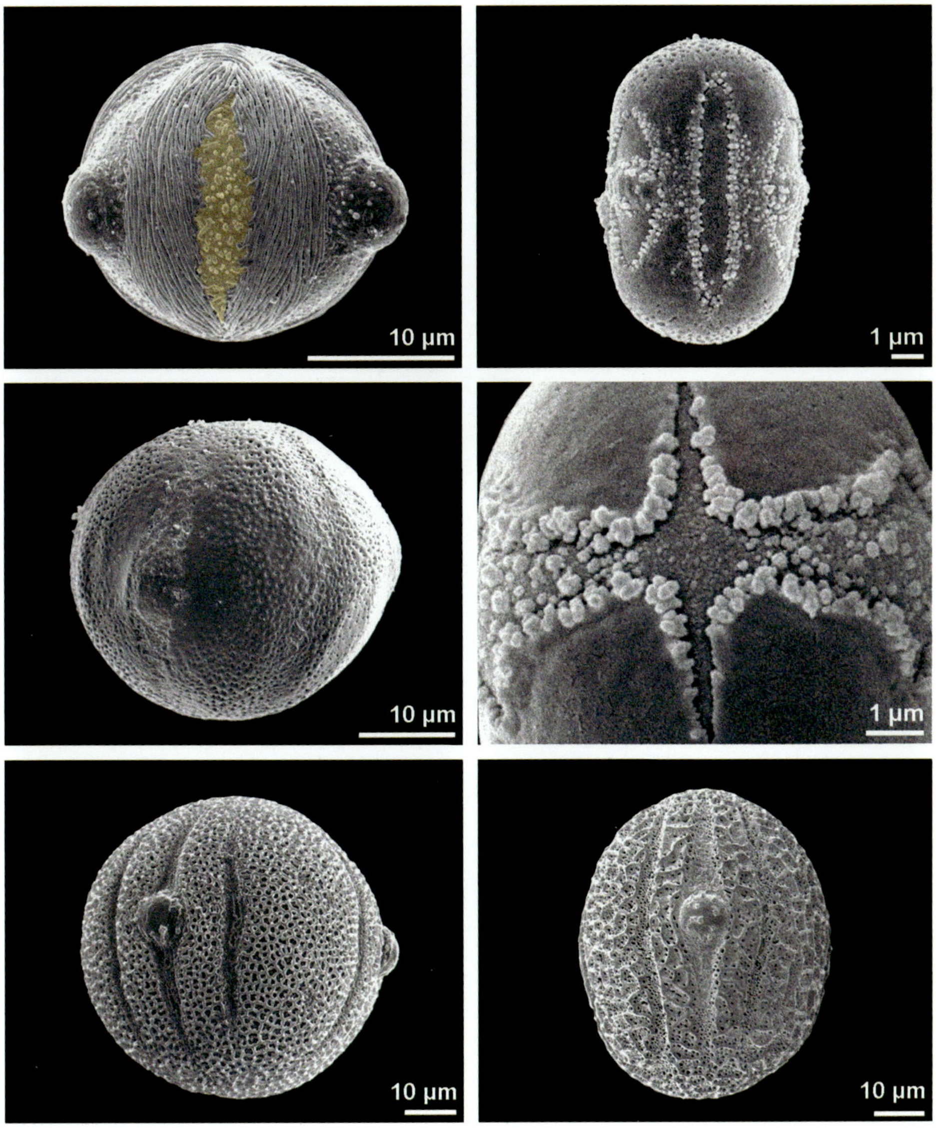

Lythrum salicaria
Lythraceae
equatorial view

Asperugo procumbens
Boraginaceae
equatorial view

Lumnitzera racemosa
Combretaceae
equatorial view

Cynoglossum officinale
Boraginaceae

Justicia furcata
Acanthaceae
equatorial view, two pseudocolpi flanking colporus

Pachystachys lutea
Acanthaceae
equatorial view, two pseudocolpi flanking colporus

LM SEM TEM mo ana fnc

colpus: elongated aperture (length : width ratio > 2) situated at the equatorial region or regularly distributed over the pollen grain.

Cruciata laevipes
Rubiaceae
stephanocolpate, equatorial view

Chimonanthus praecox
Calycanthaceae
dicolpate, polar view

Melampyrum arvense
Scrophulariaceae
tricolpate, polar view

Clinopodium vulgare
Lamiaceae
hexacolpate, equatorial view

Sherardia arvensis
Rubiaceae
stephanocolpate, oblique equatorial view

Talinum paniculatum
Portulacaceae
pantocolpate

LM SEM TEM mo ana fnc

tricolpate: pollen grain with three colpi.

Lamium maculatum
Lamiaceae

Erysimum odoratum
Brassicaceae
polar view

Nelumbo nucifera
Nelumbonaceae
polar view

Lonicera fragrantissima
Caprifoliaceae
equatorial view

Stachys palustris
Lamiaceae
dry pollen

Acer sp.
Sapindaceae, fossil
equatorial view

LM SEM TEM mo ana fnc

tricolpate: pollen grain with three colpi.

Fraxinus excelsior
Oleaceae
equatorial view

Nandina domestica
Berberidaceae
polar view

Trollius europaeus
Ranunculaceae
polar view

Odontites luteus
Scrophulariaceae
dry pollen

Corylopsis platypetala
Hamamelidaceae
equatorial view

Veronica serpyllifolia
Scrophulariaceae
equatorial view

LM SEM TEM mo ana fnc

porus: more or less circular aperture situated at the equator or regularly spread over the pollen grain.

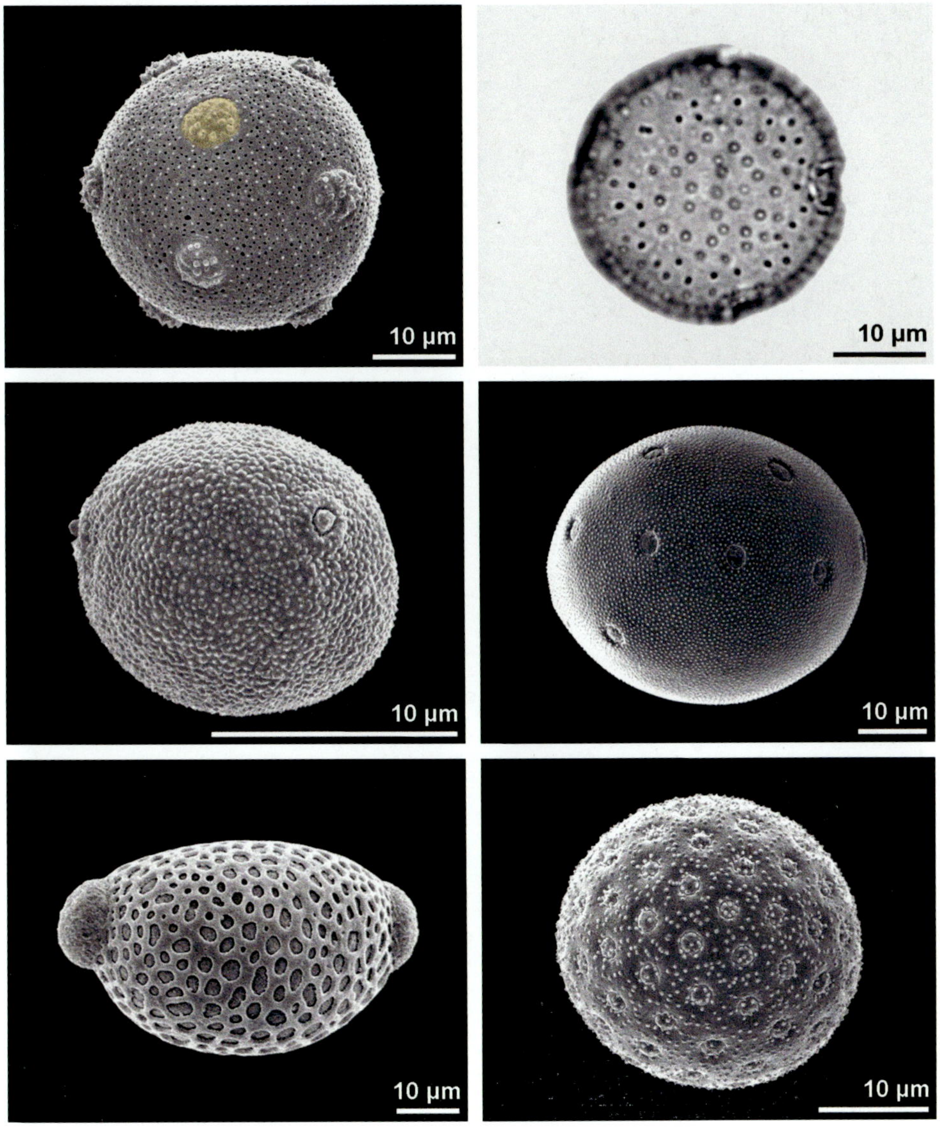

Arenaria pungens
Caryophyllaceae
pantoporate

Urtica dioica
Urticaceae
triporate, oblique equatorial view

Aechmea allenii
Bromeliaceae
diporate, equatorial view

Campanula sp.
Caryophyllaceae, fossil
stephanoporate, polar view

Juglans regia
Juglandaceae
equatorial view
pores irregularly spread (exception)

Chenopodium glaucum
Chenopodiaceae
pantoporate

LM SEM TEM mo ana fnc

diporate: pollen grains with two pori.

Colchicum autumnale
Colchicaceae

Aechmea drakeana
Bromeliaceae
equatorial view

Sanchezia nobilis
Acanthaceae

Whitfieldia lateralis
Acanthaceae
dry (left) and turgescent pollen (right)

Broussonetia papyrifera
Moraceae

Quesnelia lateralis
Bromeliaceae

triporate: pollen grain with three pori.

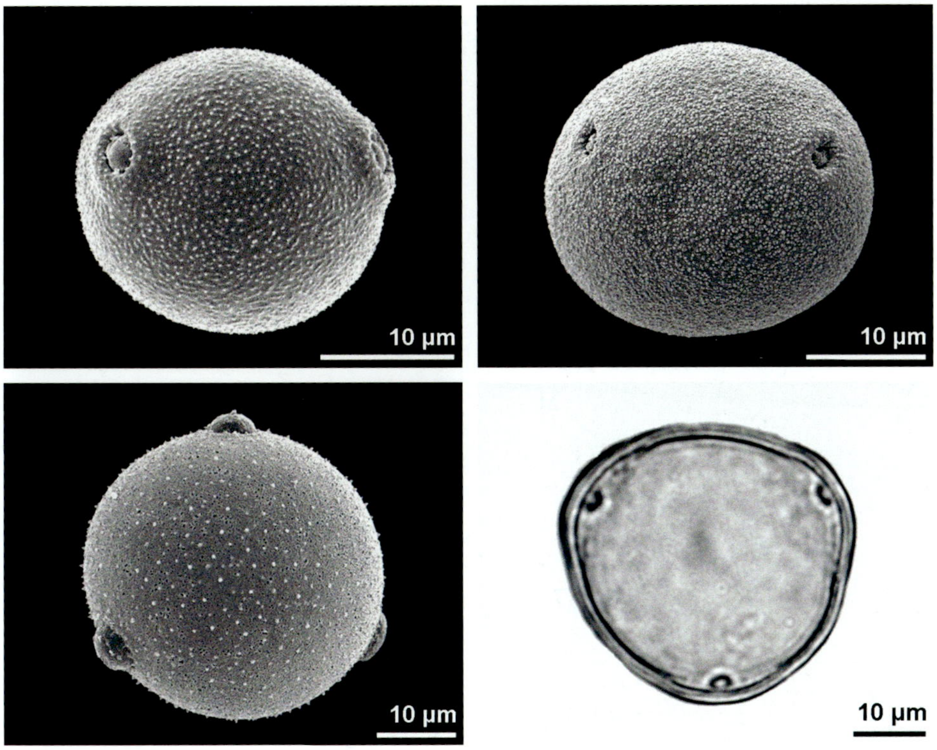

stephanoporate
stephano-: prefix meaning equatorially situated; see "Alphabetic Glossary".

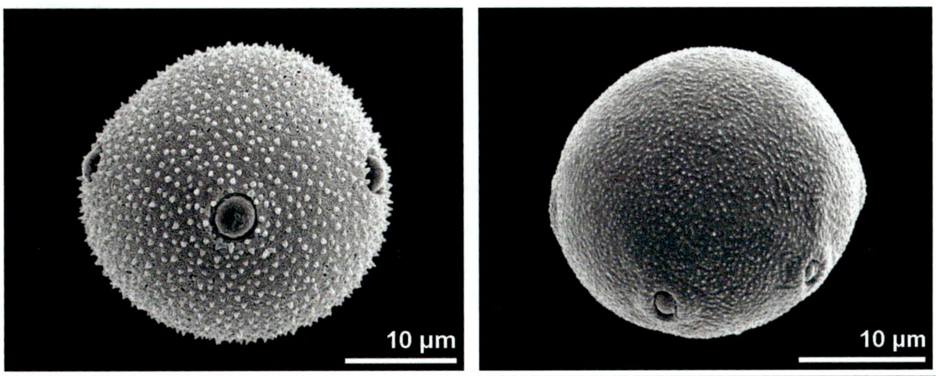

Betula pendula
Betulaceae
equatorial view

Cannabis sativa
Cannabaceae
equatorial view

Campanula saxatilis
Campanulaceae
polar view

Carya sp.
Juglandaceae, fossil
polar view

Legousia speculum-veneris
Caryophyllaceae
equatorial view

Alnus viridis
Betulaceae
oblique equatorial view

LM SEM TEM mo ana fnc

pantoporate
panto-: prefix for global; see "Alphabetic Glossary"

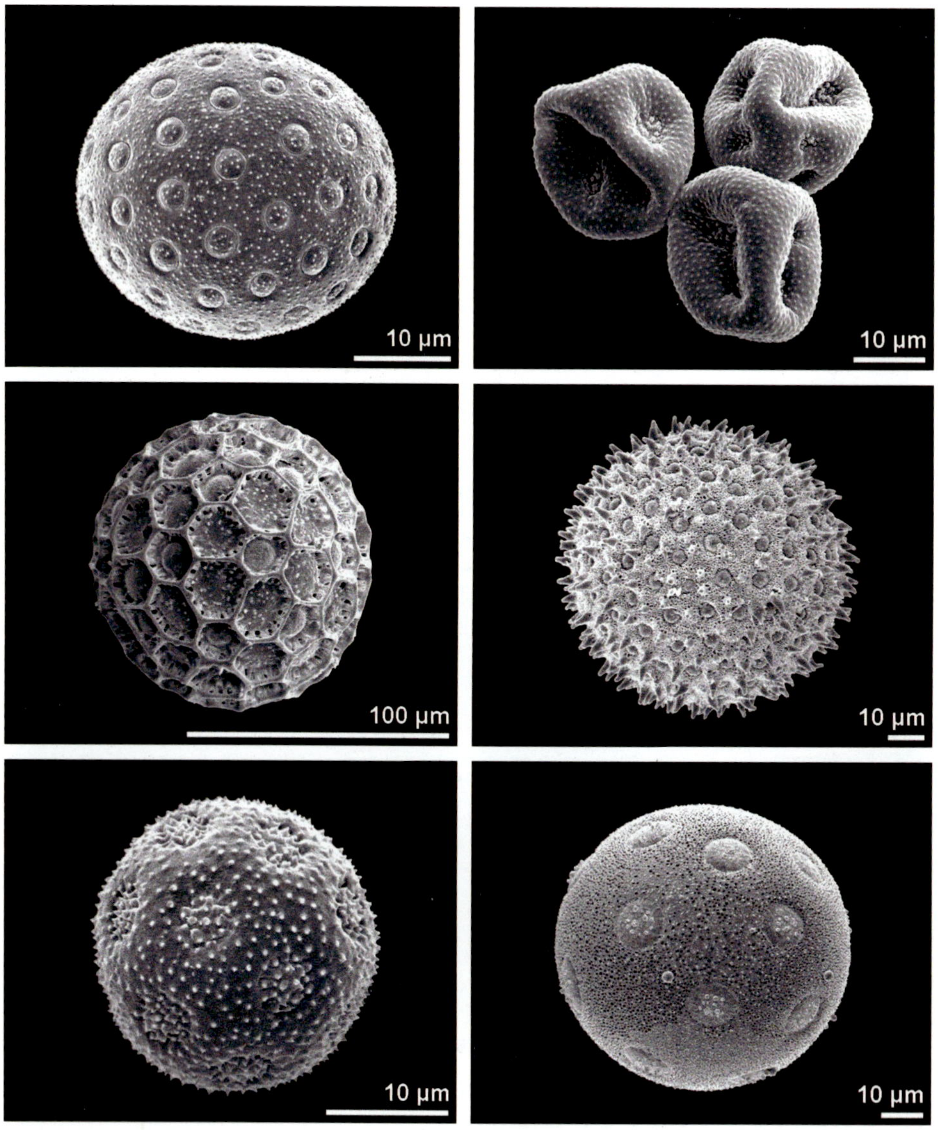

Bassia scoparia
Chenopodiaceae

Thalictrum minus
Ranunculaceae
dry pollen

Cobaea scandens
Polemoniaceae

Ipomoea batatas
Convolvulaceae

Echinodorus quadricostatus
Alismataceae

Calystegia sepium
Convolvulaceae

pantoporate
panto-: prefix for global; see "Alphabetic Glossary"

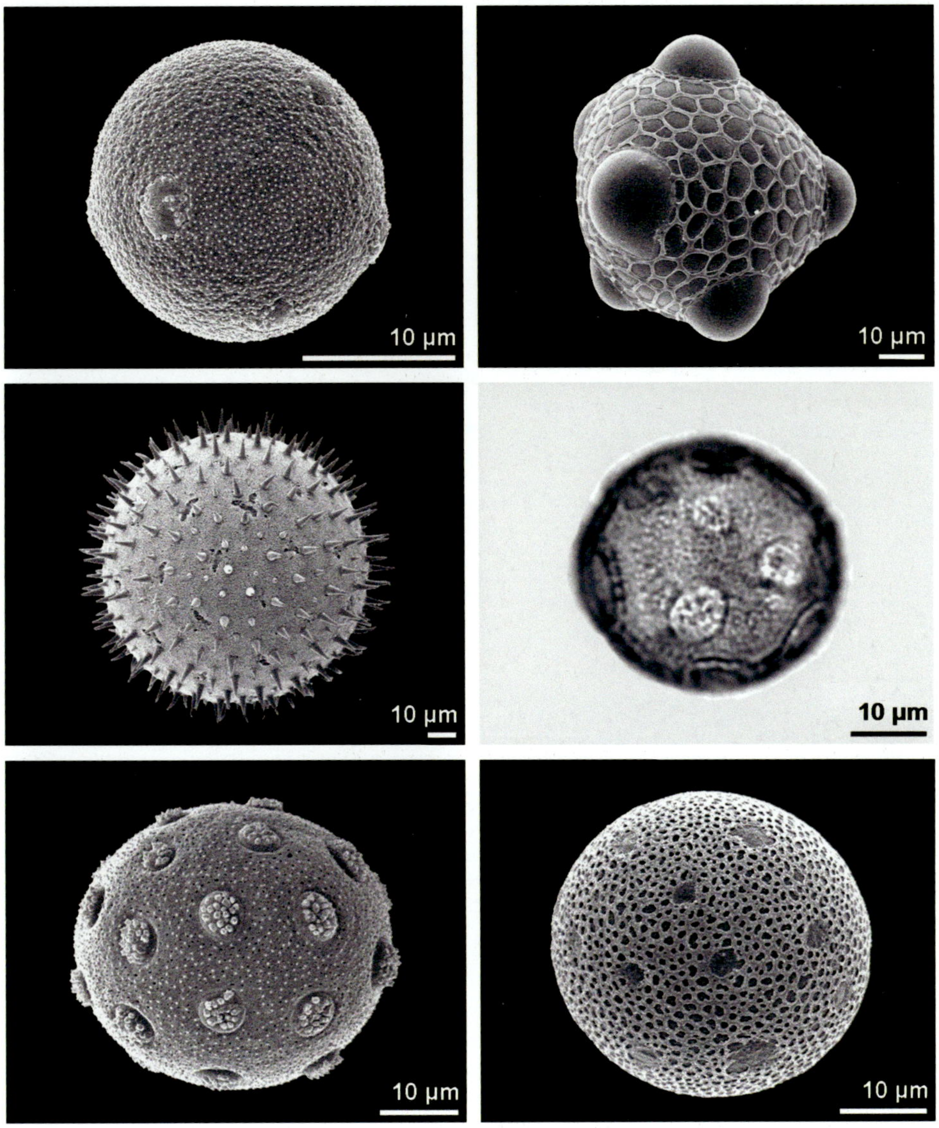

Plantago major
Plantaginaceae

Aechmea azurea
Bromeliaceae

Malva moschata
Malvaceae

Liquidambar sp.
Hamamelidaceae, fossil

Arenaria ciliata
Caryophyllaceae

Buxus sempervirens
Buxaceae

LM SEM TEM mo ana fnc

poroid: circular or elliptic aperture, with indistinct margin.

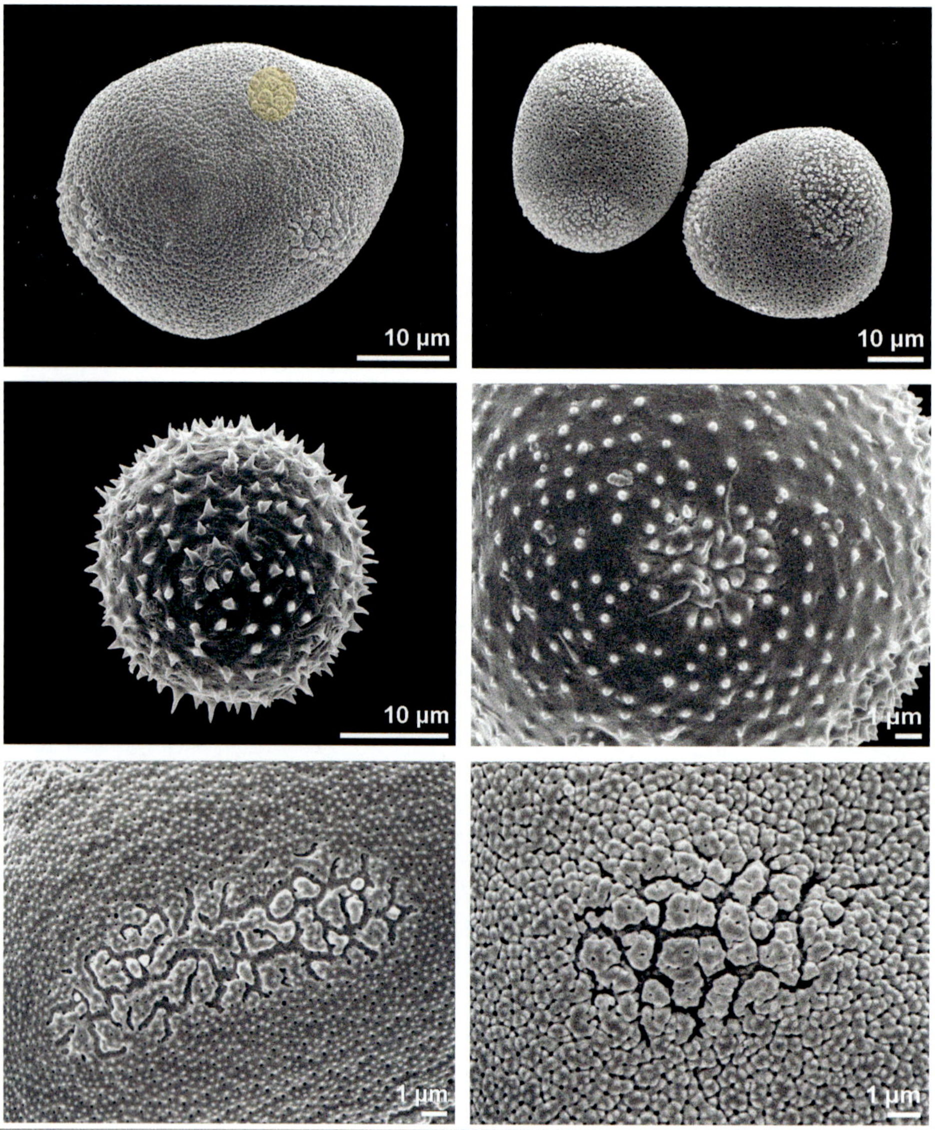

Carex remota
Cyperaceae
pseudomonad

Cercidiphyllum japonicum
Cercidiphyllaceae
triporoidate (in turgescent state)

Sagittaria sagittifolia
Alismataceae

Caldesia parnassifolia
Alismataceae

Schoenoplectus lacustris
Cyperaceae

Scirpus sylvaticus
Cyperaceae

tenuitas: general term for a thinning of the pollen wall.

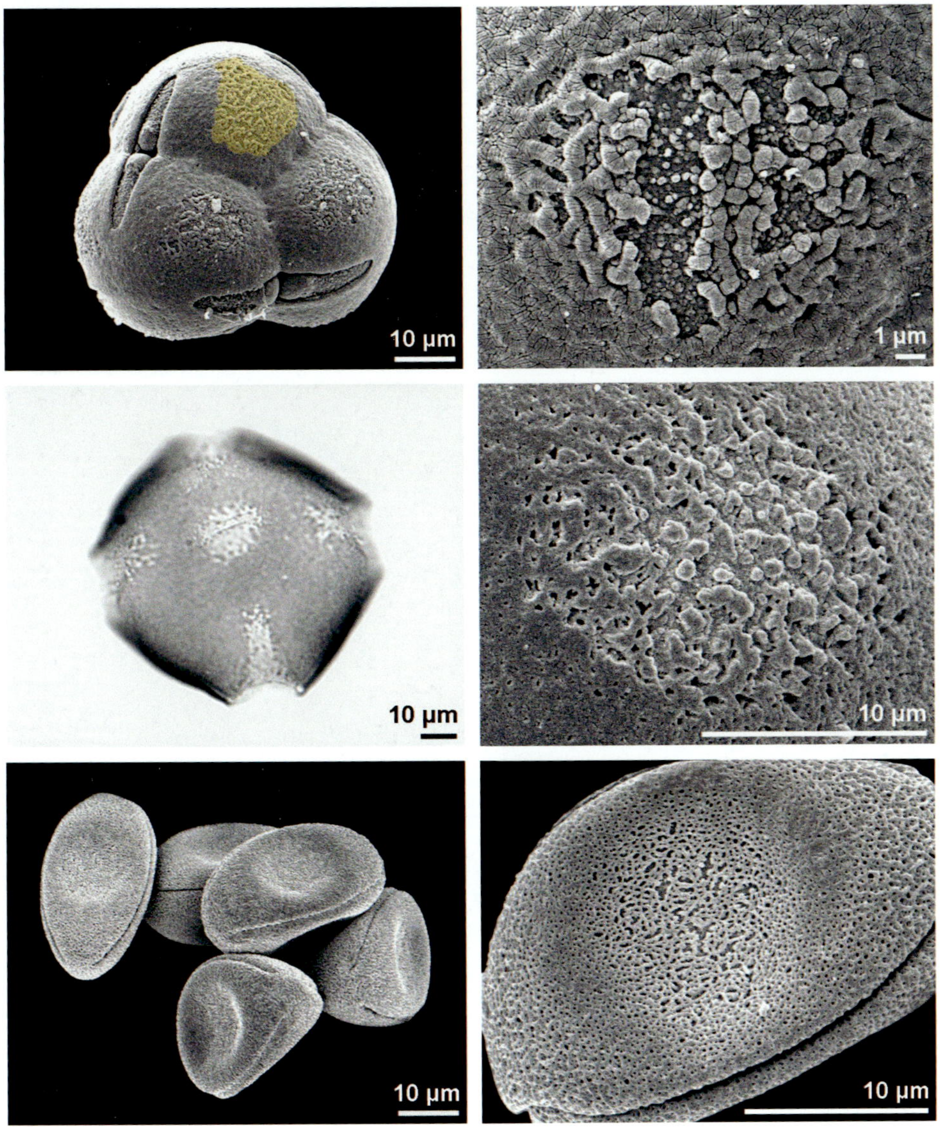

Agapetes variegata
Ericaceae
tetrad

Agapetes variegata
Ericaceae

Viola tricolor
Violaceae
polar view

Viola tricolor
Violaceae

Odontites vulgaris
Scrophulariaceae
dry pollen

Odontites vulgaris
Scrophulariaceae
dry pollen

LM SEM TEM mo ana fnc

colporus: compound aperture composed of a colpus (ektoaperture) combined with an endoaperture of variable size and shape.

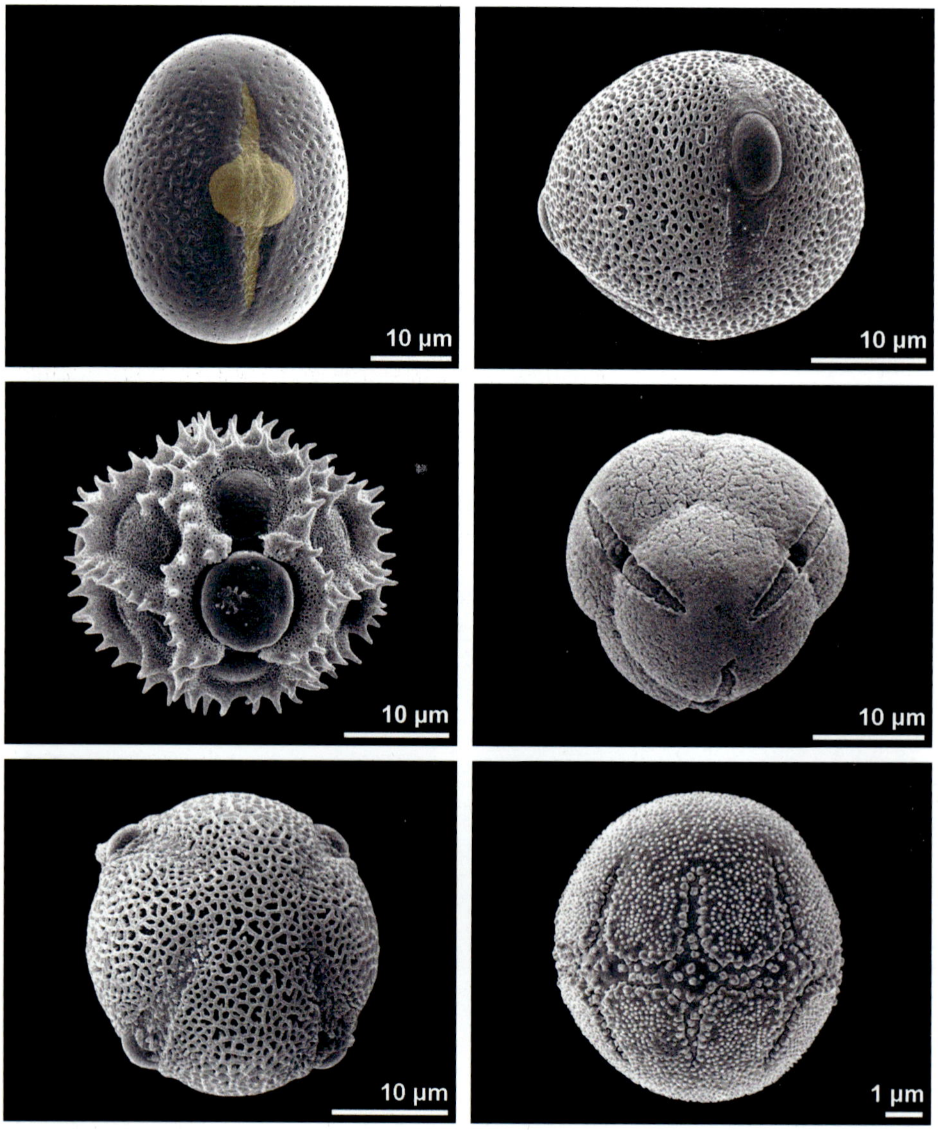

Lathyrus vernus
Fabaceae
tricolporate, equatorial view

Hieracium hoppeanum
Asteraceae
tricolporate, equatorial view

Poncirus trifoliata
Rutaceae
tetracolporate, polar view

Kraussia floribunda
Rubiaceae
tricolporate, equatorial view

Erica herbacea
Ericaceae
tricolporate, tetrad

Cerinthe minor
Boraginaceae
stephanocolporate, equatorial view

LM SEM TEM mo ana fnc

colporus: compound aperture composed of a colpus (ektoaperture) combined with an endoaperture of variable size and shape.

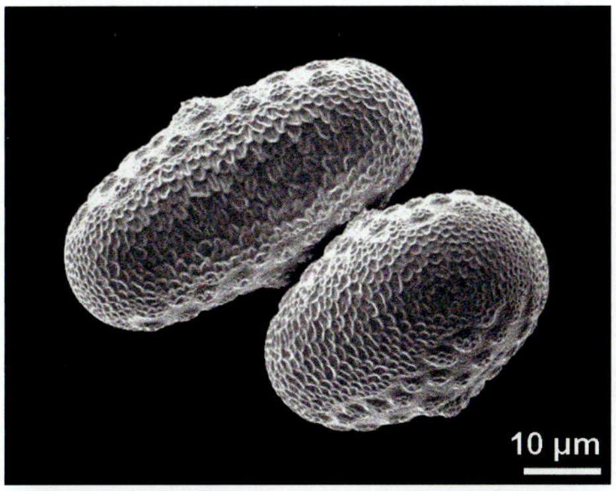

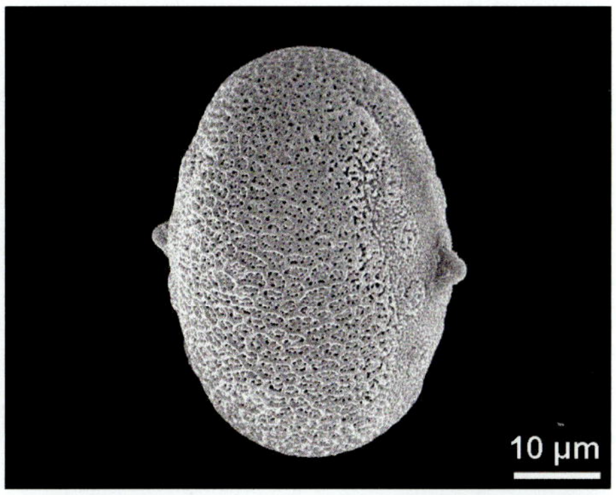

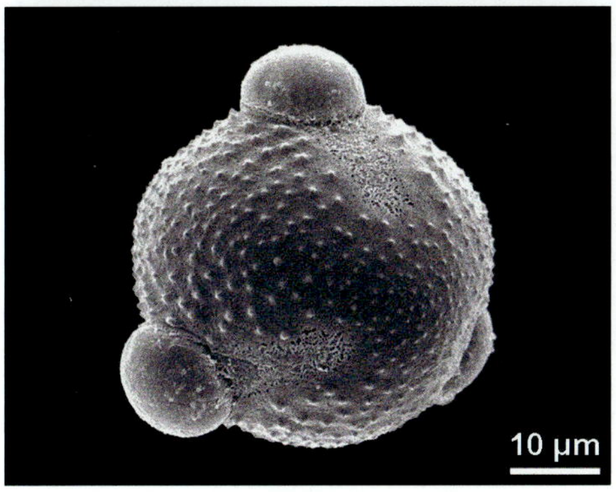

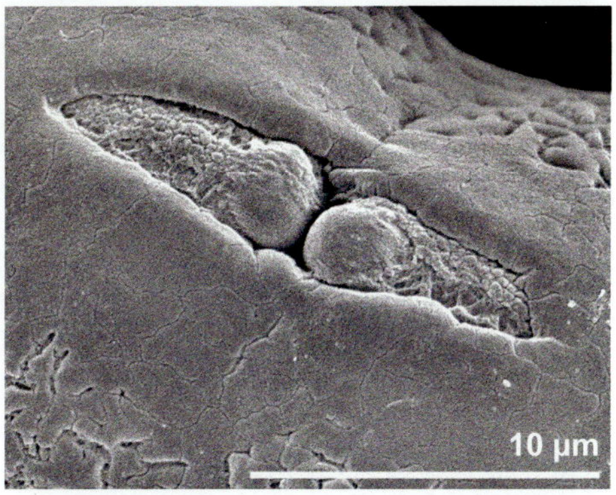

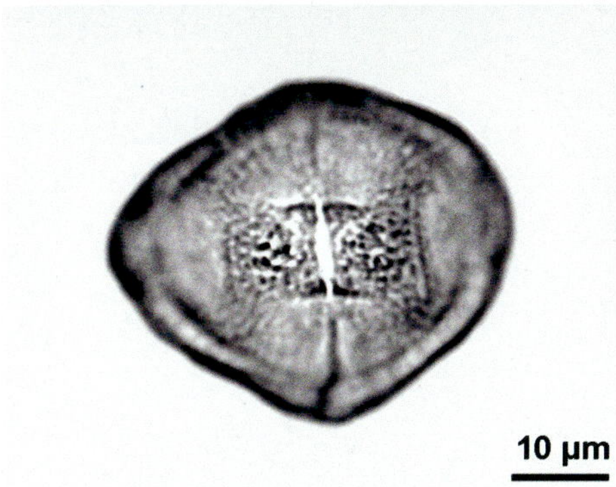

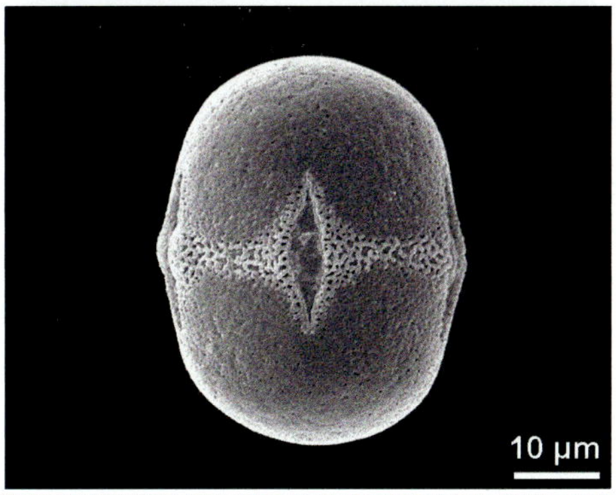

Justicia xylosteoides
Acanthaceae
dicolporate

Centaurea scabiosa
Asteraceae
tricolporate, polar view

Tricolporopollenites wackersdorfensis
Fabaceae, fossil
equatorial view

Adhatoda schimperiana
Acanthaceae
dicolporate, equatorial view

Rhododendron hirsutum
Ericaceae
tricolporate, tetrad

Pulmonaria mollissima
Boraginaceae
stephanocolporate, equatorial view

LM SEM TEM mo ana fnc

tricolporate: pollen grain with three colpori.

Fagus sp.
Fagaceae, fossil
equatorial view

Rumex acetosa
Polygonaceae
equatorial view

Parnassia palustris
Parnassiaceae
equatorial view

Verbena officinalis
Verbenaceae
polar view

Echium vulgare
Boraginaceae
equatorial view

Fatsia japonica
Araliaceae

LM SEM TEM mo ana fnc

tricolporate: pollen grain with three colpori.

Antennaria dioica
Asteraceae
equatorial view

Lycium barbarum
Solanaceae
equatorial view

Potentilla erecta
Rosaceae
equatorial view

Erica arborea
Ericaceae
tetrad

Myrrhis odorata
Apiaceae
dry pollen

Cirsium oleraceum
Asteraceae
equatorial view

LM SEM TEM mo ana fnc

stephanocolporate
stephano-: prefix meaning equatorially situated; see "Alphabetic Glossary"

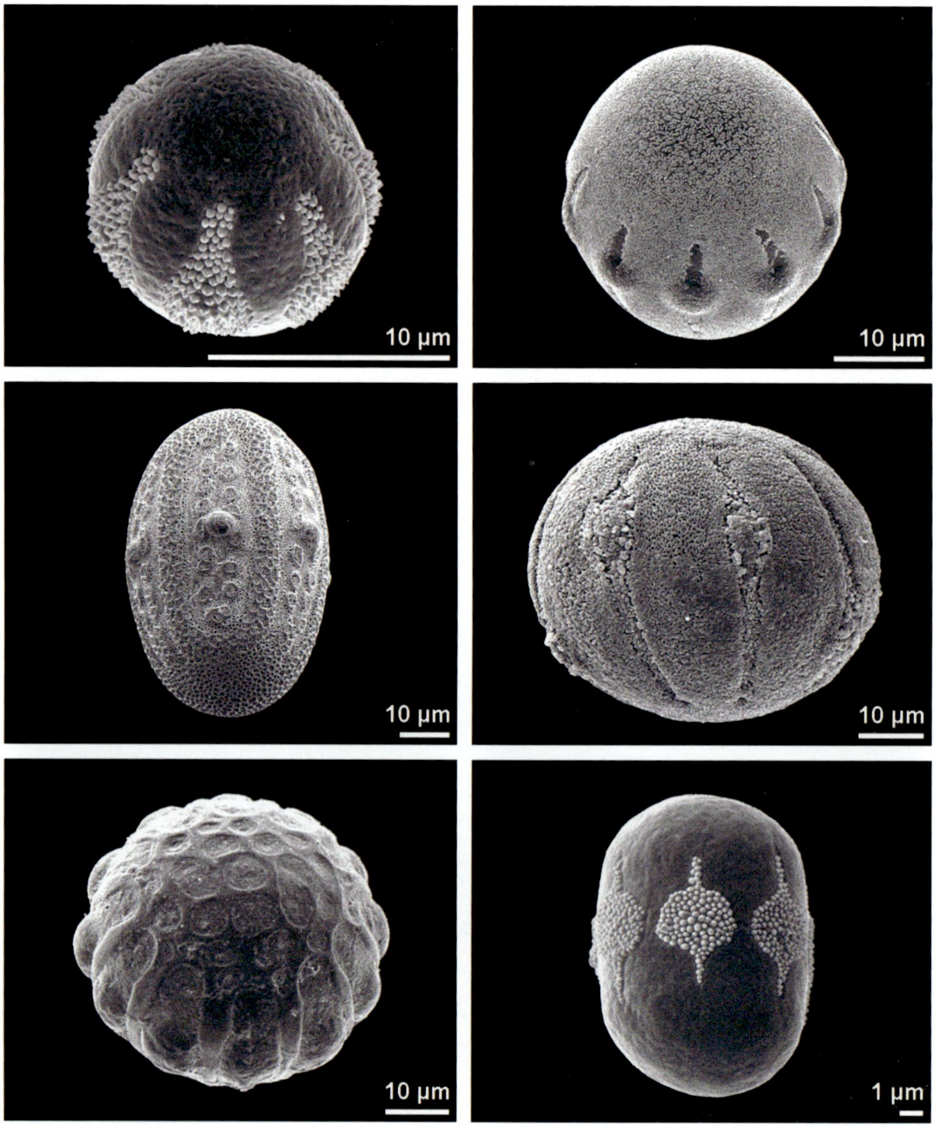

Moltkia petraea
Boraginaceae
oblique polar view

Symphytum caucasicum
Boraginaceae
oblique polar view

Justicia menesii
Acanthaceae
equatorial view

Echinopepon wrightii
Cucurbitaceae
equatorial view

Polygala major
Polygalaceae
oblique polar view

Buglossoides arvensis
Boraginaceae
equatorial view

LM SEM TEM mo ana fnc

ulcus: more or less circular aperture situated distally.

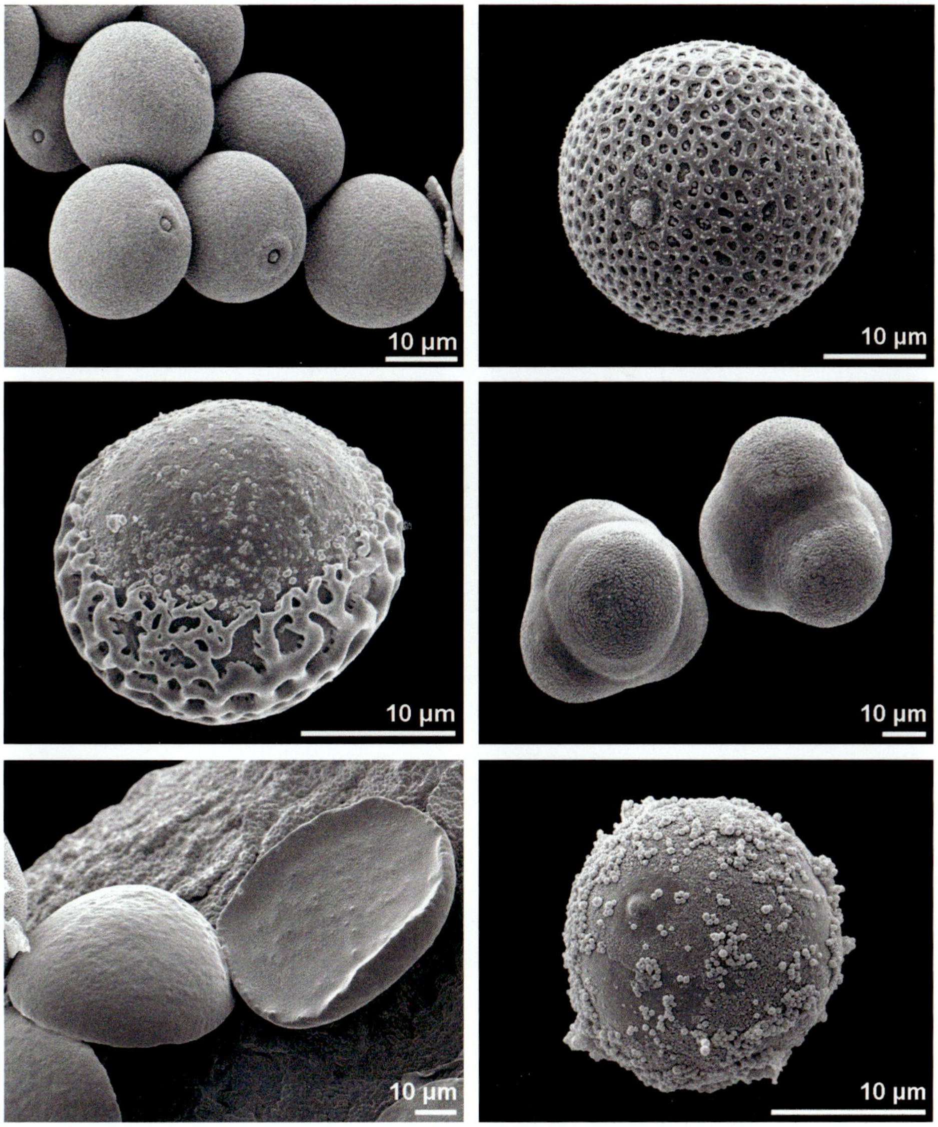

Poa annua
Poaceae

Cephalanthera longifolia
Orchidaceae
oblique distal polar view

Sansevieria dooneri
Dracaenaceae
dry pollen

Sparganium erectum
Sparganiaceae
distal polar view

Luzula luzuloides
Juncaceae
tetrads

Juniperus communis
Cupressaceae
distal polar view

LM SEM TEM mo ana fnc

ulcus: more or less circular aperture situated distally.

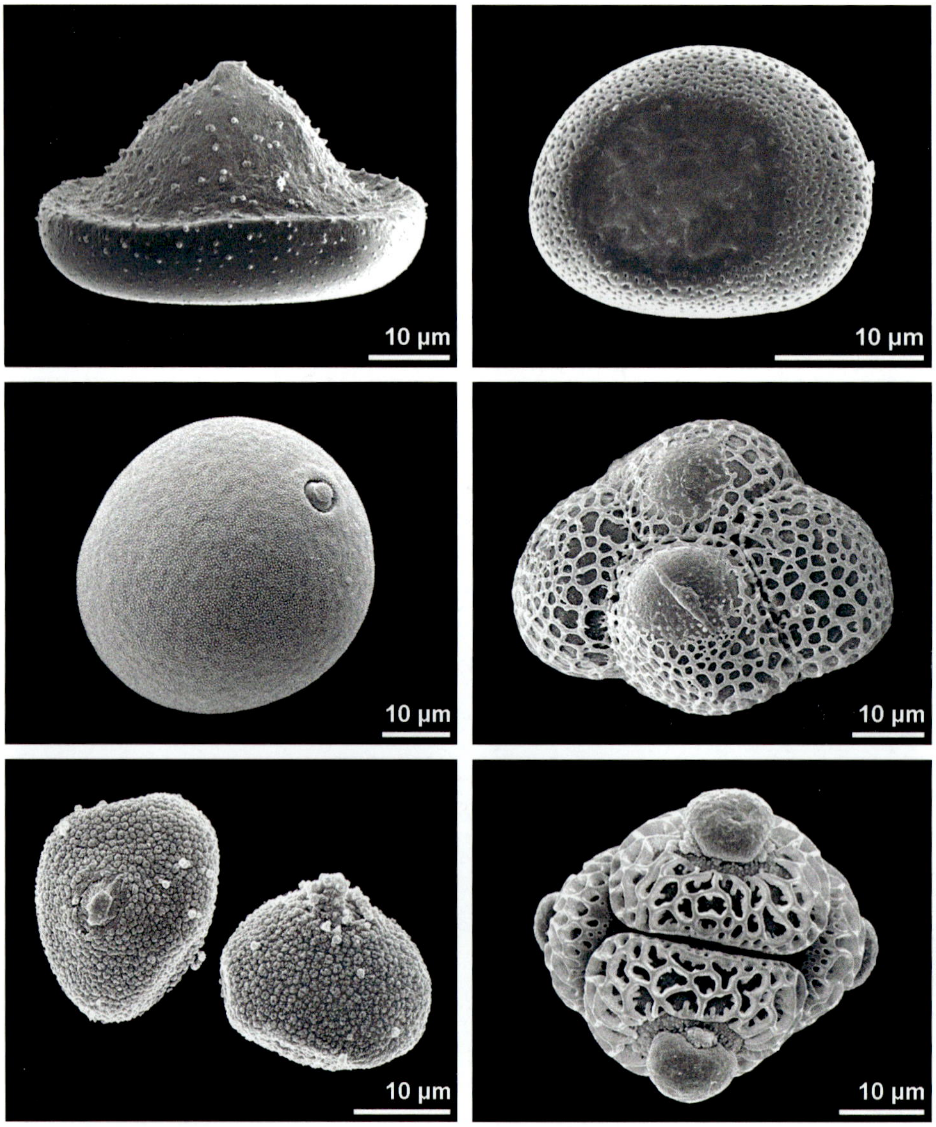

Heliconia sp.
Heliconiaceae
equatorial view, dry pollen

Bromus erectus
Poaceae
oblique distal polar view

Amborella trichopoda
Amborellaceae

Cyrtosperma beccarianum
Araceae
distal polar view

Neottia nidus-avis
Orchidaceae
tetrad

Drimys granatensis
Winteraceae
tetrad

sulcus: elongated aperture situated distally.

Lilium martagon
Liliaceae
distal polar view

Galanthus nivalis
Amaryllidaceae
distal polar view

Doryanthes palmeri
Agavaceae
dry pollen

Allium ursinum
Alliaceae
distal polar view

Cabomba palaeformis
Cabombaceae
oblique distal polar view

Asphodeline lutea
Liliaceae

LM SEM TEM mo ana fnc

sulcus: elongated aperture situated distally.

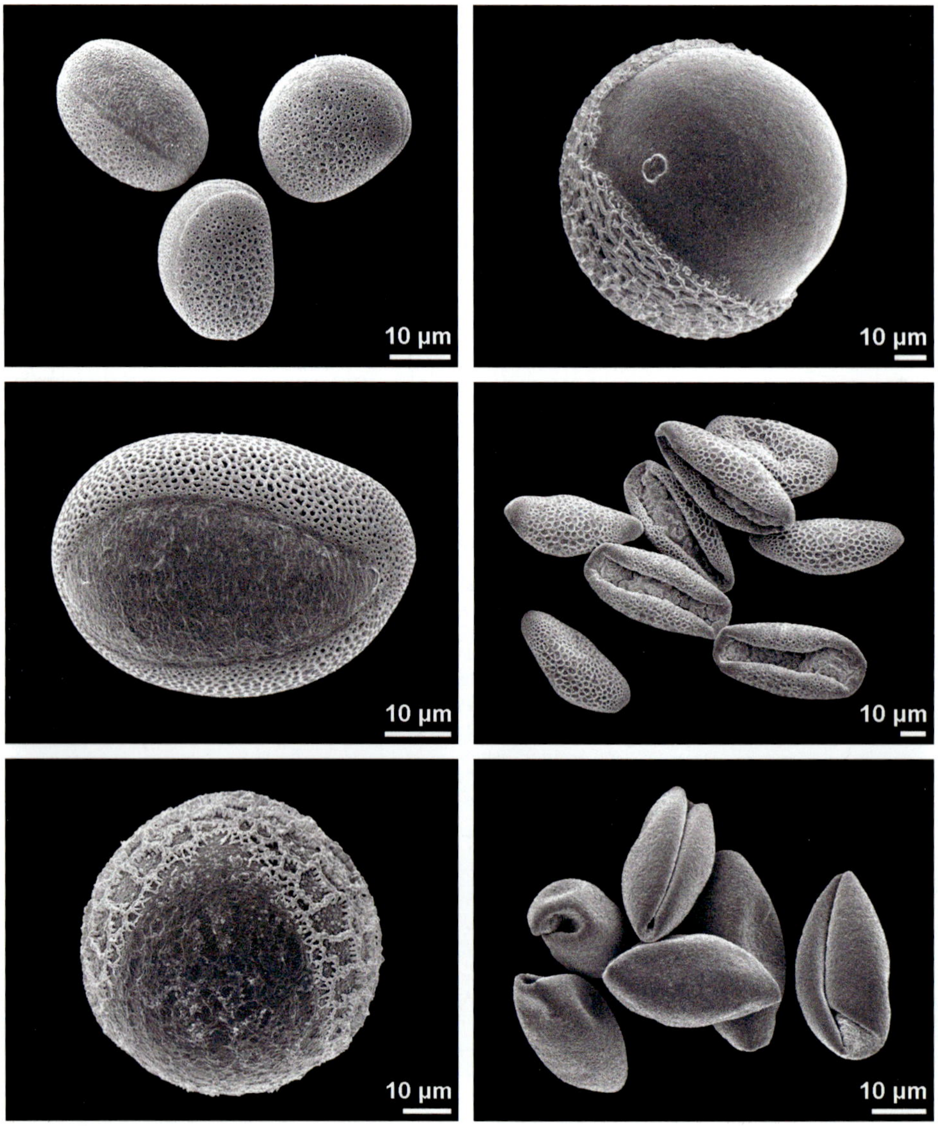

Lachenalia aloides
Hyacinthaceae

Iris reichenbachii
Iridaceae
oblique distal polar view

Catopsis floribunda
Bromeliaceae
distal polar view

Vriesea neoglutinosa
Bromeliaceae
dry pollen

Paradisea liliastrum
Anthericaceae
equatorial view

Chamaedorea microspadix
Arecaceae
dry pollen

sulcus: elongated aperture situated distally.

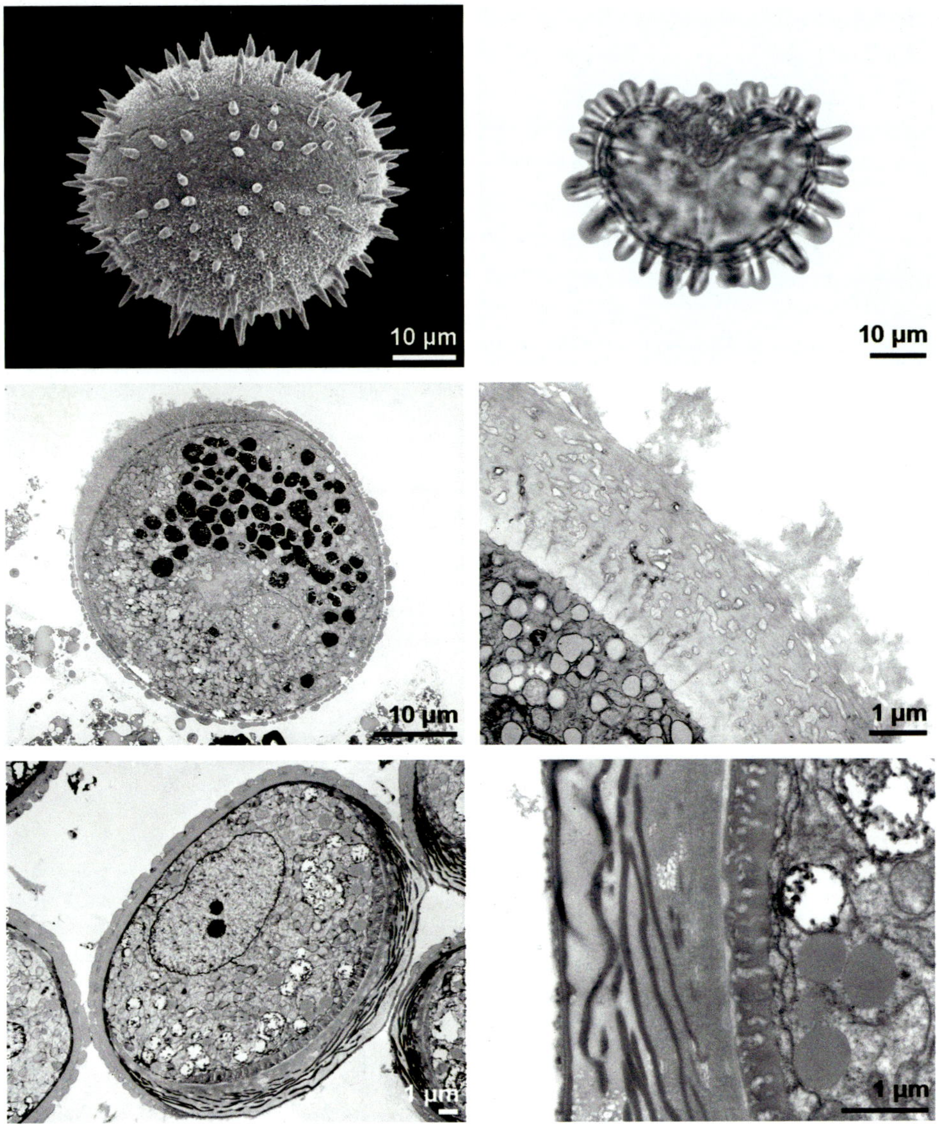

Nuphar luteum
Nymphaeaceae
oblique distal polar view

Nuphar luteum
Nymphaeaceae
equatorial view

Liriodendron tulipifera
Magnoliaceae
PA+TCH+SP

Liriodendron tulipifera
Magnoliaceae
PA+TCH+SP
detail of sulcus

Anaphyllopsis americana
Araceae
U+Pb

Anaphyllopsis americana
Araceae
U+Pb
detail of sulcus

LM SEM TEM mo ana fnc

disulcate: pollen grain with two sulci.

Tofieldia calyculata
Tofieldiaceae
equatorial view

Tofieldia calyculata
Tofieldiaceae
dry pollen

Uvularia grandiflora
Convallariaceae
equatorial view

Uvularia grandiflora
Convallariaceae
dry pollen

Eichhornia crassipes
Pontederiaceae
dry pollen

Crinum augustum
Amaryllidaceae
dry pollen

LM SEM TEM mo ana fnc

trichotomosulcus: three-radiate sulcus.

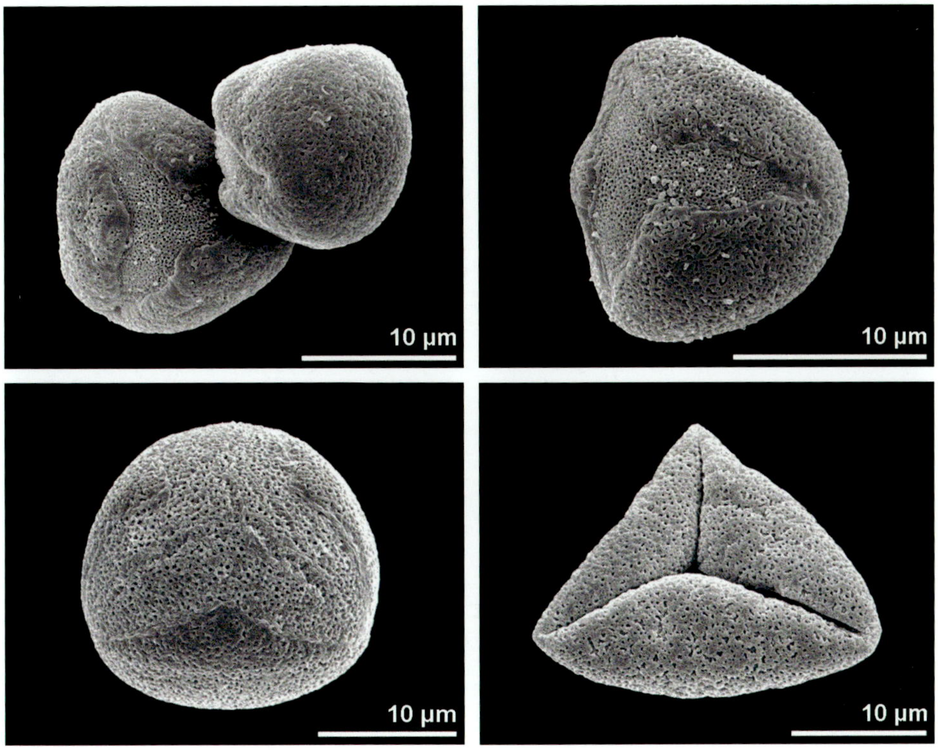

polychotomosulcus: sulcus with more than three arms.

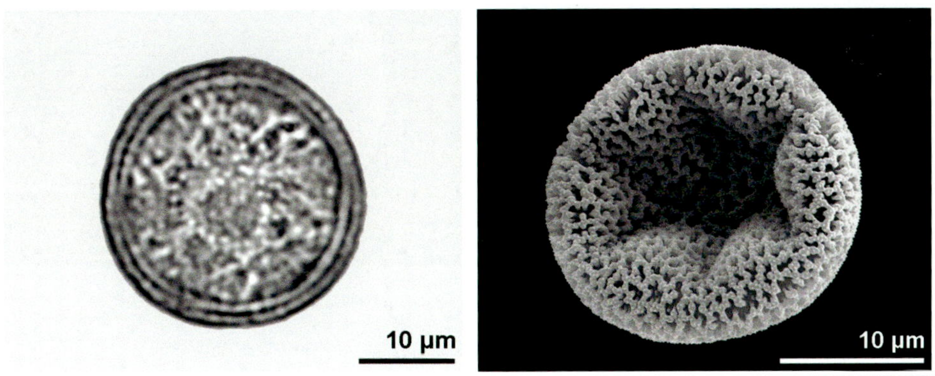

Dianella intermedia
Phormiaceae

Dianella intermedia
Phormiaceae
oblique distal polar view

Dianella tasmanica
Phormiaceae
distal polar view

Dianella tasmanica
Phormiaceae
dry pollen

Hedyosmum goudotianum
Chloranthaceae
polar view, acetolyzed

Hedyosmum goudotianum
Chloranthaceae
polar view, acetolyzed

LM SEM TEM mo ana fnc

ring-like aperture: circumferential aperture (situated more or less equatorially or, rarely, meridionally).

Limnanthes douglasii
Limnanthaceae
ring-like aperture running equatorially

Iris histrioides
Iridaceae
dry pollen, ring-like aperture running equatorially

Pedicularis palustris
Scrophulariaceae
dry pollen, ring-like aperture running meridionally

Cephalostemon riedelianus
Rapateaceae
oblique equatorial view, ring-like aperture running equatorially

Pedicularis rostrato–capitata
Scrophulariaceae
dry pollen, ring-like aperture running meridionally

Pedicularis palustris
Scrophulariaceae
ring-like aperture running meridionally

ring-like aperture: circumferential aperture (situated more or less equatorially or, rarely, meridionally).

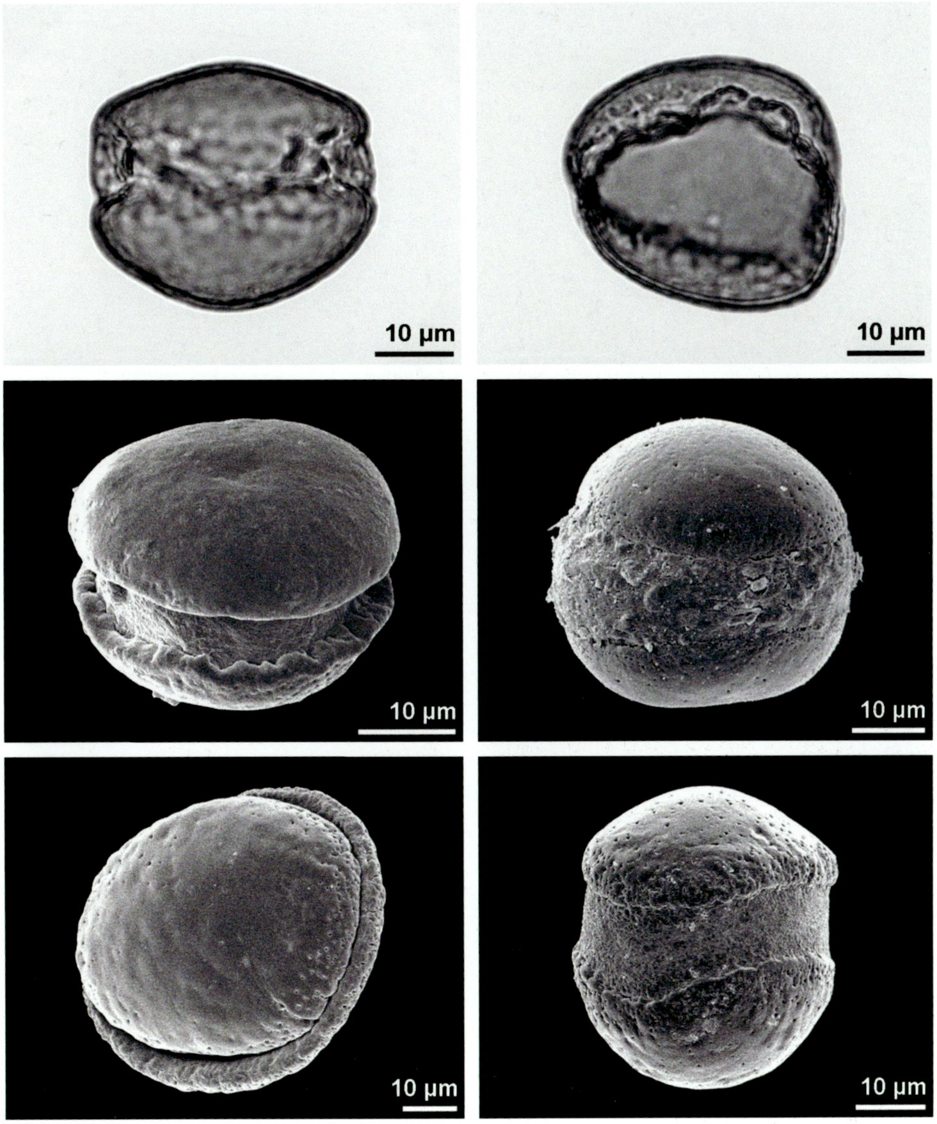

Zamioculcas zamiifolia
Araceae
equatorial view, acetolyzed

Zamioculcas zamiifolia
Araceae
oblique polar view, artificially ruptured, acetolyzed

Zamioculcas zamiifolia
Araceae
equatorial view

Monstera deliciosa
Araceae
equatorial view

Gonatopus angustus
Araceae
oblique polar view, dry pollen

Gonatopus angustus
Araceae
equatorial view

LM SEM TEM mo ana fnc

margo: exine area surrounding an aperture and differentiated in ornamentation.

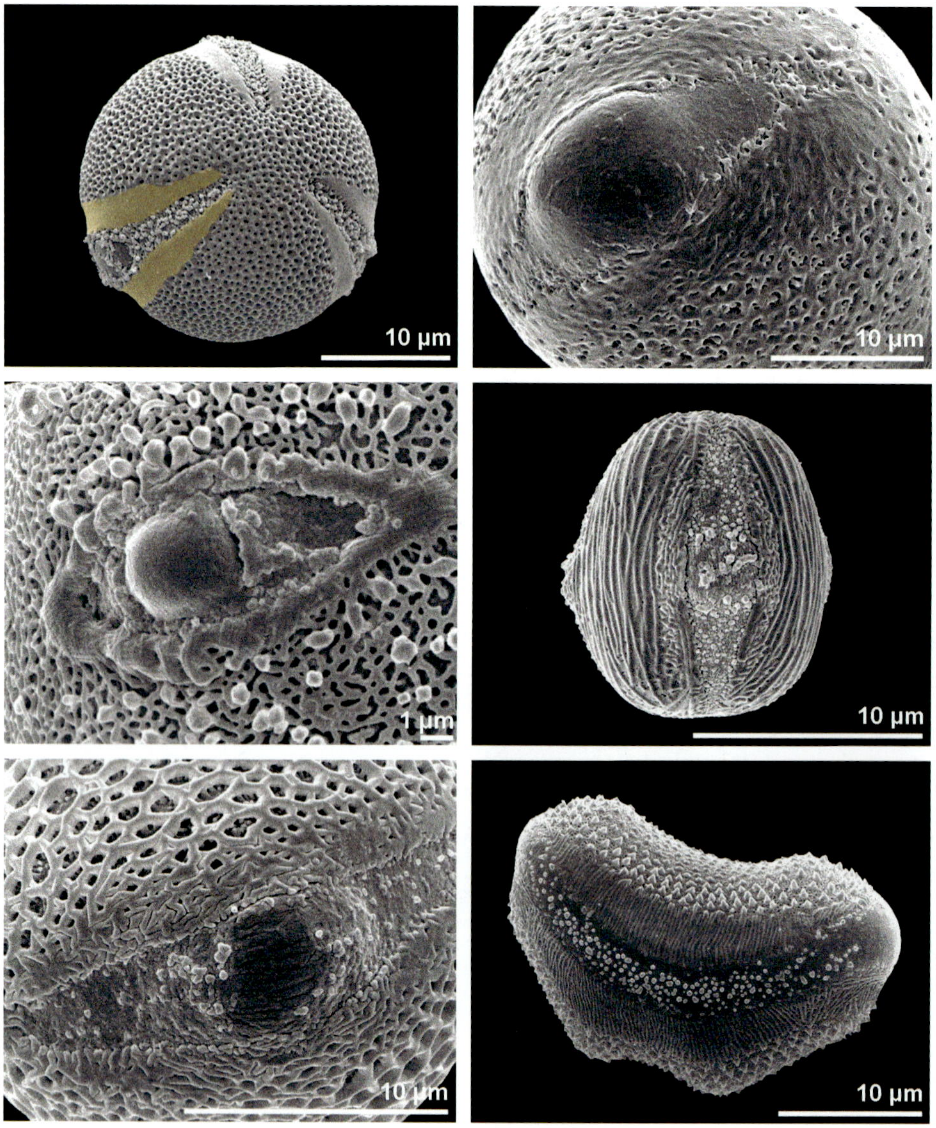

Discocleidion rufescens
Euphorbiaceae
tricolporate, polar view

Fatsia japonica
Araliaceae
tricolporate

Lysimachia vulgaris
Primulaceae
tricolporate

Medicago minima
Fabaceae
tricolporate

Begonia heracleifolia
Begoniaceae
tricolporate, equatorial view

Limnanthes douglasii
Limnanthaceae
ring-like aperture, equatorial view

LM SEM TEM mo ana fnc

annulus: ring-like thickening of the pollen wall surrounding a porus or ulcus.

Comment:
"anulus" orthographical variant of "annulus".

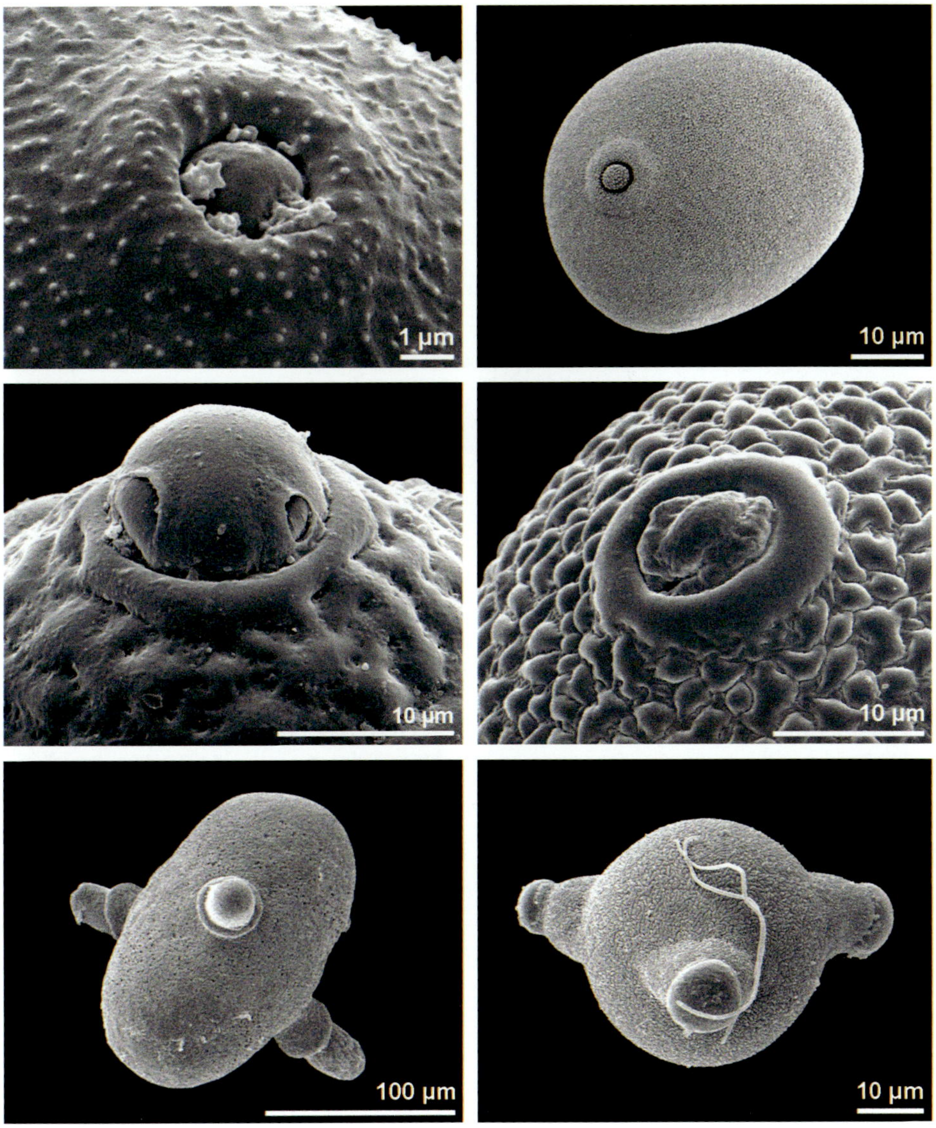

Betula humilis
Betulaceae

Secale cereale
Poaceae
distal polar view

Fumaria officinalis
Fumariaceae

Trichosanthes anguina
Cucurbitaceae

Morina longifolia
Morinaceae
equatorial view

Circaea lutetiana
Onagraceae
equatorial view, viscin threads

LM SEM TEM mo ana fnc

annulus: ring-like thickening of the pollen wall surrounding a porus or ulcus.

Comment:
"anulus" orthographical variant of "annulus".

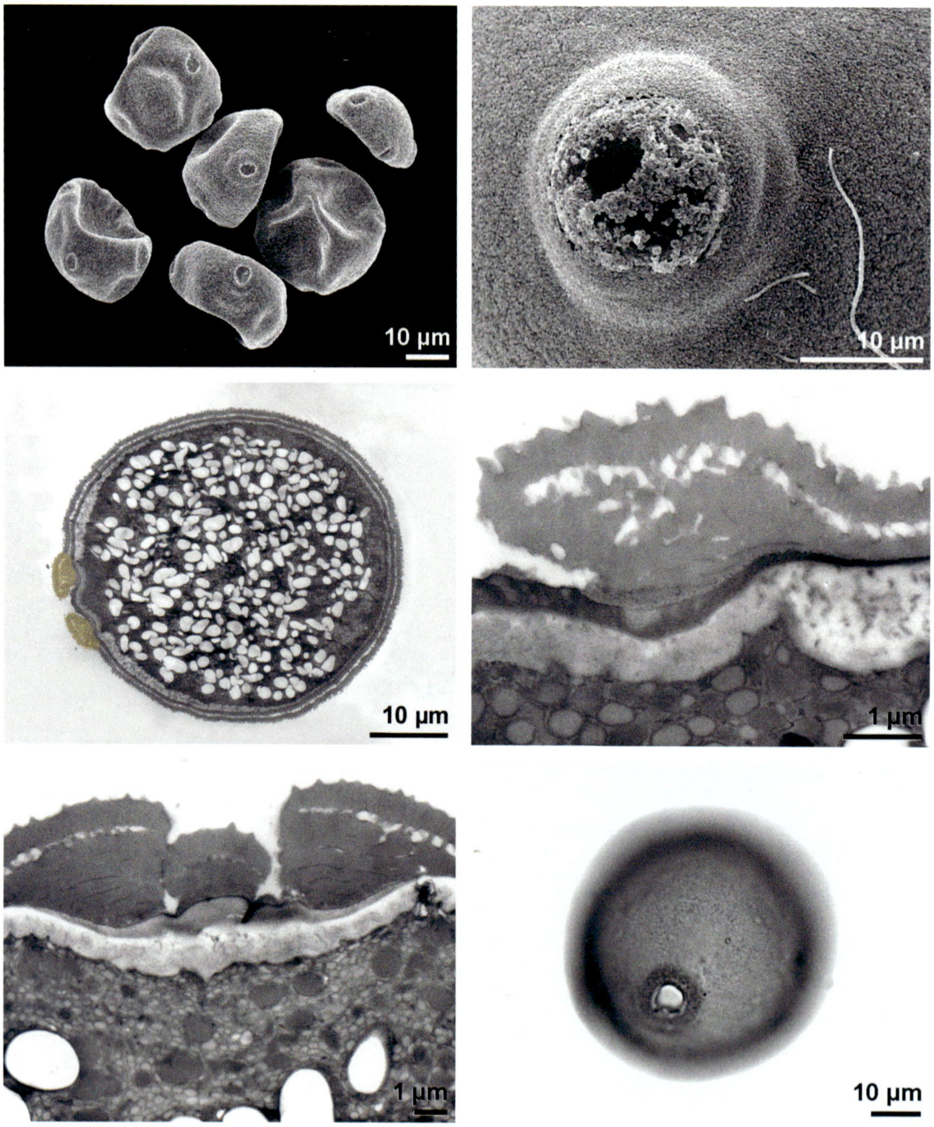

Myriophyllum spicatum
Haloragaceae
dry pollen

Epilobium angustifolium
Onagraceae
viscin threads

Triticum aestivum
Poaceae
U+Pb

Triticum aestivum
Poaceae
U+Pb

Zea mays
Poaceae
U+Pb

indet.
Poaceae
acetolyzed

aperture membrane: exine layer covering an aperture; aperture membrane can be smooth or ornamented.

Comment:
the terms "smooth" and "ornamented" should be used when the feature is remarkably expressed.

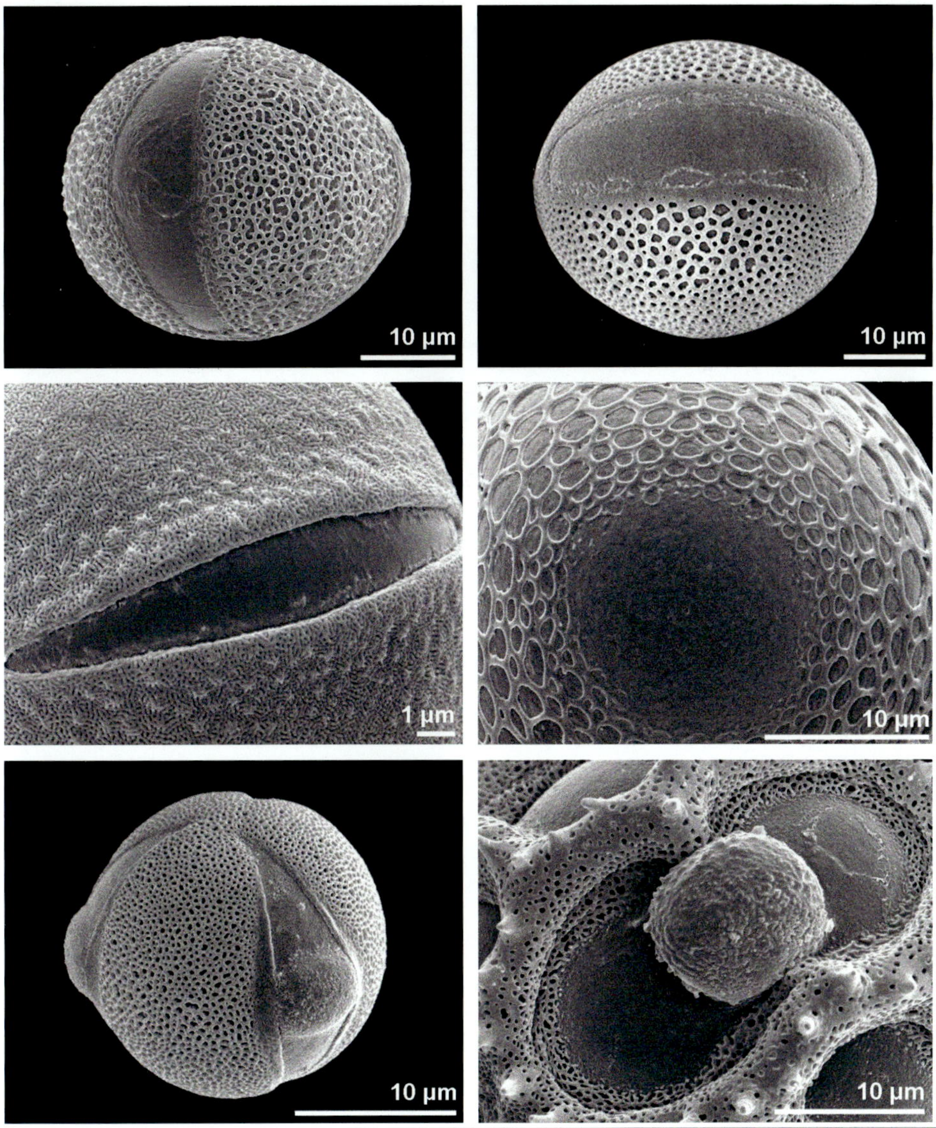

Scrophularia vernalis
Scrophulariaceae
tricolporate, equatorial view

Melampyrum subalpinum
Scrophulariaceae
tricolpate

Kickxia spuria
Scrophulariaceae
tricolporate, equatorial view

Doryanthes palmeri
Agavaceae
sulcate, distal polar view

Aechmea caesia
Bromeliaceae
diporate

Tragopogon dubius
Asteraceae
tricolporate

LM SEM TEM mo ana fnc

aperture membrane: exine layer covering an aperture; aperture membrane can be smooth or ornamented.

Comment:
the terms "smooth" and "ornamented" should be used when the feature is remarkably expressed.

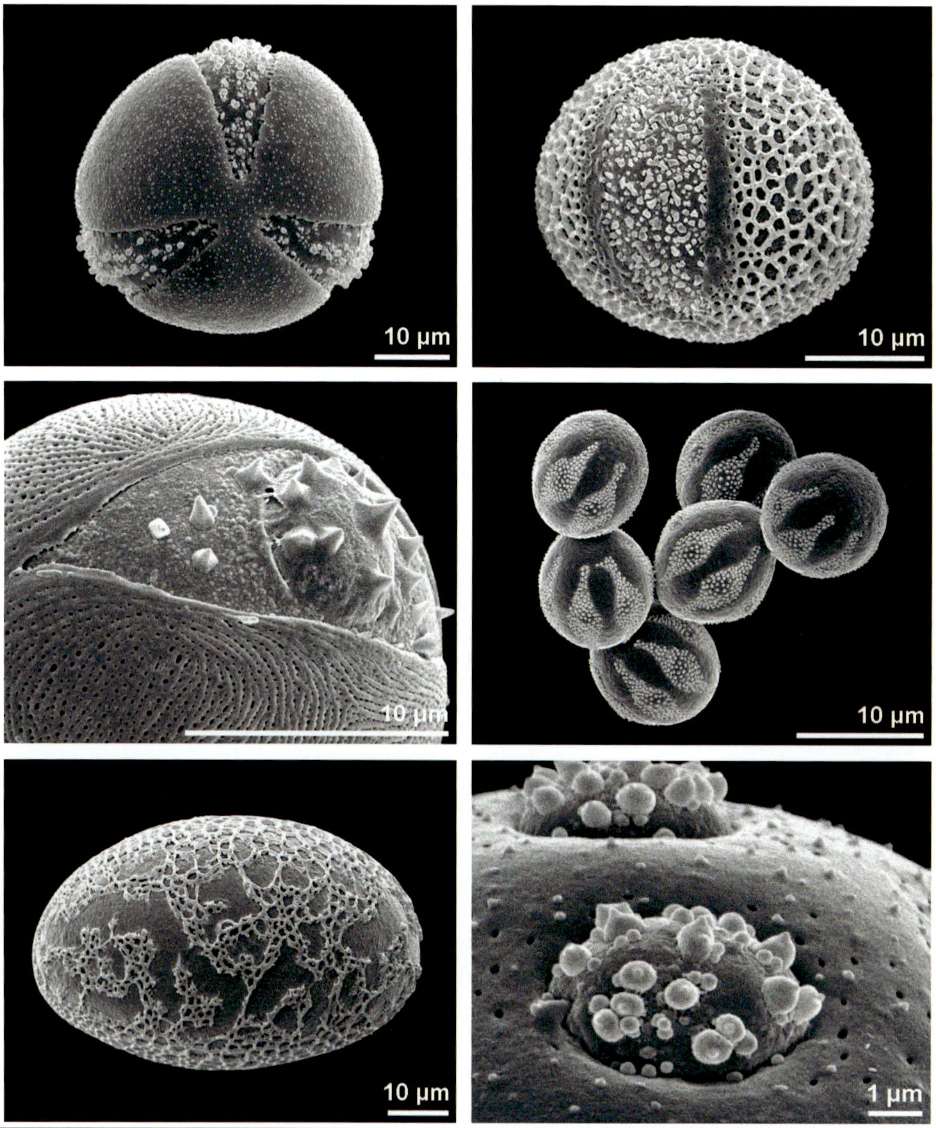

Convolvulus tricolor
Convolvulaceae
tricolpate, polar view

Salix alba
Salicaceae
tricolpate, equatorial view

Aesculus hippocastanum
Hippocastanaceae
tricolporate

Moltkia petraea
Boraginaceae
stephanocolporate

Billbergia macrocalyx
Bromeliaceae
sulcate, distal polar view

Arenaria serpyllifolia
Caryophyllaceae
pantoporate

aperture membrane: exine layer covering an aperture; aperture membrane can be smooth or ornamented.

Comment:
the terms "smooth" and "ornamented" should be used when the feature is remarkably expressed.

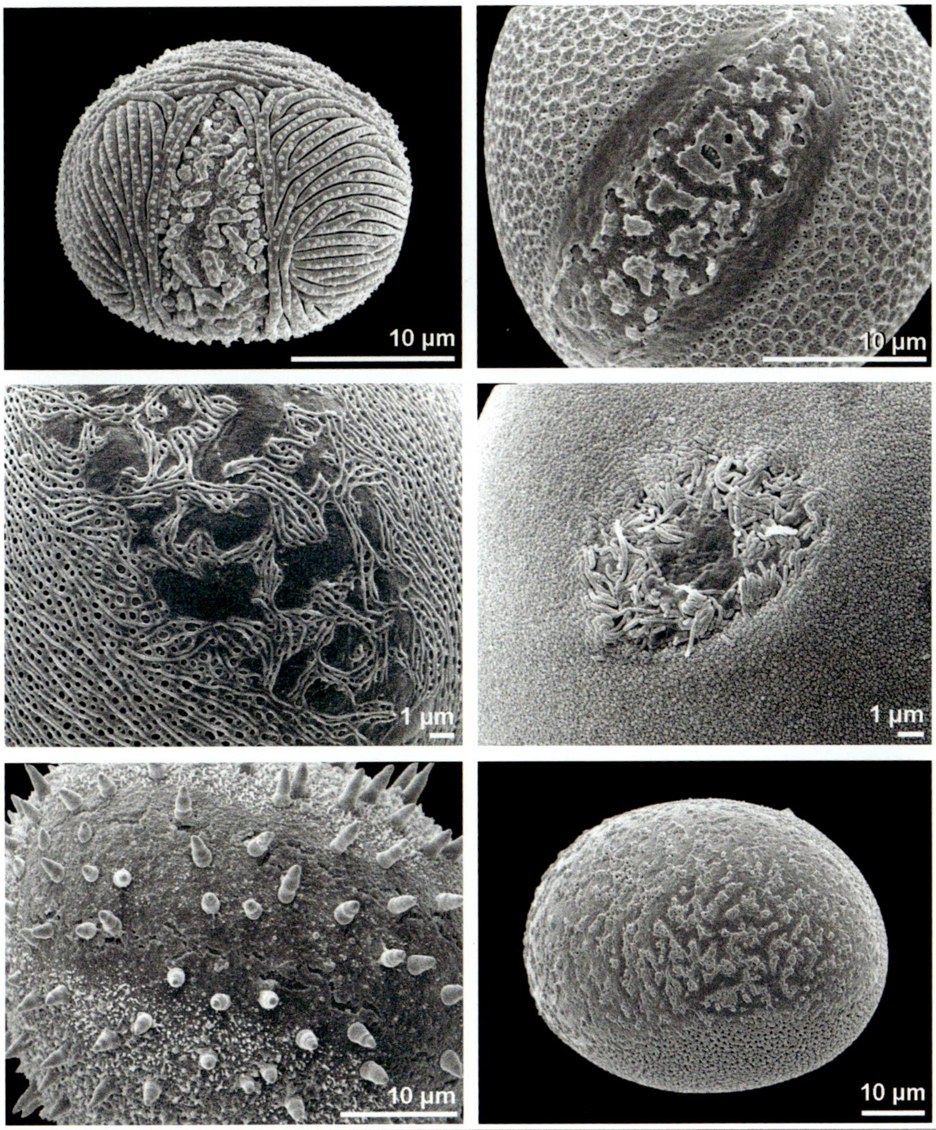

Saxifraga vandellii
Saxifragaceae
tricolpate, equatorial view

Veronica cinerea
Scrophulariaceae
tricolpate

Nuphar luteum
Nymphaeaceae
sulcate

Galeopsis tetrahit
Lamiaceae
tricolpate

Clarkia pulchella
Onagraceae
triporate

Gagea villosa
Liliaceae
sulcate

LM SEM TEM mo ana fnc

aperture membrane: exine layer covering an aperture; aperture membrane can be smooth or ornamented.

Comment:
the terms "smooth" and "ornamented" should be used when the feature is remarkably expressed.

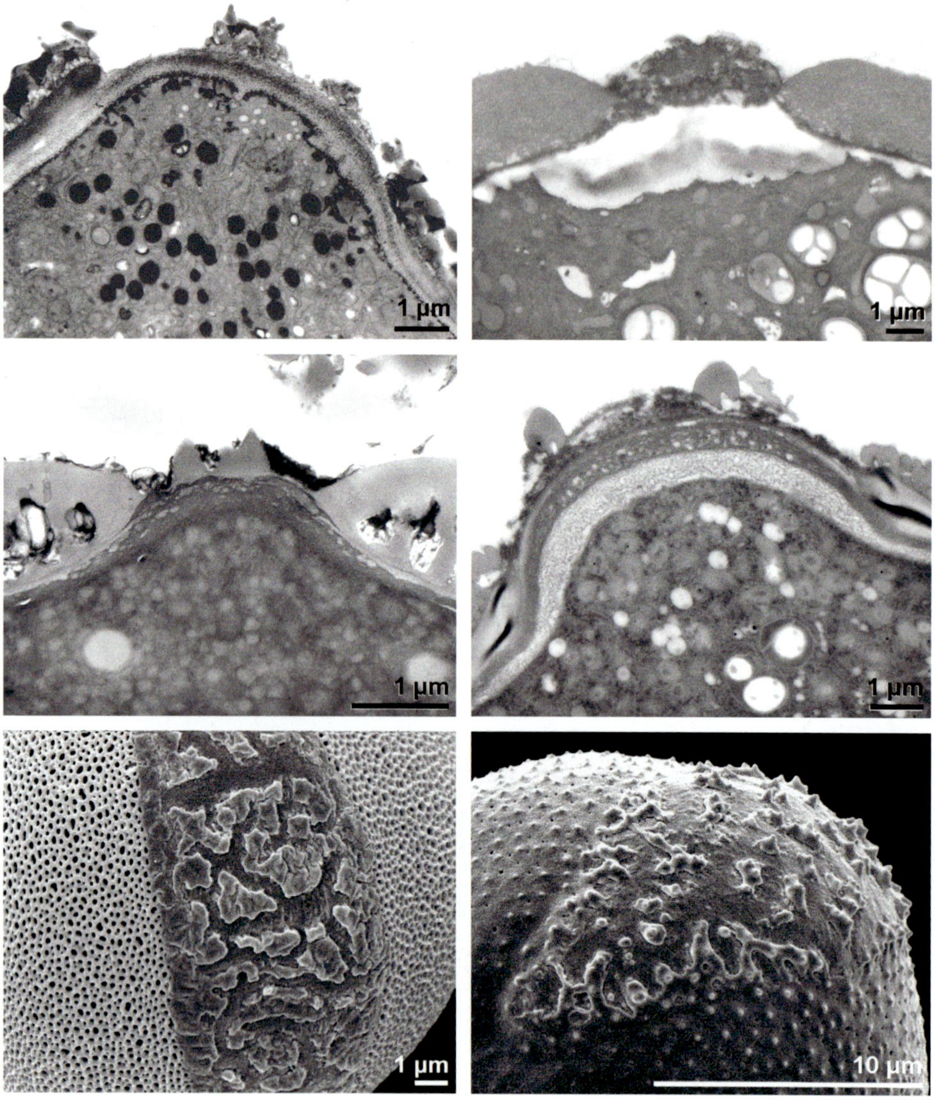

Chenopodium hybridum
Chenopodiaceae
U+Pb

Aesculus carnea
Hippocastanaceae
U+Pb

Mercurialis perennis
Euphorbiaceae
PA+TCH+SP (short)

Ulmus minor
Ulmaceae
U+Pb

Lamiastrum montanum
Lamiaceae

Aconitum lycoctonum
Ranunculaceae

LM SEM TEM mo ana fnc

operculum: coherent exine structure covering an aperture.

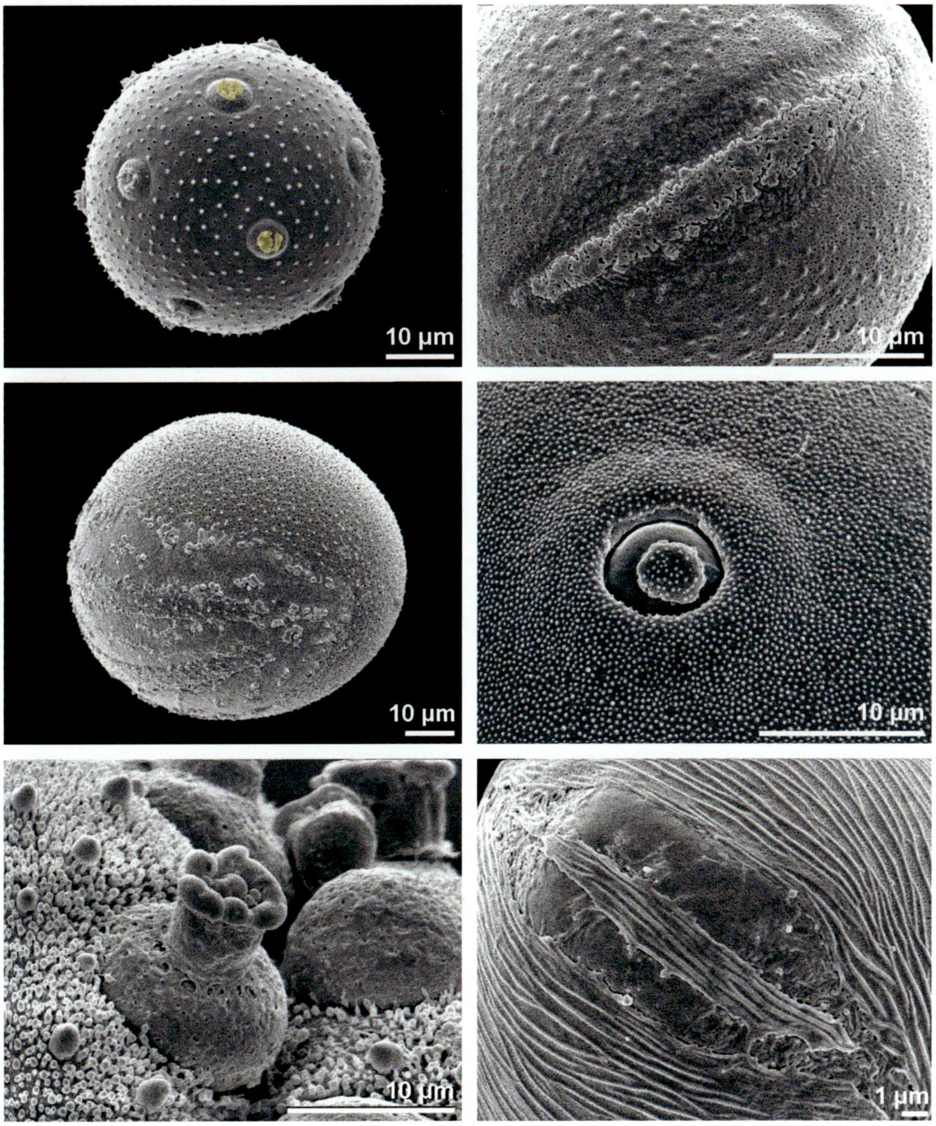

Dianthus carthusianorum
Caryophyllaceae

Teucrium pyrenaicum
Lamiaceae

Babiana velutina
Iridaceae
sulcate, two opercula
oblique equatorial view

Zea mays
Poaceae

Dionaea muscipula
Droseraceae

Potentilla arenaria
Rosaceae

LM SEM TEM mo ana fnc

operculum: coherent exine structure covering an aperture.

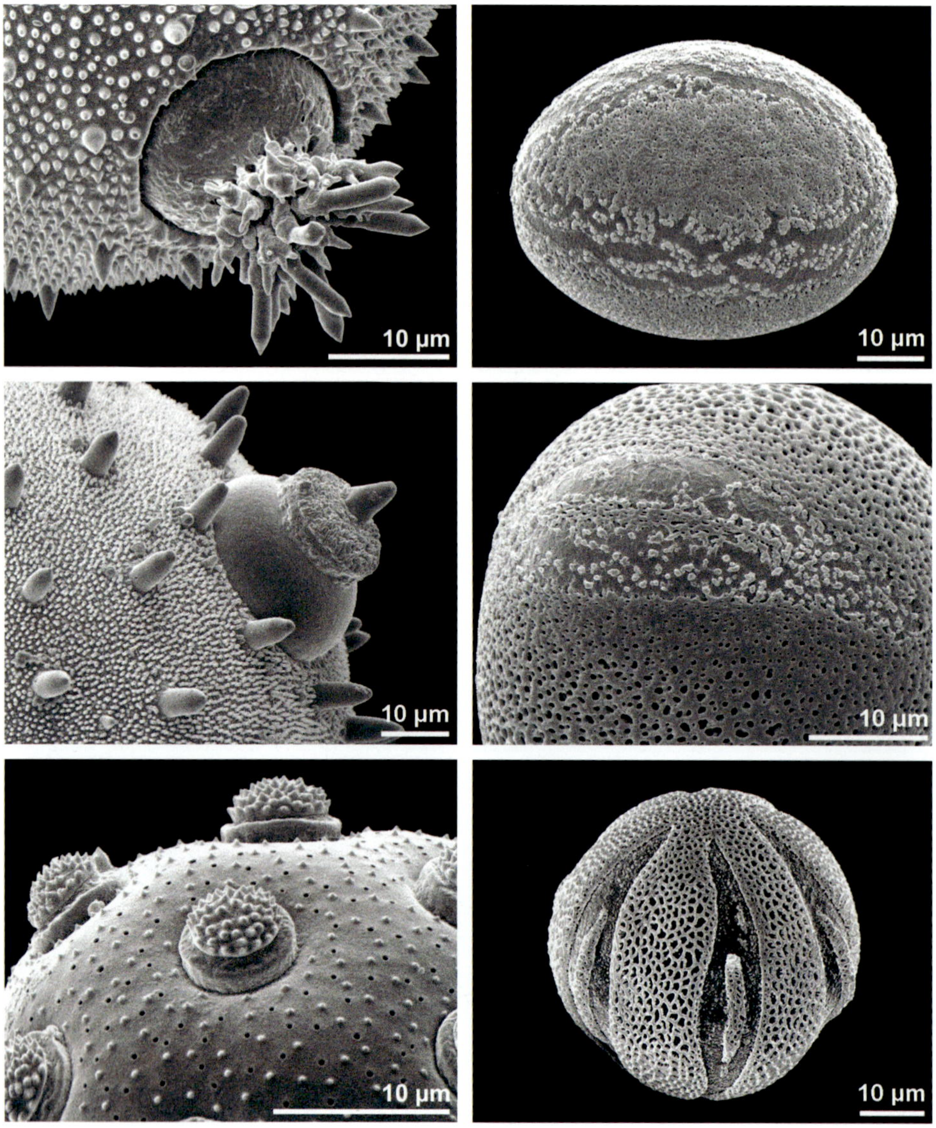

Knautia drymeia
Dipsacaceae

Tulipa sylvestris
Liliaceae
distal polar view

Cucurbita pepo
Cucurbitaceae

Camellia japonica
Theaceae

Agrostemma githago
Caryophyllaceae

Passiflora citrina
Passifloraceae
oblique equatorial view

LM SEM TEM mo ana fnc

operculum: coherent exine structure covering an aperture.

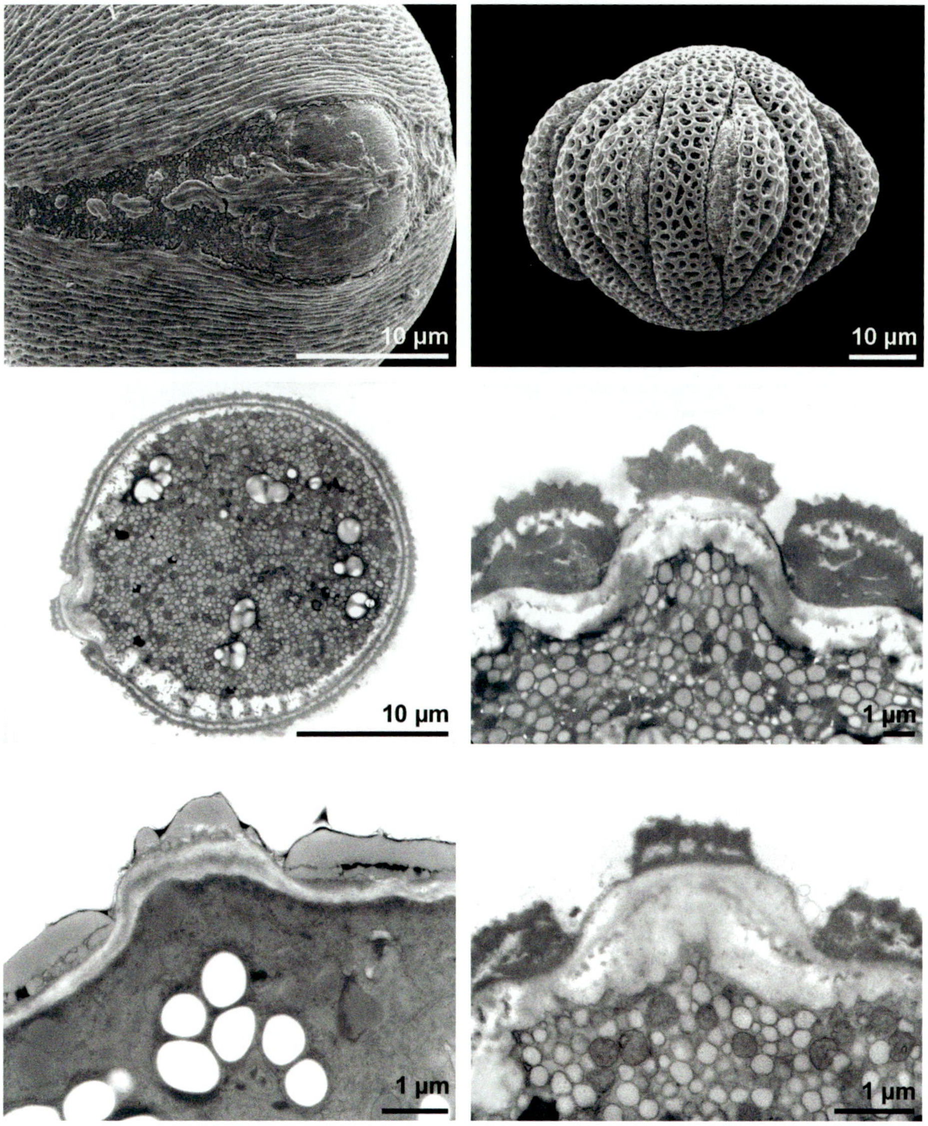

Rosa pendulina Rosaceae	*Passiflora suberosa* Passifloraceae equatorial view
Poa pratensis Poaceae PA+TCH+SP (short)	*Triticum aestivum* Poaceae PA+TCH+SP (short)
Plantago lanceolata Plantaginaceae U+Pb	*Poa angustifolia* Poaceae PA+TCH+SP (short)

LM SEM TEM mo ana fnc

pontoperculum: operculum covering a colpus, not completely isolated from the remainder of the sexine.

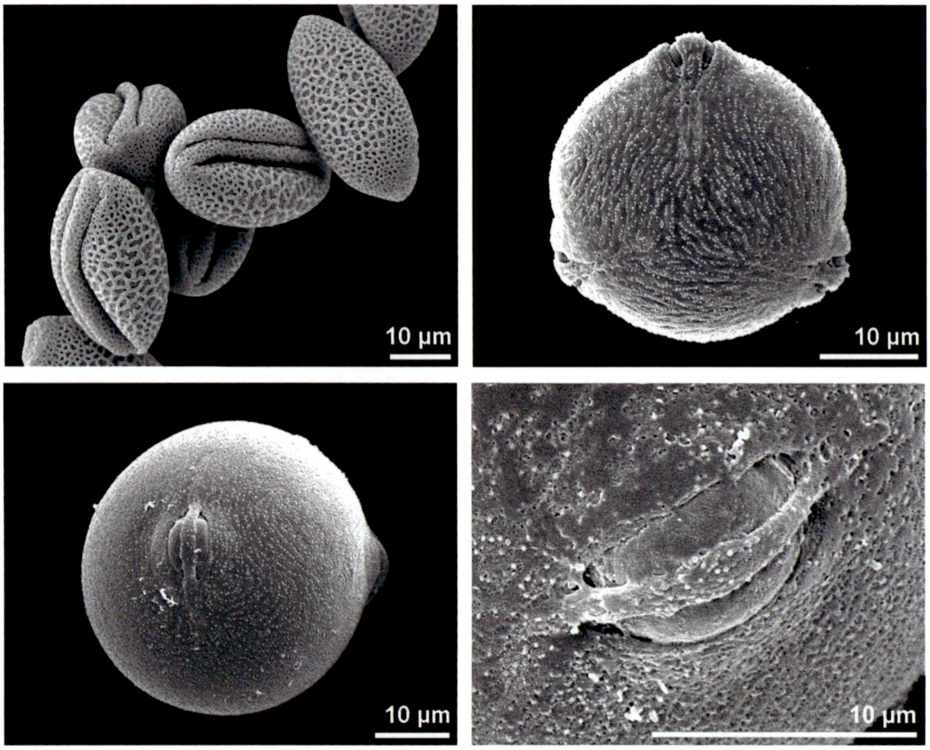

Veratrum nigrum
Melanthiaceae
dry pollen

Sanguisorba minor
Rosaceae
polar view

Sanguisorba cretica
Rosaceae
equatorial view

Sanguisorba cretica
Rosaceae

LM SEM TEM mo ana fnc

bridge: exine connection between the margins of a colpus in the equatorial region.

Comment:
the term is often used in a more general context, e.g., for exine connections within tetrads.

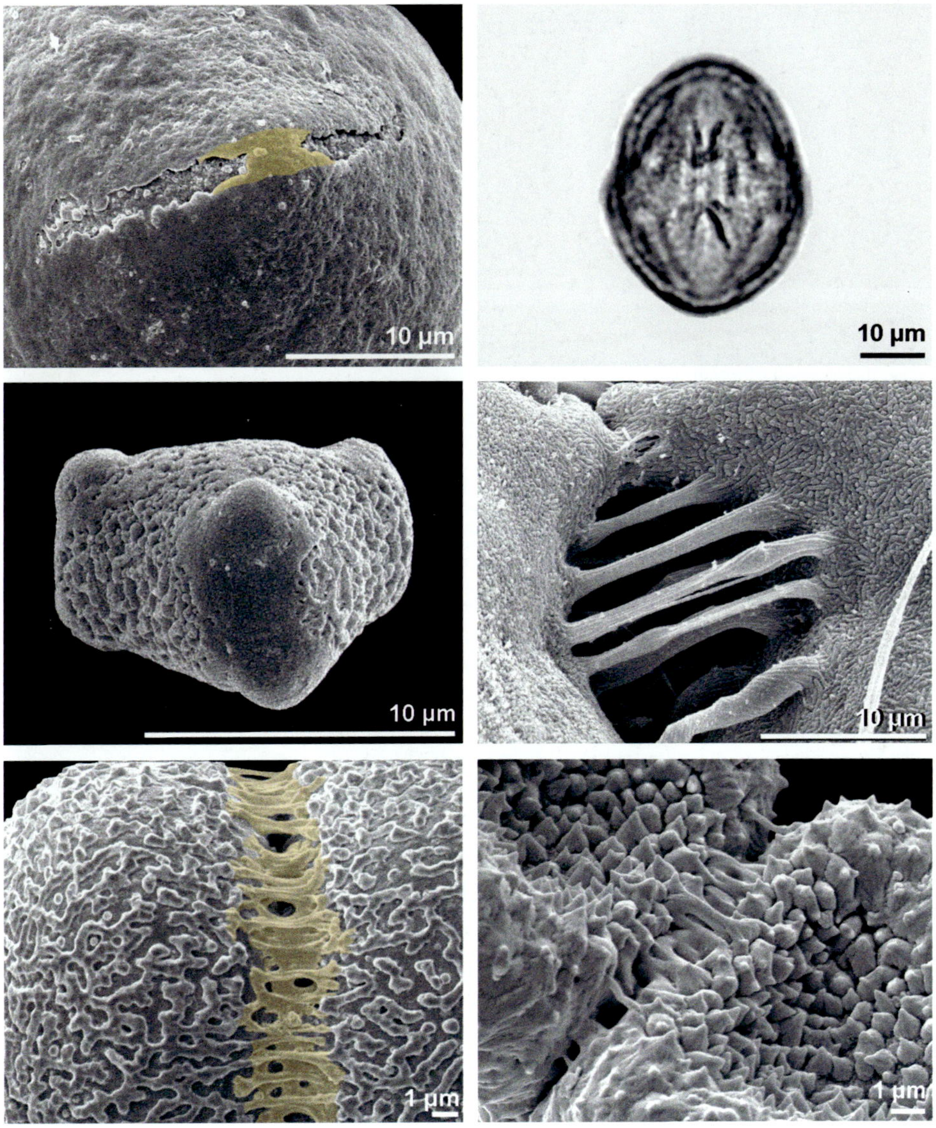

Elaeagnus angustifolia
Elaeagnaceae

Rehderodendron sp.
Styracaceae, fossil
equatorial view

Cunonia capensis
Cunoniaceae
equatorial view

Epilobium hirsutum
Onagraceae
tetrad

Typha latifolia
Typhaceae
tetrad

Thelethylax minutiflora
Podostemaceae
dyad

LM SEM TEM mo ana inc

papilla: small protuberance typical for Taxoidoideae-pollen (Cupressaceae).

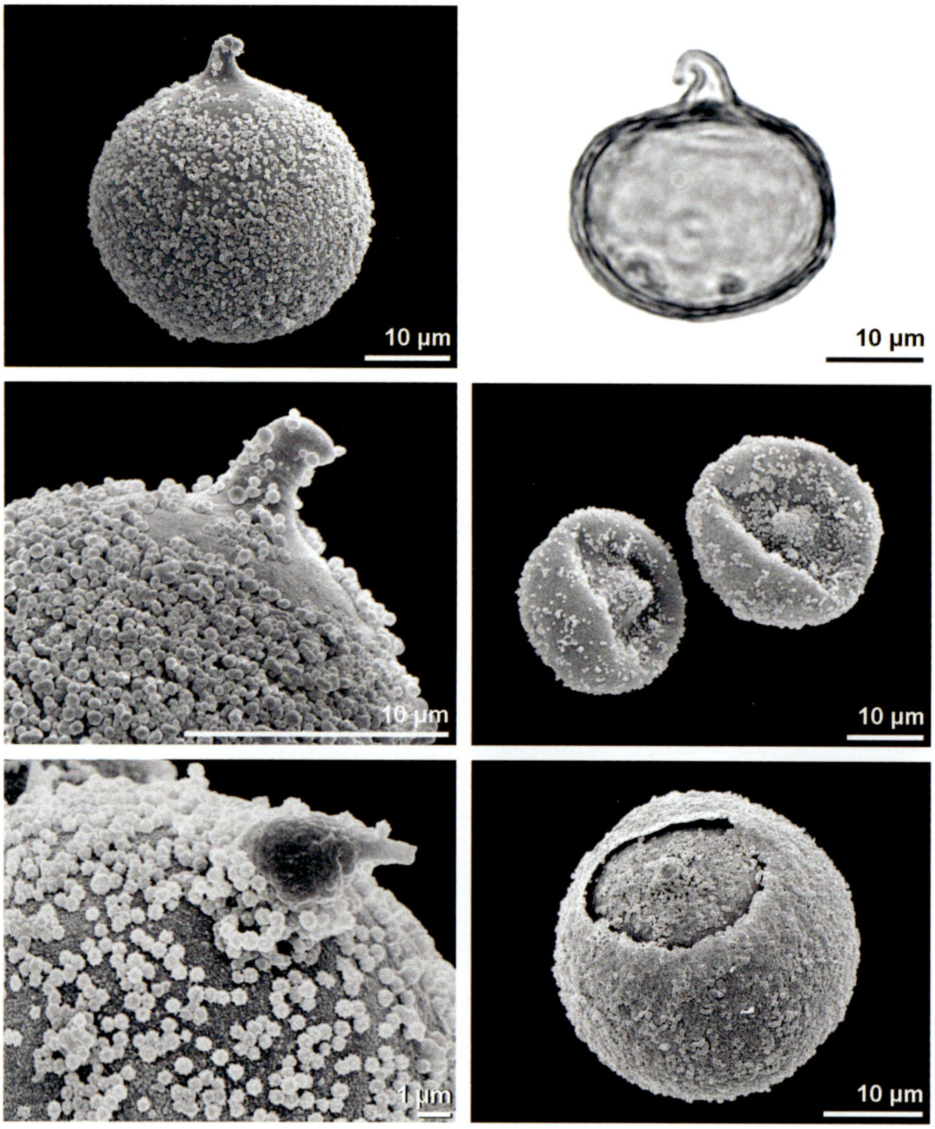

Cryptomeria japonica
Cupressaceae
equatorial view

Cryptomeria japonica
Cupressaceae

Metasequoia glyptostroboides
Cupressaceae

Cryptomeria sp.
Cupressaceae, fossil
equatorial view

Metasequoia glyptostroboides
Cupressaceae
dry pollen

Cunninghamia lanceolata
Cupressaceae
oblique distal polar view

LM SEM TEM mo ana fnc

reticulum: network-like pattern formed by exine elements (muri), where the lumina are wider than 1 µm.

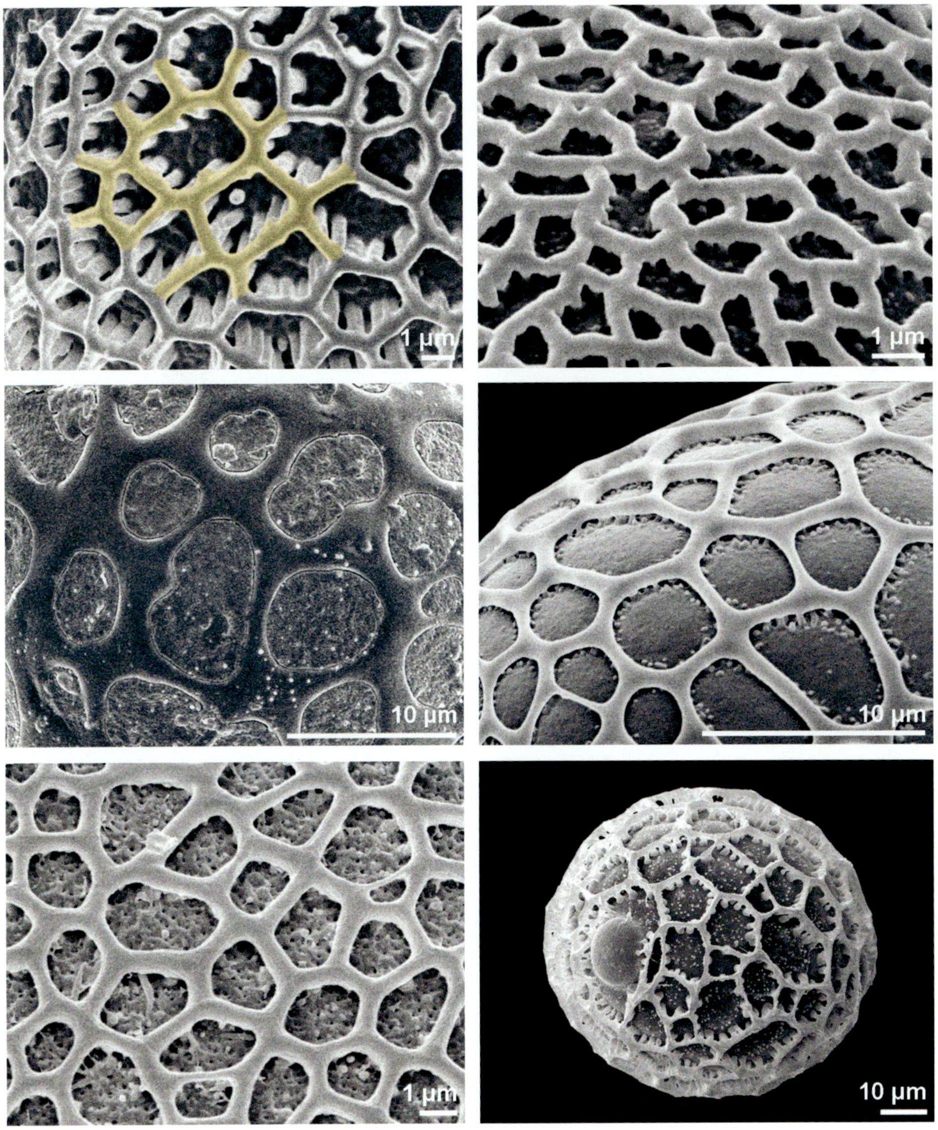

Cardamine pratensis
Brassicaceae

Polygala major
Polygalaceae

Luffa cylindrica
Cucurbitaceae

Veratrum album
Melanthiaceae

Aechmea azurea
Bromeliaceae

Razisea citrina
Acanthaceae
equatorial view

LM SEM TEM mo ana fnc

reticulum: network-like pattern formed by exine elements (muri), where the lumina are wider than 1 μm.

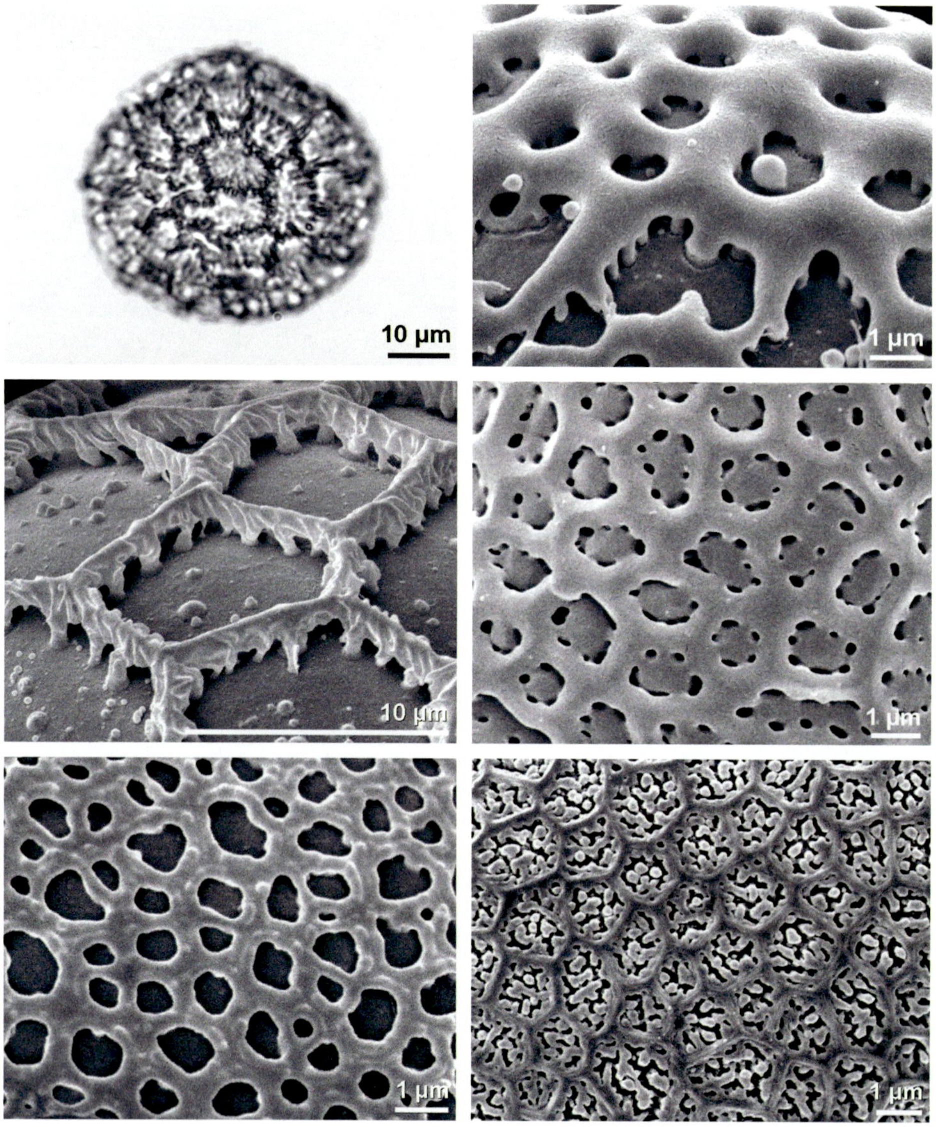

Persicaria sp.
Polygonaceae, fossil
equatorial view

Ruellia brevifolia
Acanthaceae

Buxus sempervirens
Buxaceae

Cephalanthera longifolia
Orchidaceae

Beloperone guttata
Acanthaceae
reticulate, perforate

Impatiens glandulifera
Balsaminaceae

LM SEM TEM mo ana fnc

reticulum: network-like pattern formed by exine elements (muri), where the lumina are wider than 1 µm.

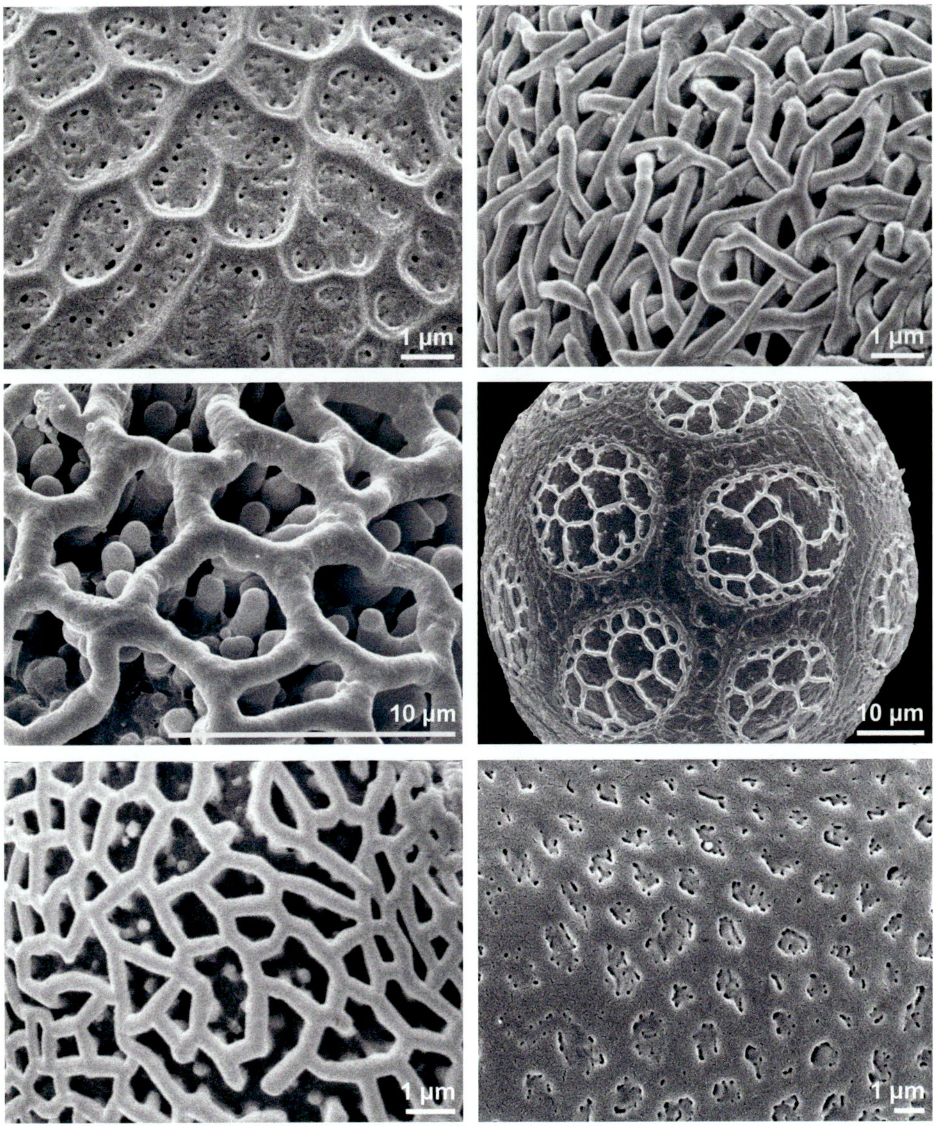

Lupinus polyphyllus
Fabaceae
reticulate, perforate

Sollya heterophylla
Pittosporaceae
reticulate to rugulate

Opuntia paraguayensis
Cactaceae
free standing columellae

Ibicella lutea
Martyniaceae
clypeate, reticulate

Poncirus trifoliata
Rutaceae

Lathyrus vernus
Fabaceae
reticulate, perforate

LM SEM TEM mo ana fnc

reticulum: network-like pattern formed by exine elements (muri), where the lumina are wider than 1 µm.

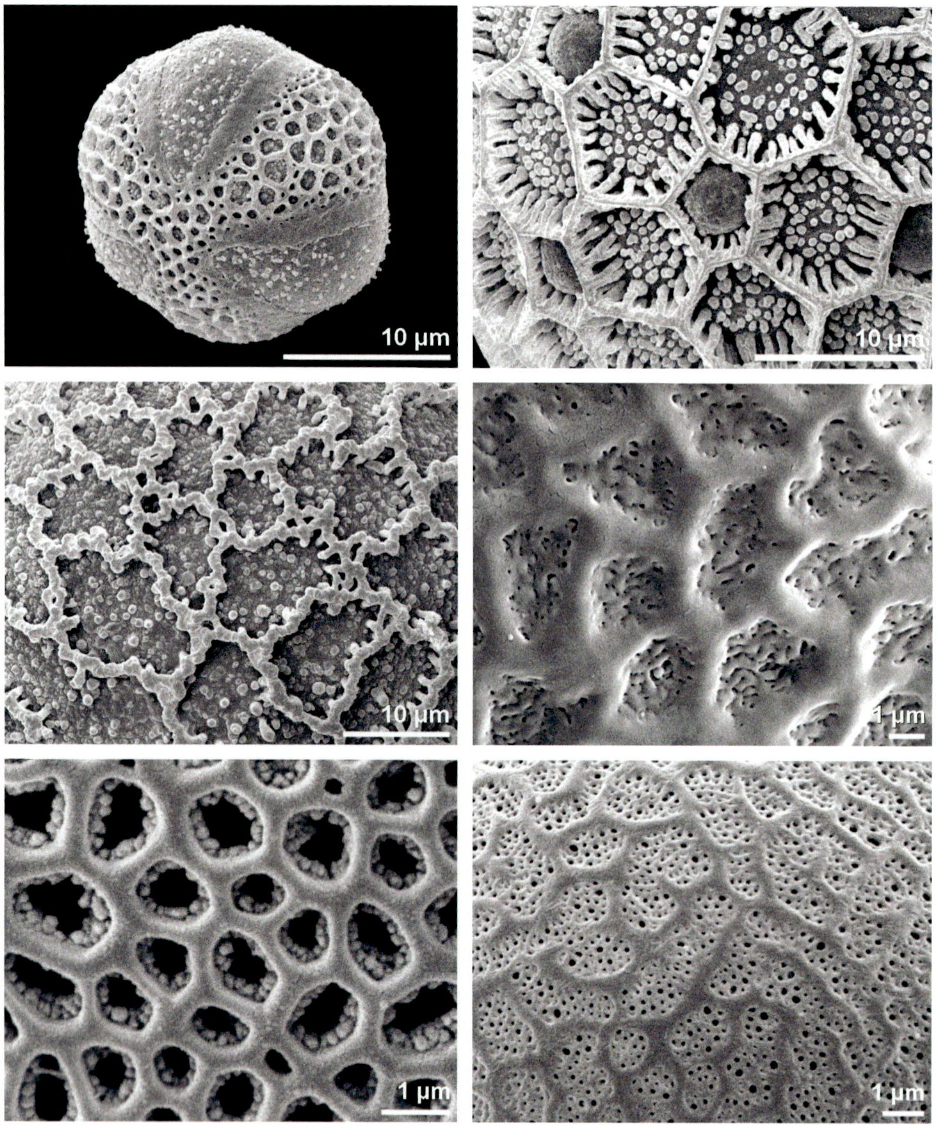

Salix daphnoides
Salicaceae
polar view

Persicaria mitis
Polygonaceae
pantoporate, free standing columellae

Paradisea liliastrum
Liliaceae

Trifolium rubens
Fabaceae
reticulate, perforate

Ecballium elaterium
Cucurbitaceae

Ajuga genevensis
Lamiaceae
reticulate, perforate

LM SEM TEM mo ana fnc

reticulum: network-like pattern formed by exine elements (muri), where the lumina are wider than 1 µm.

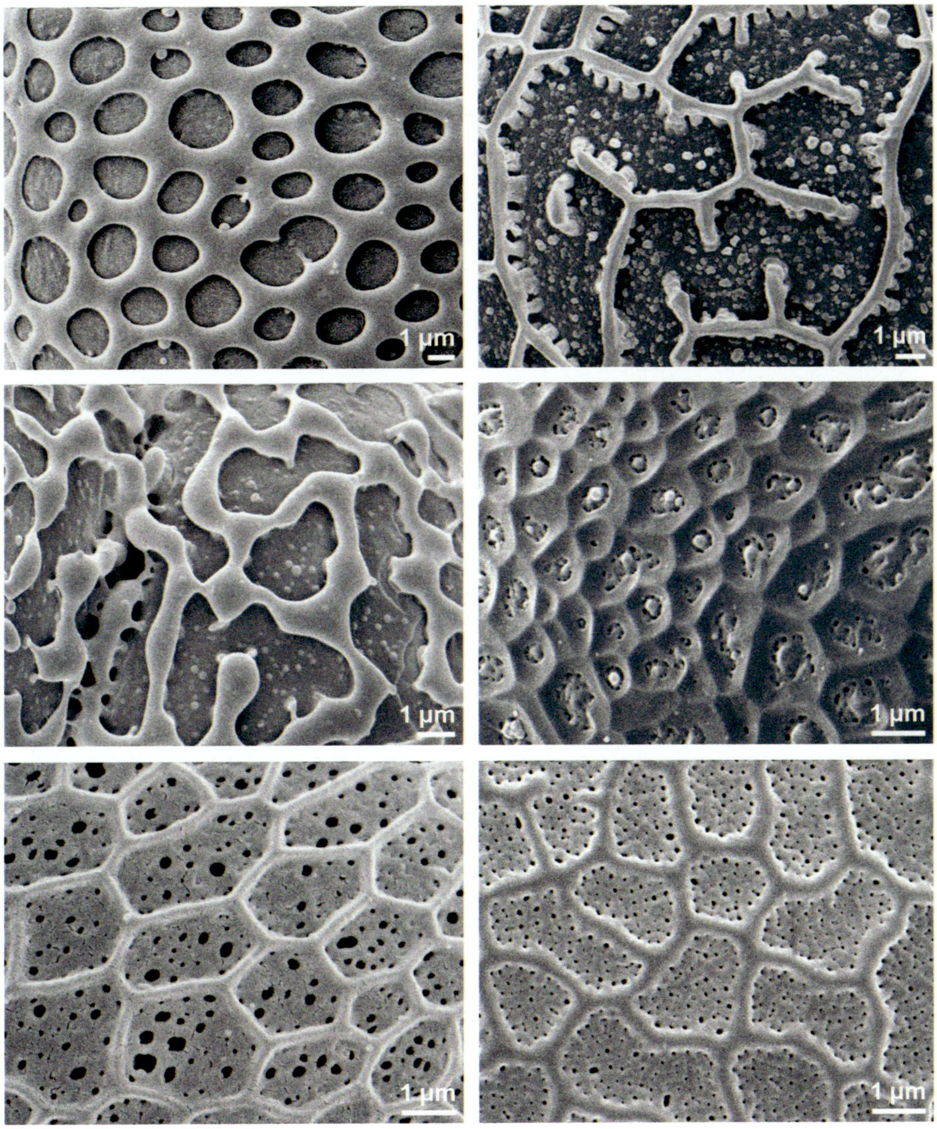

Aechmea allenii
Bromeliaceae
reticulate to foveolate

Epipactis helleborine
Orchidaceae
incomplete reticulum

Tropaeolum majus
Tropaeolaceae
reticulate, perforate

Adenia fruticosa
Passifloraceae
incomplete reticulum

Melilotus officinalis
Fabaceae

Physostegia virginiana
Lamiaceae
reticulate, perforate

LM SEM TEM mo ana fnc

reticulum: network like pattern formed by exine elements (muri), where the lumina are wider than 1 µm.

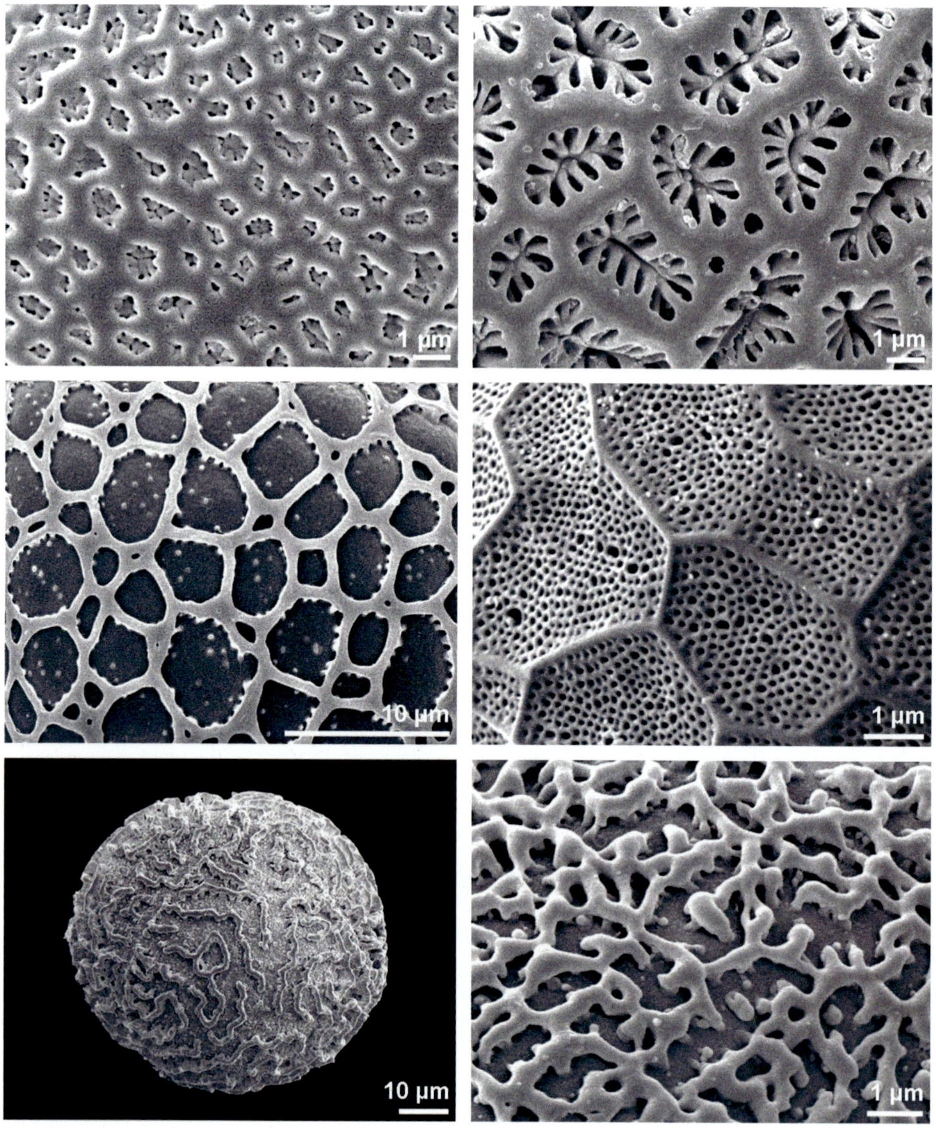

Pisum sativum
Fabaceae
reticulate, perforate

Harpochilus neesianus
Acanthaceae
reticulate, brochi with inclined columellae

Werauhia tarmaensis
Bromeliaceae

Plectranthus ornatus
Lamiaceae
bireticulate

Thladiantha hookeri
Cucurbitaceae
oblique equatorial view, incomplete reticulum

Pinguicula alpina
Lentibulariaceae
incomplete reticulum

LM SEM TEM mo ana fnc

bireticulate: special type of reticulate ornamentation, where the brochi of the large-meshed reticulum are filled by a small-meshed reticulum.

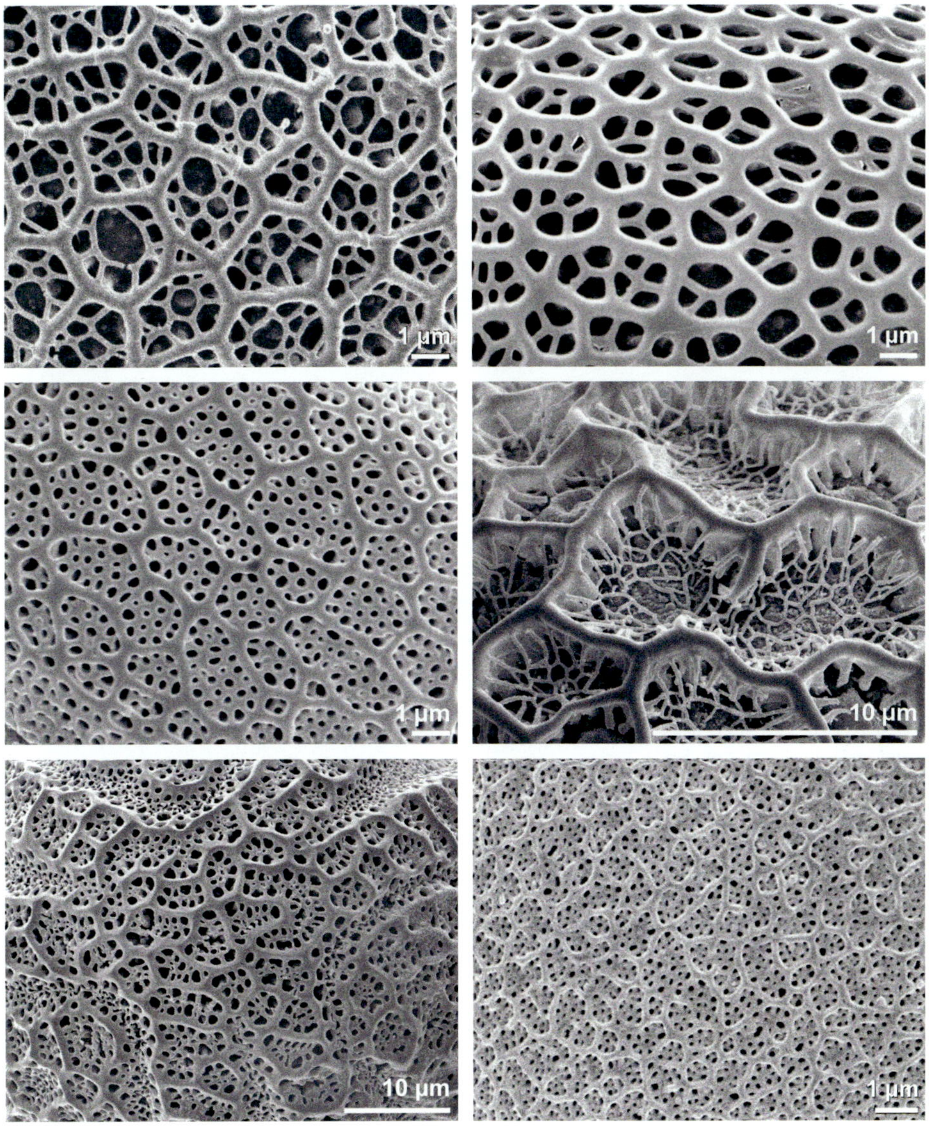

Salvia argentea
Lamiaceae

Salvia glutinosa
Lamiaceae

Prunella grandiflora
Lamiaceae

Phlox paniculata
Polemoniaceae

Pachystachys lutea
Acanthaceae

Melittis melissophyllum
Lamiaceae

LM SEM TEM mo ana fnc

brochus: mesh of a reticulum consisting of one lumen and the adjoining half of the muri.
homobrochate: reticulate pollen wall with lumina of uniform size.

Comment:
the term "homobrochate" should be used when the feature is remarkably expressed.

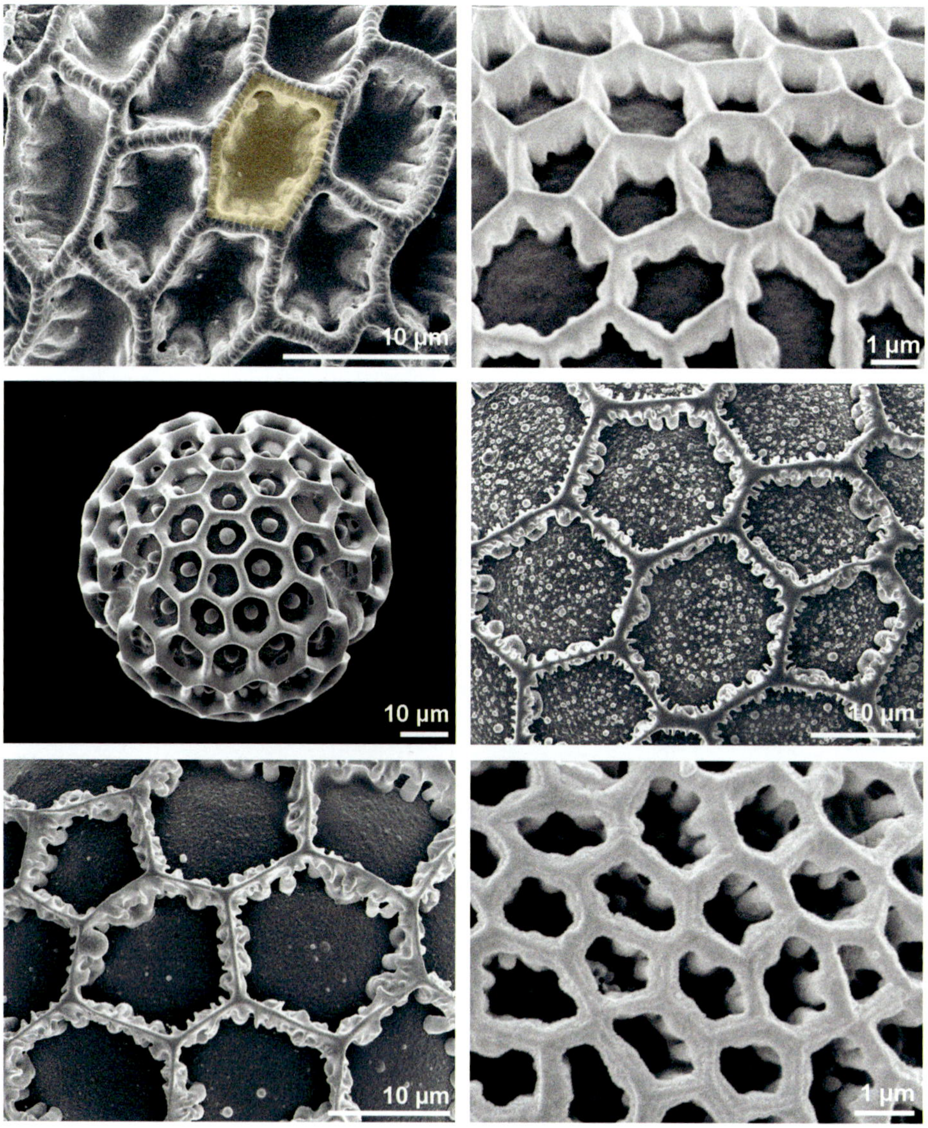

Acantholimon glumaceum
Plumbaginaceae

Abeliophyllum distichum
Hamamelidaceae

Eranthemum wattii
Acanthaceae
polar view

Strobilanthes roseus
Acanthaceae

Ruellia graecizans
Acanthaceae

Thlaspi montanum
Brassicaceae

LM SEM TEM mo ana fnc

brochus: mesh of a reticulum consisting of one lumen and the adjoining half of the muri.
heterobrochate: reticulate pollen wall with lumina of different sizes.

Comment:
the term "heterobrochate" should be used when the feature is remarkably expressed.

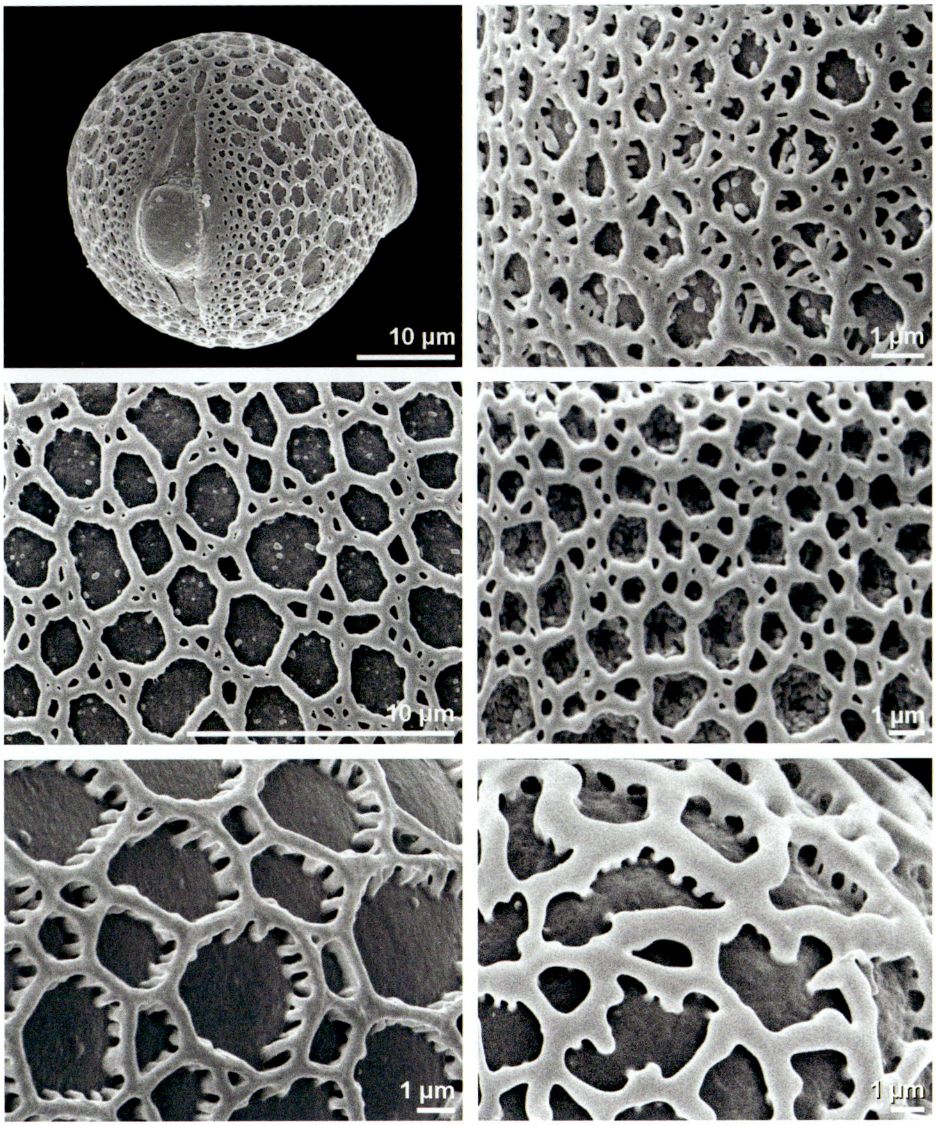

Hedera helix
Araliaceae
equatorial view

Fritillaria meleagris
Liliaceae

Billbergia seidelii
Bromeliaceae

Lachenalia aloides
Hyacinthaceae

Anthericum ramosum
Anthericaceae

Limodorum abortivum
Orchidaceae

LM SEM TEM mo ana fnc

microreticulate
micro-: prefix for small; features smaller as 1 µm; see "Alphabetic Glossary".

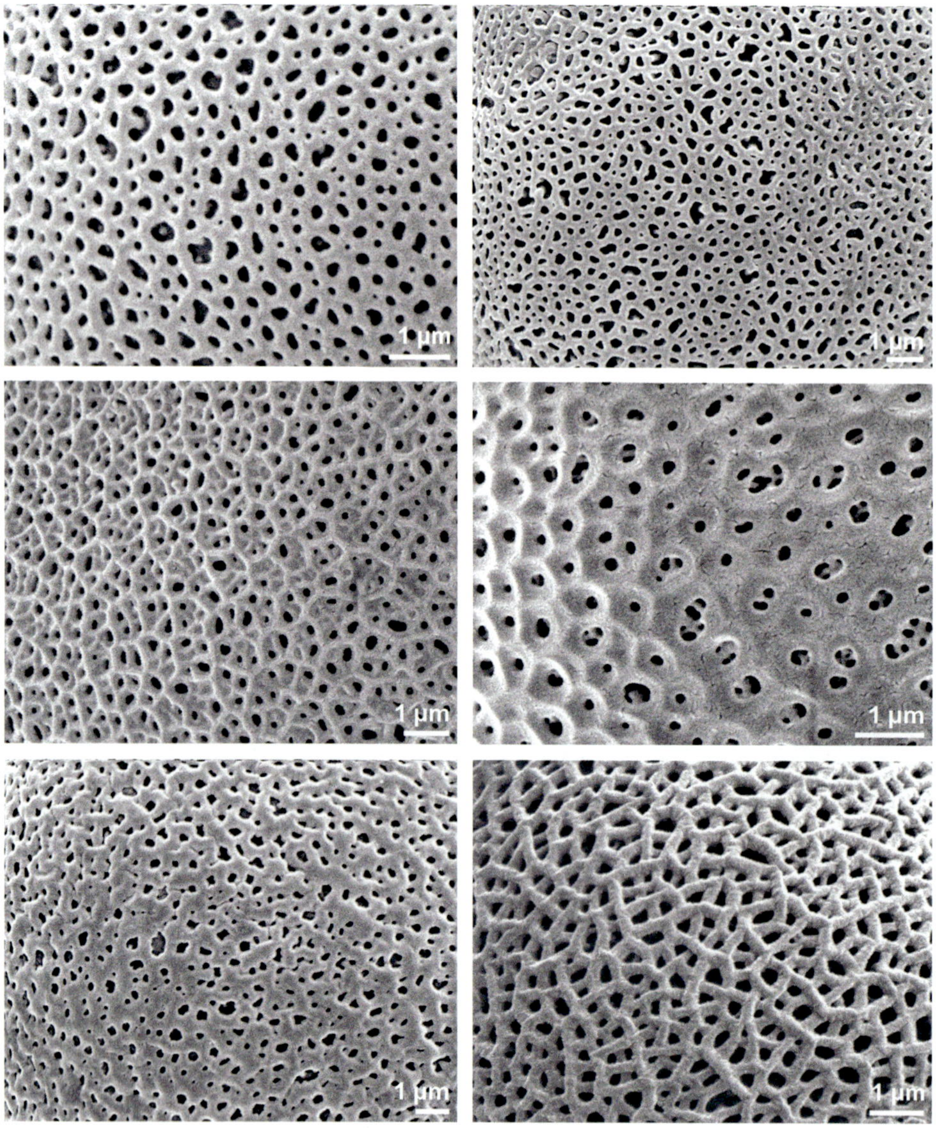

Kickxia spuria
Scrophulariaceae

Pseudolysimachion longifolium
Scrophulariaceae

Lamium purpureum
Lamiaceae

Cytisus nigricans
Fabaceae

Dianella tasmanica
Phormiaceae

Reseda luteola
Resedaceae

reticulum cristatum: special type of reticulum; muri with prominent sculpture elements.

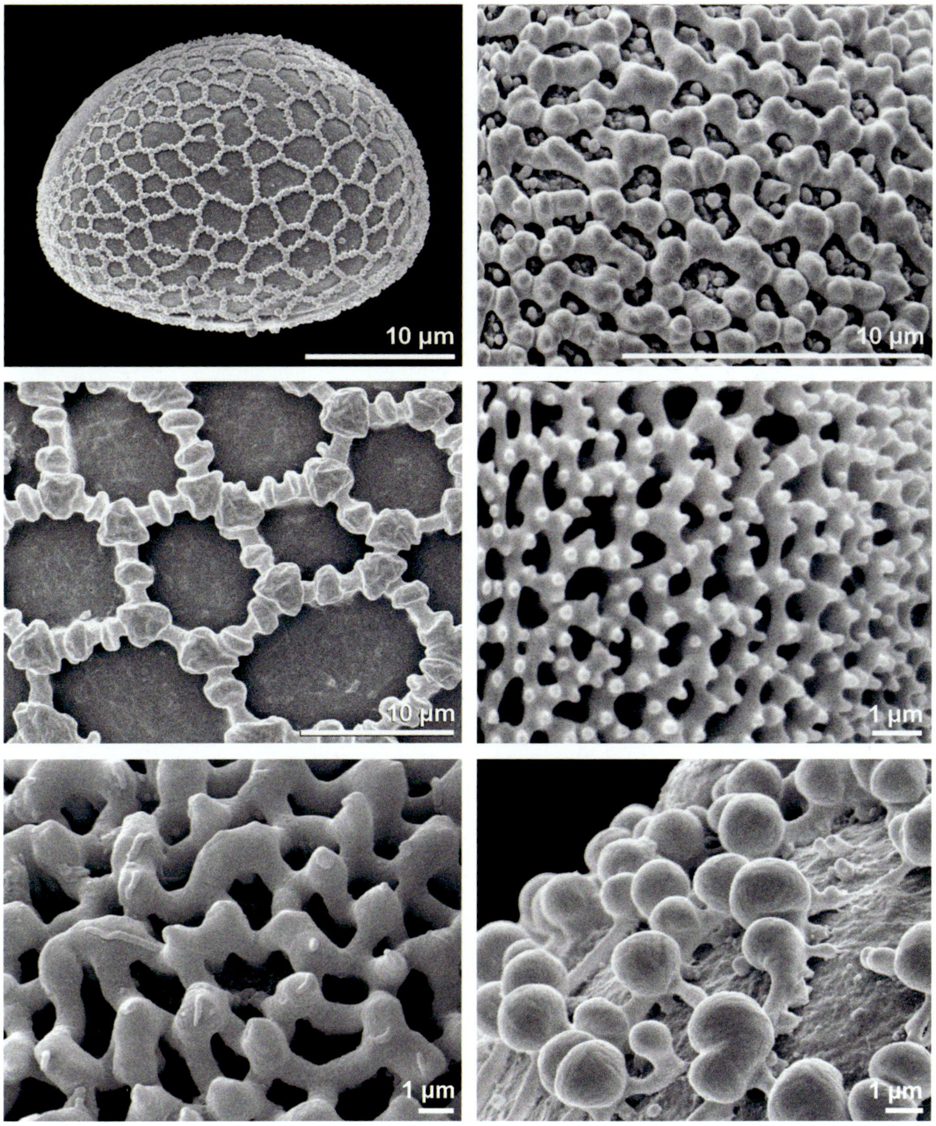

Lilium martagon
Liliaceae
equatorial view

Fumana procumbens
Cistaceae

Lilium candidum
Liliaceae

Mercurialis perennis
Euphorbiaceae
reticulum with microechini

Pachira aquatica
Bombacaceae

Phaleria capitata
Thymelaeaceae
reticulum with gemmae or clavae

LM SEM TEM mo ana fnc

reticulum cristatum: special type of reticulum; muri with prominent sculpture elements.

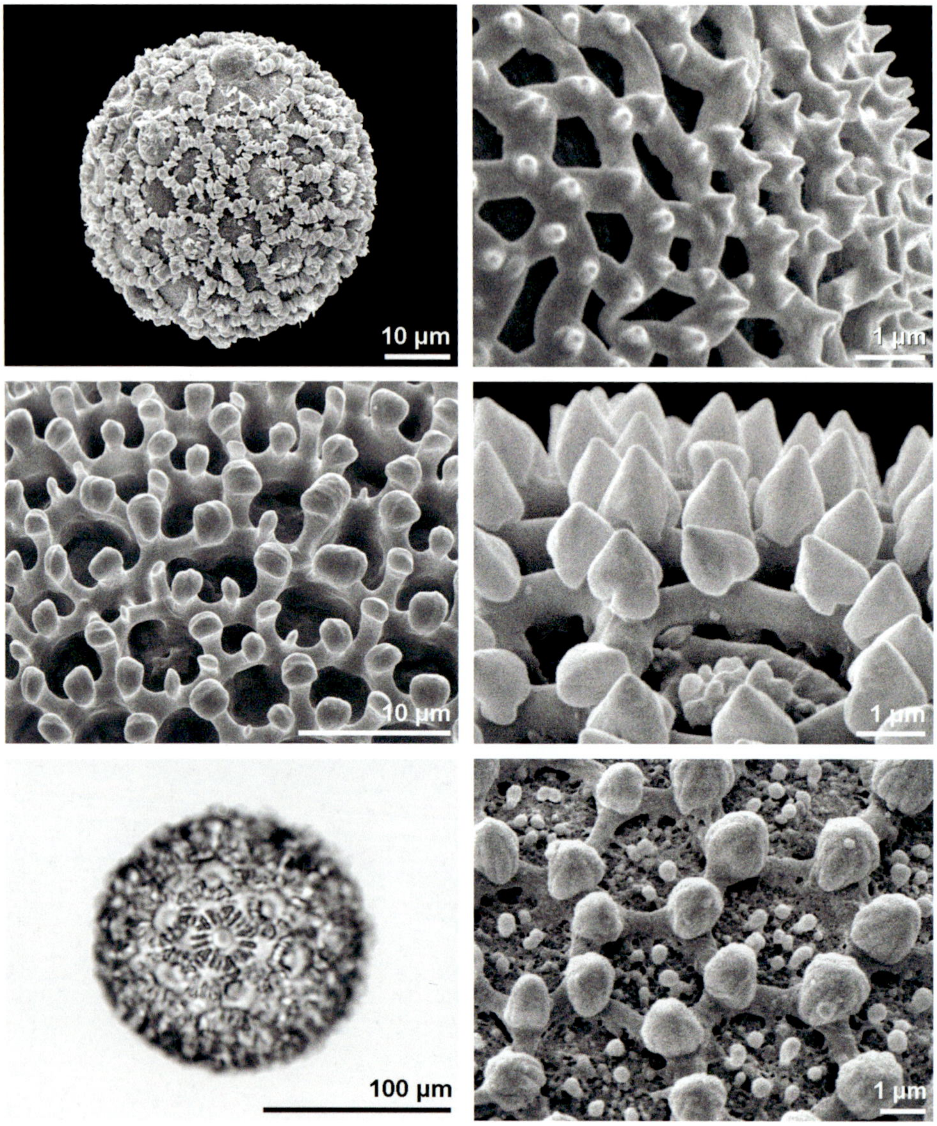

Pachysandra terminalis
Buxaceae

Aponogeton masoalaensis
Aponogetonaceae
reticulum with microechini

Geranium canariense
Geraniaceae
reticulum with clavae

Thymelaea passerina
Thymelaeaceae
reticulum with echini, croton pattern

Erdtmanipollis sp.
Buxaceae, fossil

Garcia nutans
Euphorbiaceae
croton pattern

LM SEM TEM mo ana fnc

croton pattern: special type of reticulum cristatum formed by regularly arranged sculpture elements on muri.

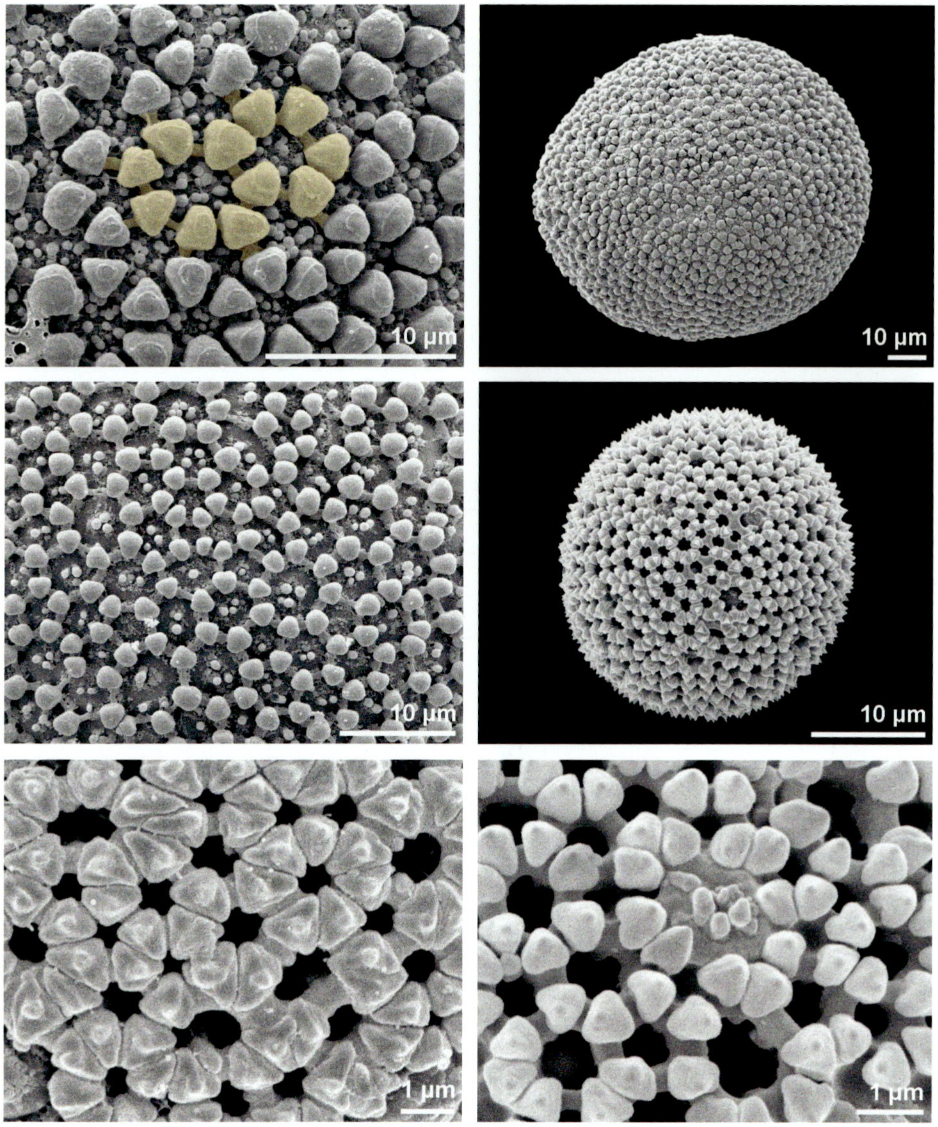

Croton triqueter
Euphorbiaceae

Croton triqueter
Euphorbiaceae

Jatropha podagrica
Euphorbiaceae

Daphne laureola
Thymelaeaceae

Daphne cneorum
Thymelaeaceae

Thymelaea passerina
Thymelaeaceae
surface detail with porus

LM SEM TEM mo ana fnc

plicae: cirumferential, parallel ridge-like folds.

Ephedra distachya
Ephedraceae

Ephedra sp.
Ephedraceae, fossil
equatorial view

Hemigraphis primulaefolia
Acanthaceae
polar (left) and equatorial view (right)

Pistia stratiotes
Araceae

Pistia stratiotes
Araceae
PA+TCH+SP
cross section of pollen grain and plicae

Pseudodracontium siamense
Araceae

striae: elongated exine elements separated by grooves predominantly parallel arranged.

Comment:
the term "striae" is used inconsistently in the literature. We use the term for the elevated elements and not for the grooves.

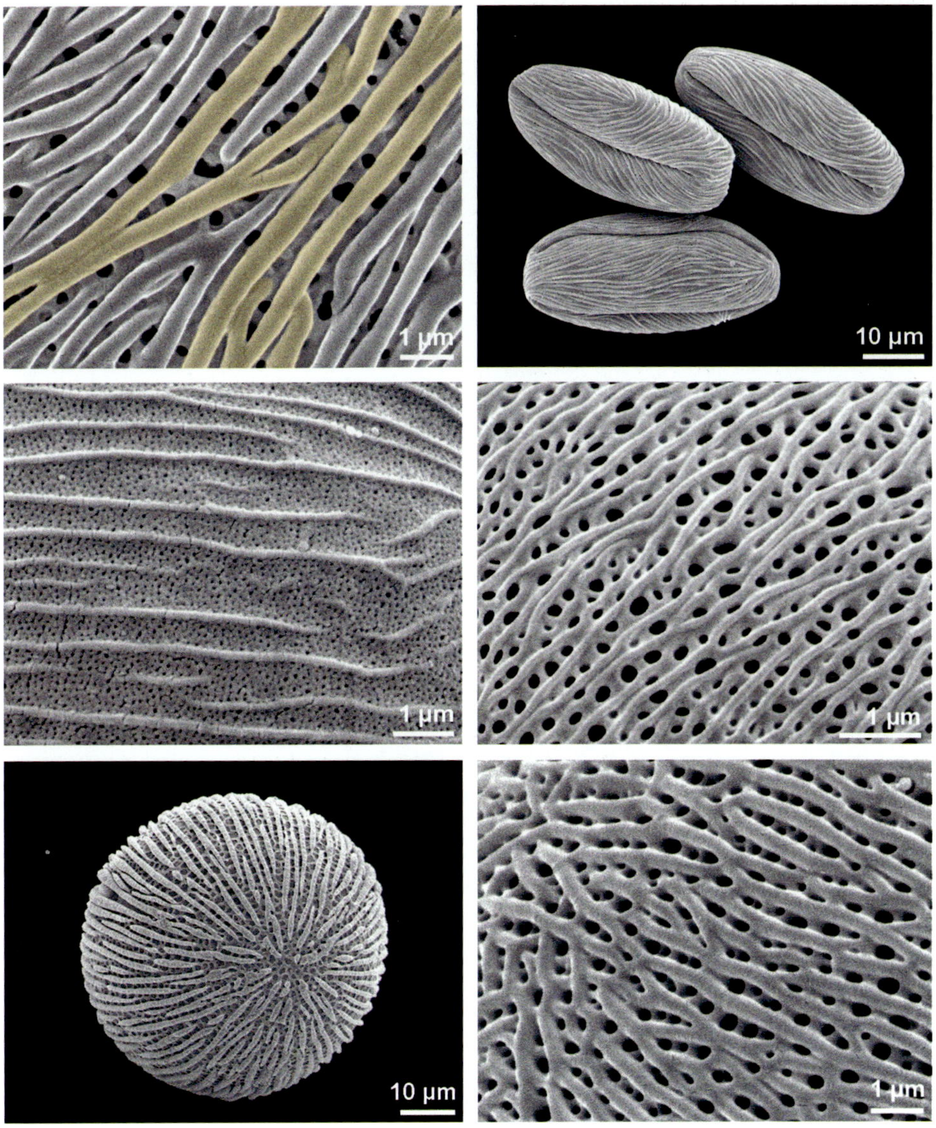

Acer pseudoplatanus
Sapindaceae
striate, perforate

Prunus avium
Rosaceae
dry pollen

Potentilla inclinata
Rosaceae
striate, perforate

Veronica cinerea
Scrophulariaceae
striate, perforate

Datura suaveolens
Solanaceae
polar view, striate, foveolate

Gentiana lutea
Gentianaceae
striate, perforate

LM SEM TEM mo ana fnc

striae: elongated exine elements separated by grooves predominantly parallel arranged.

Comment:
the term "striae" is used inconsistently in the literature. We use the term for the elevated elements and not for the grooves.

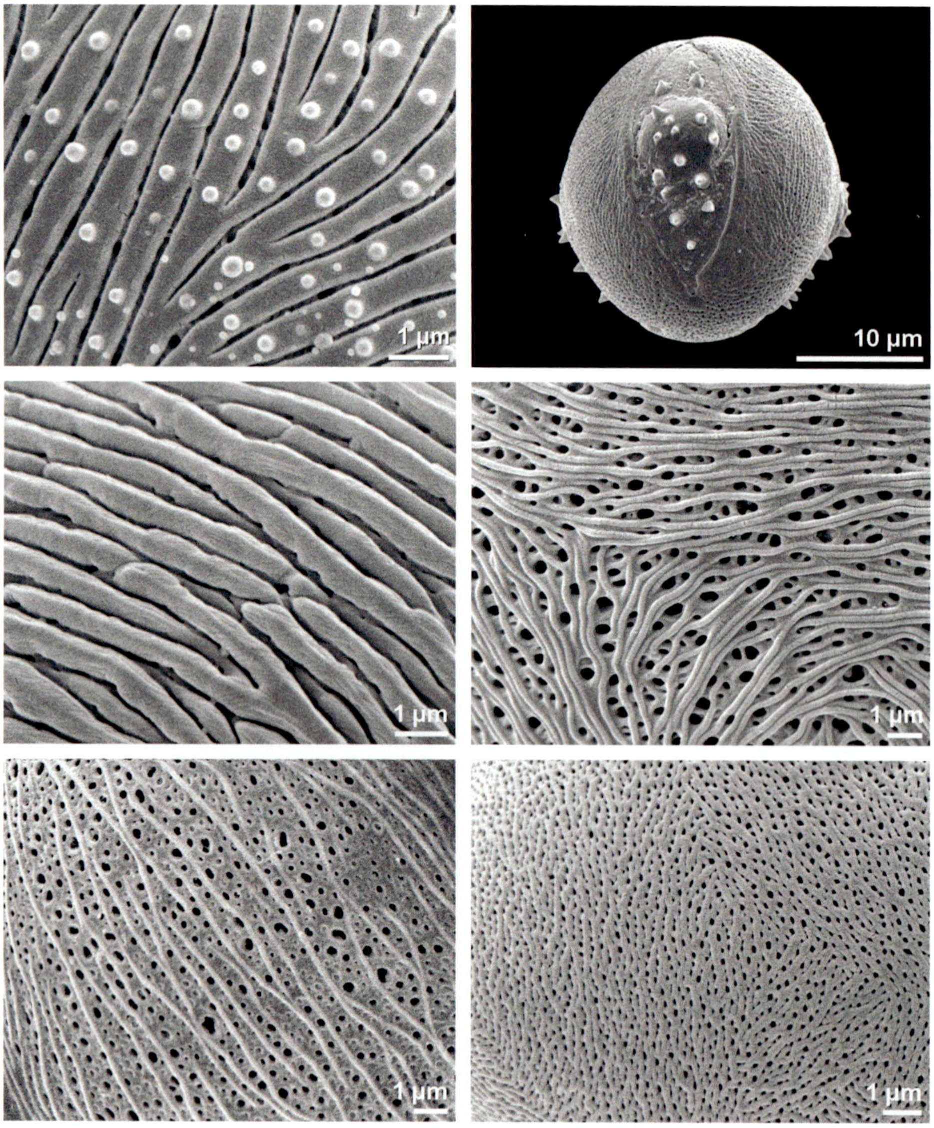

Saxifraga rotundifolia
Saxifragaceae
striate, microgemmate

Aesculus hippocastanum
Hippocastanaceae
equatorial view, striate, perforate

Lycium barbarum
Solanaceae

Cydonia sinensis
Rosaceae
striate, perforate

Rubus caesius
Rosaceae
striate, perforate

Allium flavum
Alliaceae
striate, perforate

striae: elongated exine elements separated by grooves predominantly parallel arranged.

Comment:
the term "striae" is used inconsistently in the literature. We use the term for the elevated elements and not for the grooves.

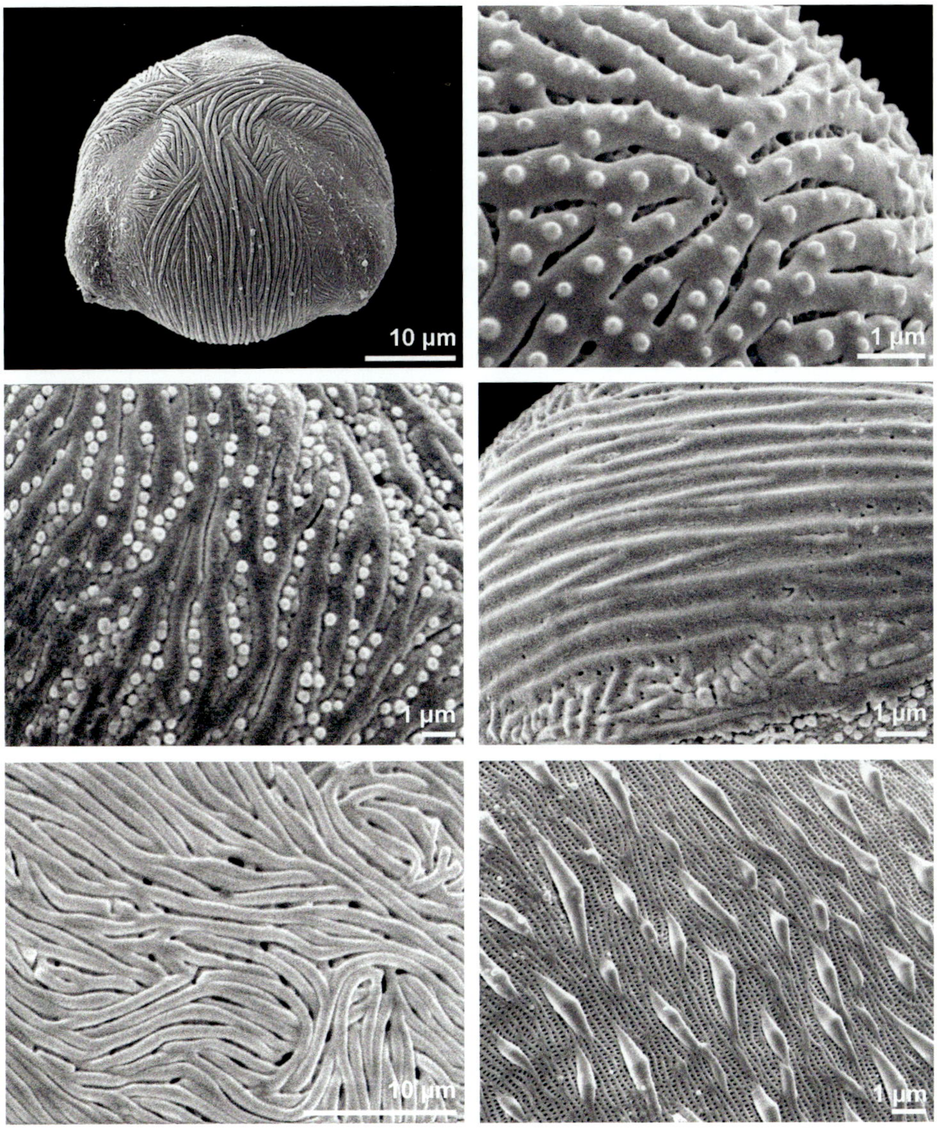

Menyanthes trifoliata
Menyanthaceae
oblique polar view

Sanguisorba minor
Rosaceae
striate, granulate

Crataegus laevigata
Rosaceae

Saxifraga tridactylites
Saxifragaceae
striate, microechinate

Begonia heracleifolia
Begoniaceae

Cabomba palaeformis
Cabombaceae
striate, perforate

LM SEM TEM mo ana fnc

striae: elongated exine elements separated by grooves predominantly parallel arranged.

Comment:
the term "striae" is used inconsistently in the literature. We use the term for the elevated elements and not for the grooves.

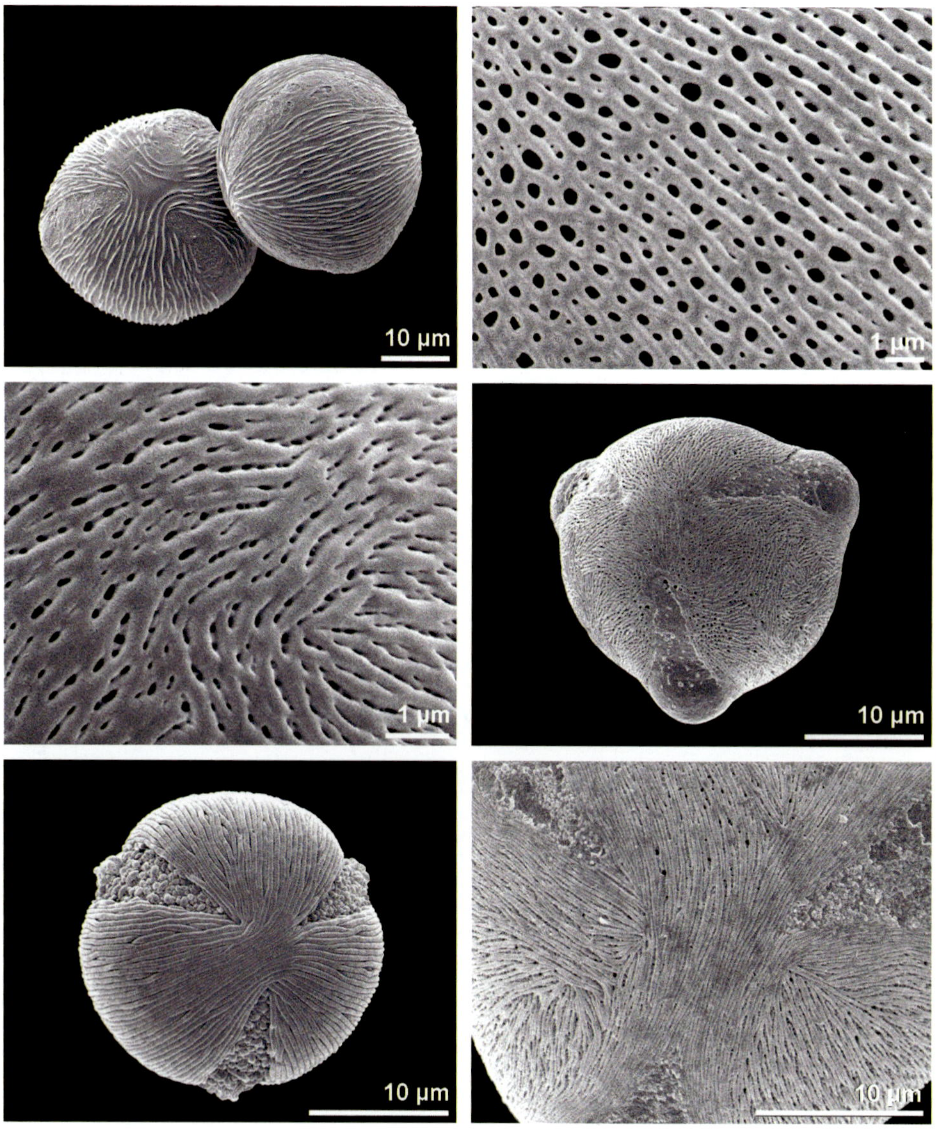

Geum reptans
Rosaceae
polar (left) and equatorial (right) view

Helianthemum nummularium
Cistaceae
striate, perforate

Prunus laurocerasus
Rosaceae
striate, perforate

Ruta graveolens
Rutaceae
polar view

Neoalsomitra sarcophylla
Cucurbitaceae
polar view

Malus sylvestris
Rosaceae
striate, perforate, polar area

LM SEM TEM mo ana fnc

striato-reticulate: ornamentation intermediate between striate and reticulate.

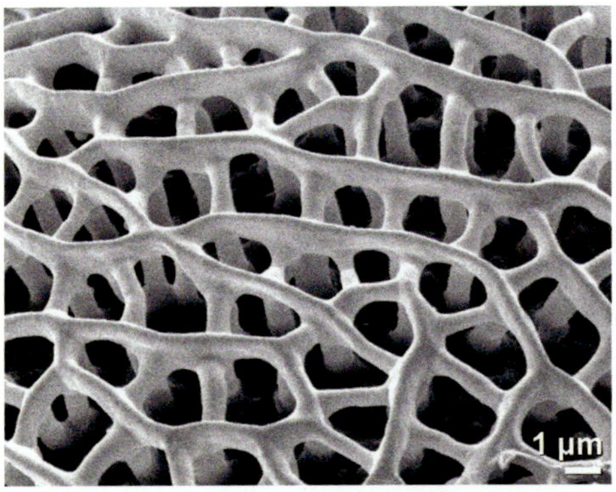

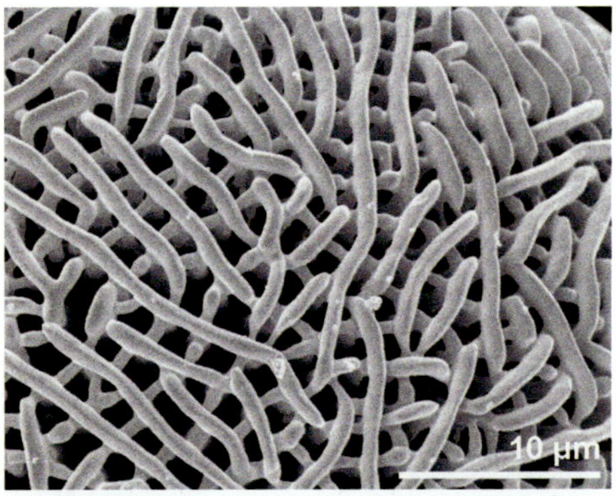

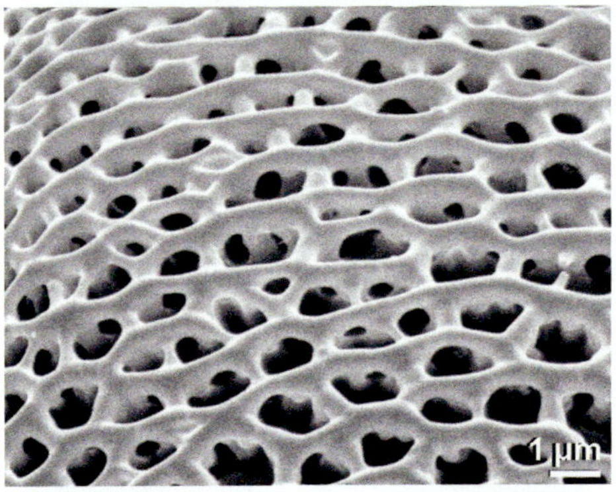

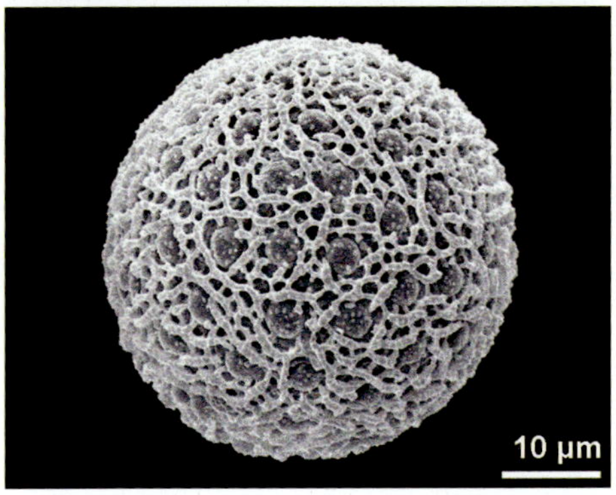

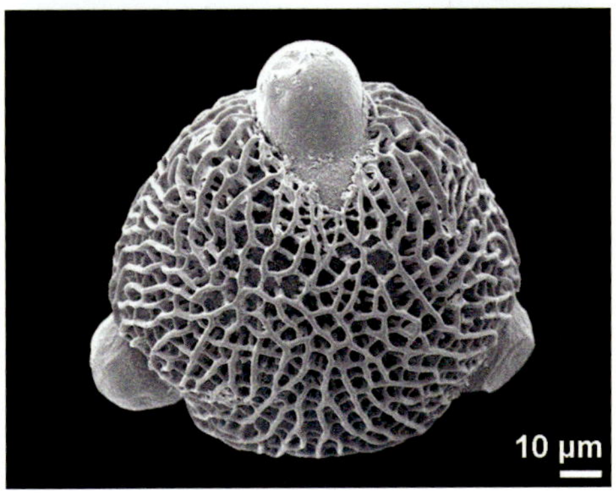

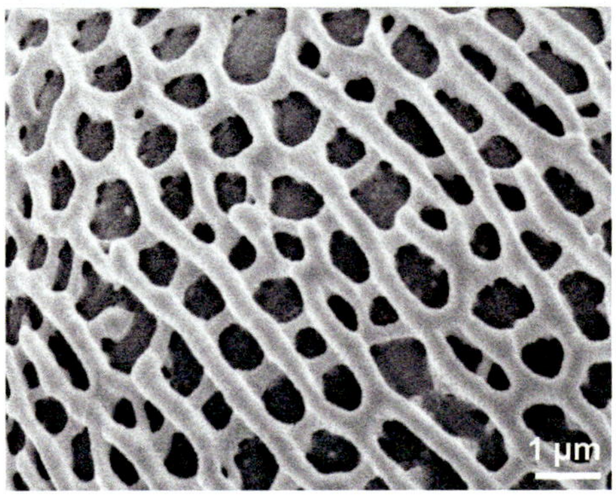

Pelargonium ferulaceum
Geraniaceae

Erodium cicutarium
Geraniaceae

Gentianella austriaca
Gentianaceae

Polemonium caeruleum
Polemoniaceae

Pelargonium tetragonum
Geraniaceae
polar view

Solandra longiflora
Solanaceae

striato-reticulate: ornamentation intermediate between striate and reticulate.

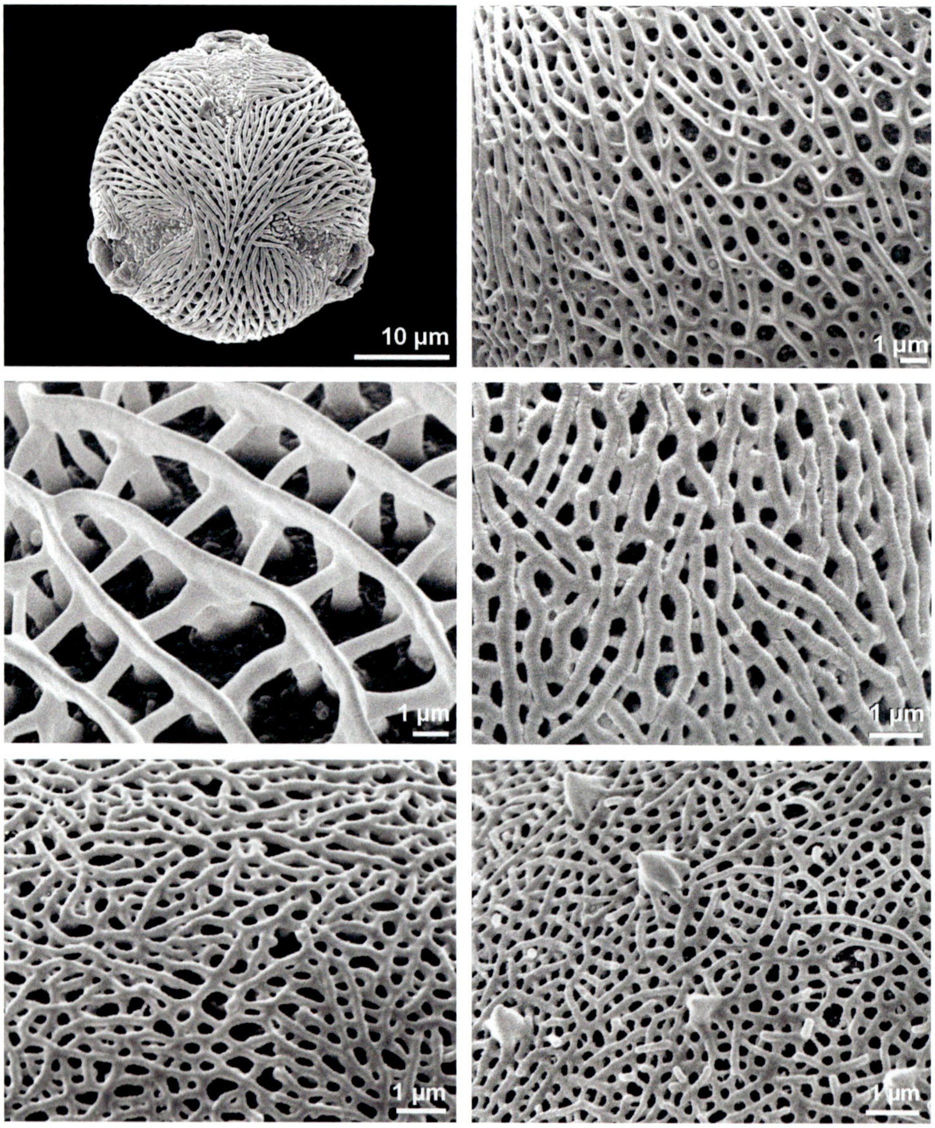

Ailanthus altissima
Simaroubaceae
polar view

Gentiana acaulis
Gentianaceae

Pelargonium carnosum
Geraniaceae

Ptelea trifoliata
Rutaceae
striato-microreticulate

Veronica prostrata
Scrophulariaceae
striato-microreticulate

Campanula persicifolia
Campanulaceae
microechinate, striato-microreticulate

LM SEM TEM mo ana fnc

rugulae: elongated exine elements longer than 1 µm; irregularly arranged.

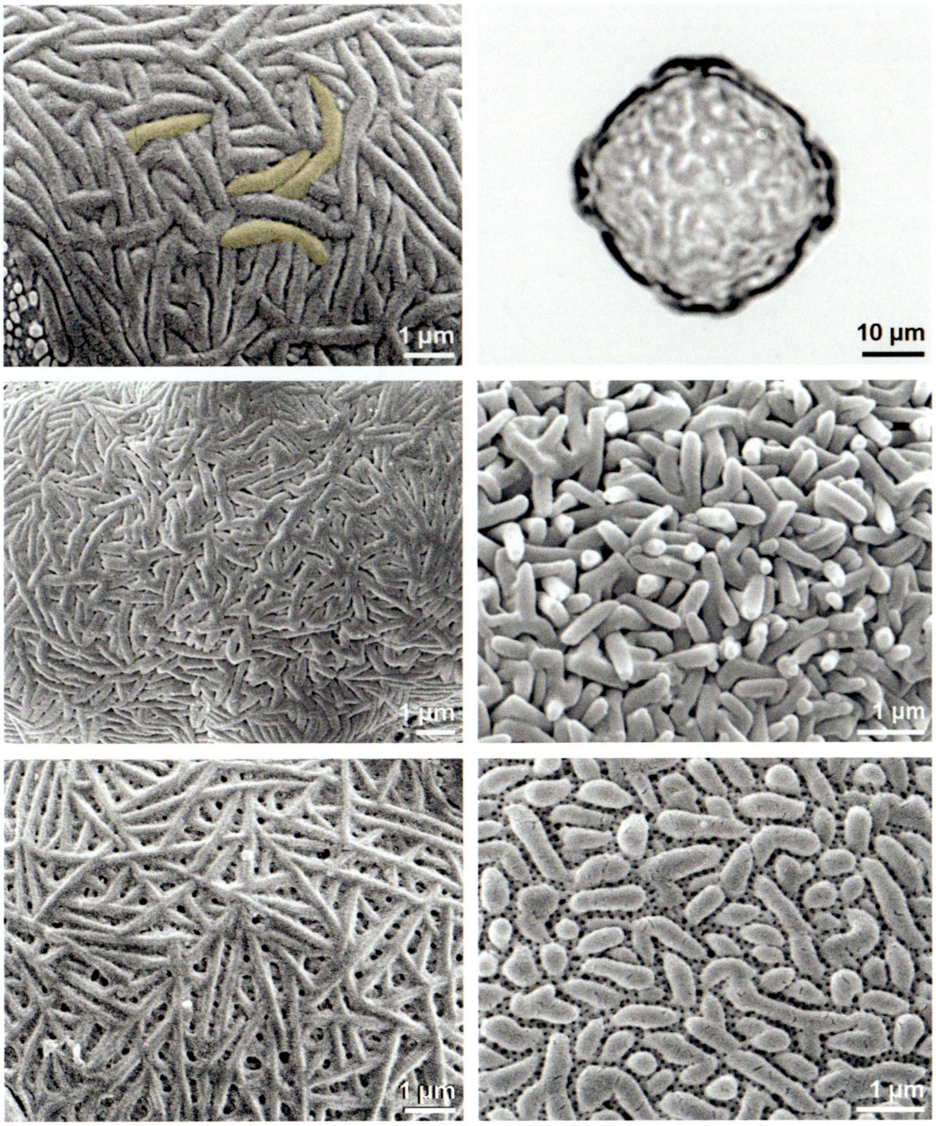

Securigera varia
Fabaceae

Peucedanum cervaria
Apiaceae

Sedum acre
Crassulaceae
rugulate, perforate

Zelkova sp.
Ulmaceae, fossil
polar view

Fagus sp.
Fagaceae, fossil

Circaea lutetiana
Onagraceae
rugulate, perforate

LM SEM TEM mo ana fnc

rugulae: elongated exine elements longer than 1 µm; irregularly arranged.

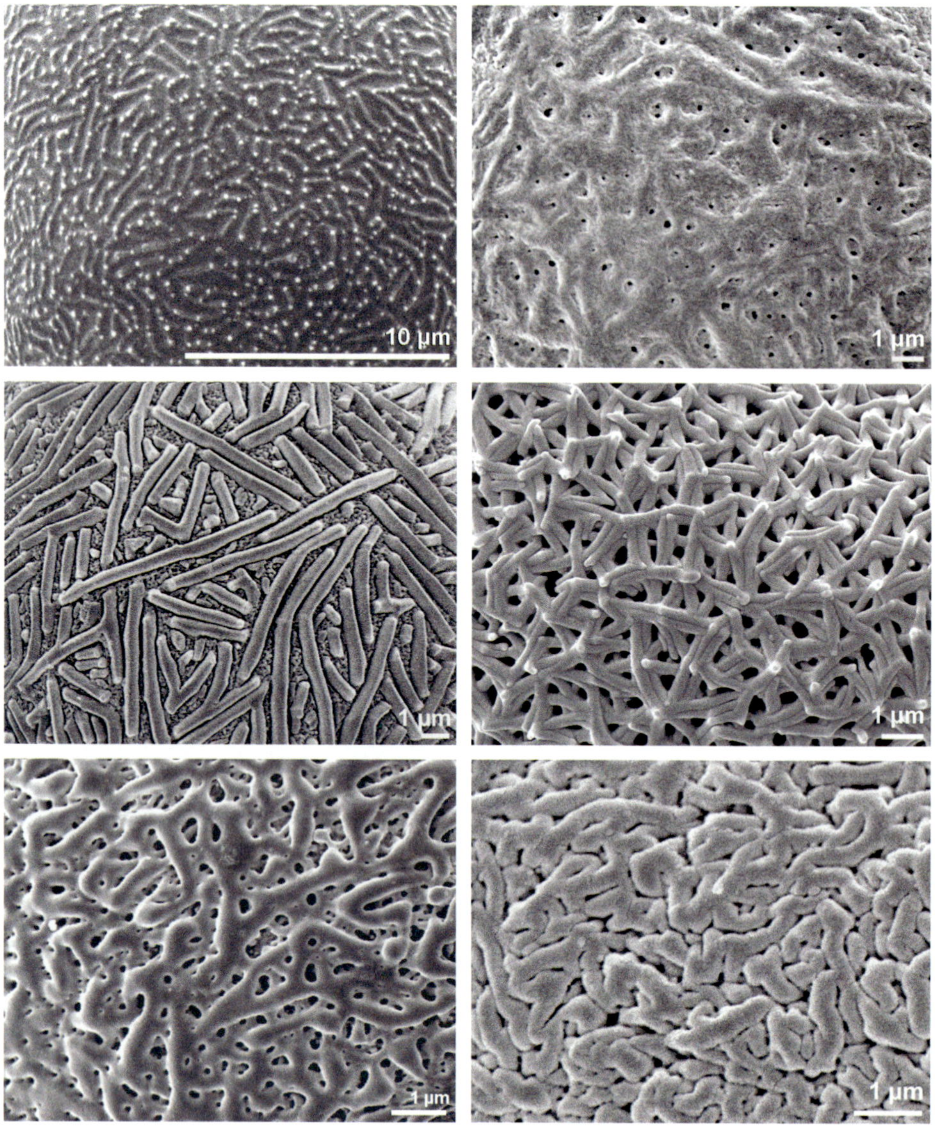

Carpinus betulus
Betulaceae
rugulate, granulate

Acer negundo
Sapindaceae
rugulate, perforate

Nymphoides peltata
Menyanthaceae

Leucadendron discolor
Proteaceae
rugulate, perforate

Nicotiana tabacum
Solanaceae
rugulate, perforate

Myrrhis odorata
Apiaceae

LM SEM TEM mo ana fnc

granulum: structure- or sculpture element of different size and shape; smaller than 1 µm.

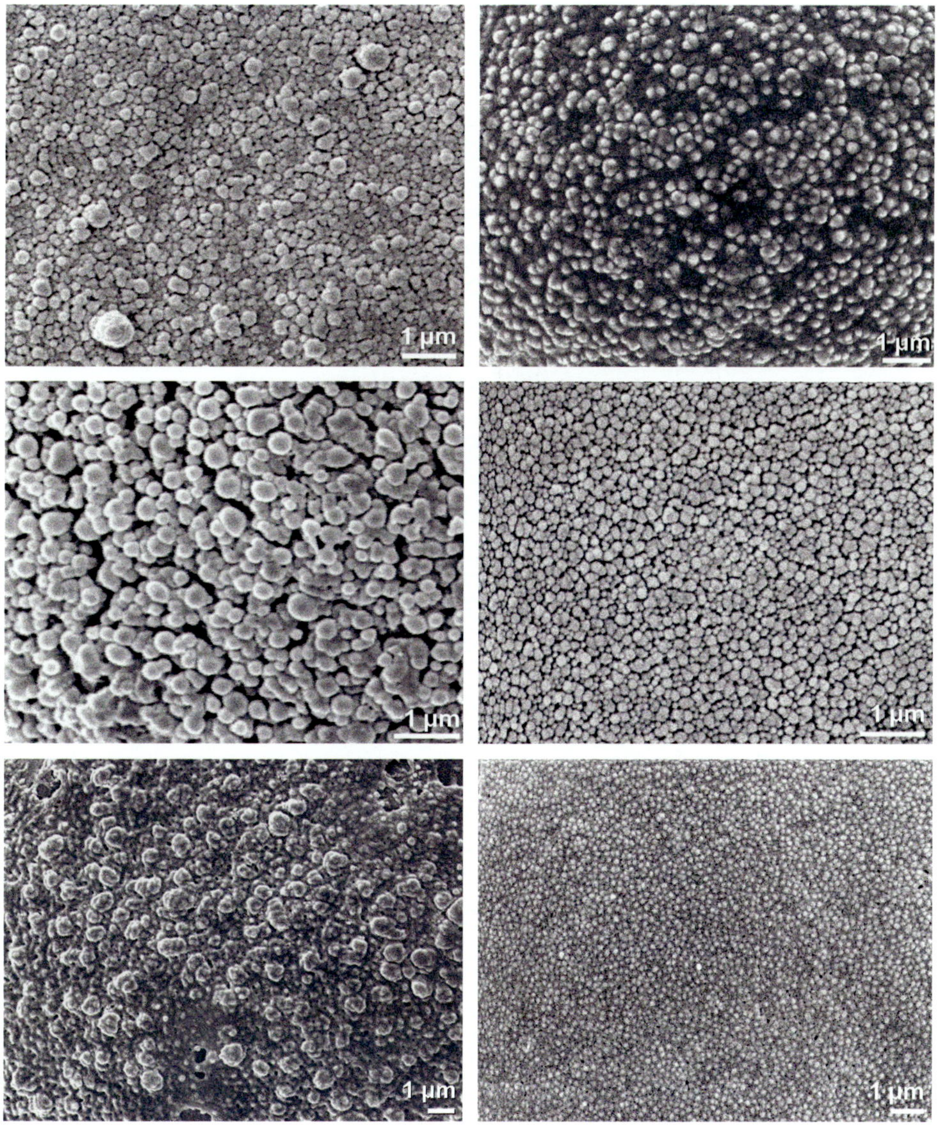

Larix decidua
Pinaceae

Humulus lupulus
Cannabaceae

Rhaphidophora africana
Araceae

Clarkia pulchella
Onagraceae

Quercus robur
Fagaceae

Luzula campestris
Juncaceae

psilate: pollen wall with smooth surface.

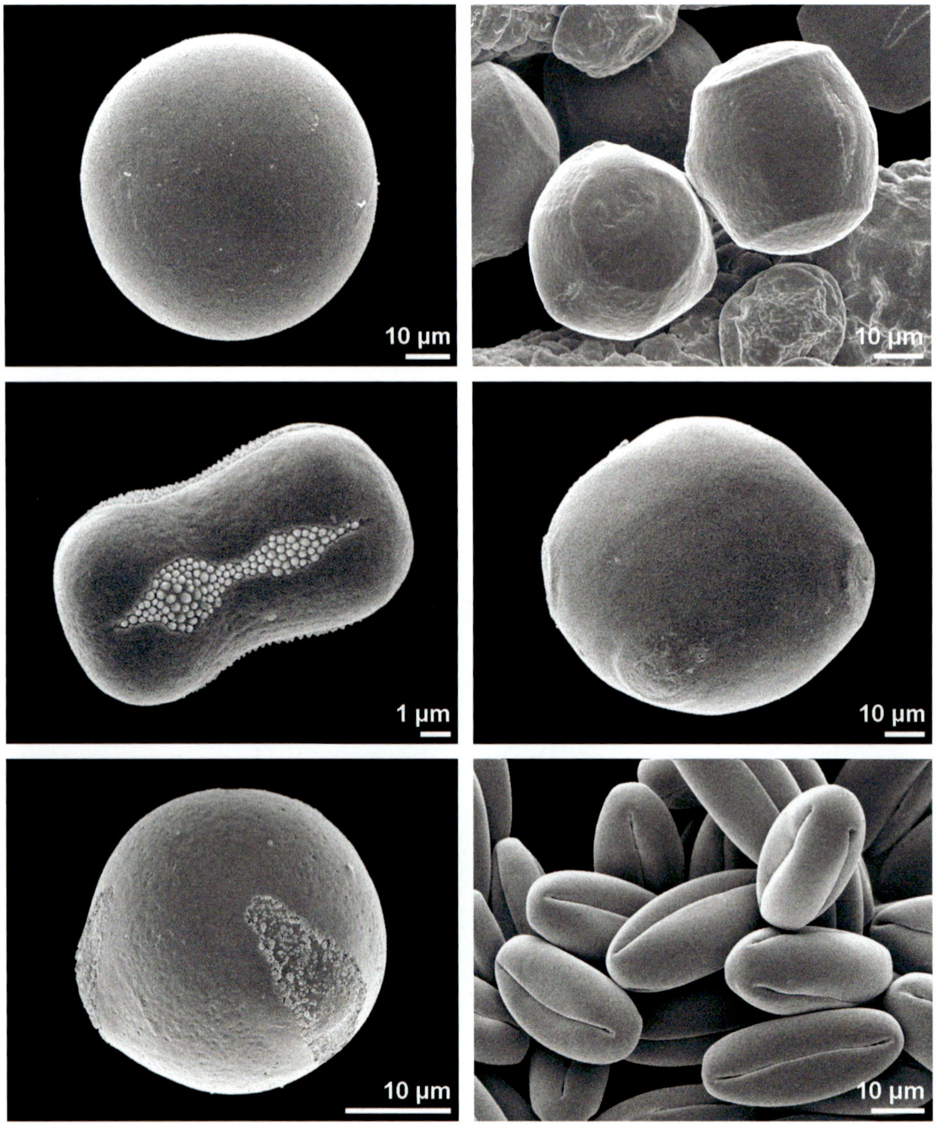

Hedychium gardnerianum
Zingiberaceae
inaperturate

Anthyllis vulneraria
Fabaceae

Lithospermum officinale
Boraginaceae
equatorial view

Vinca minor
Apocynaceae
oblique polar view

Dorycnium germanicum
Fabaceae
oblique view

Jovibarba hirta
Crassulaceae
dry pollen

psilate: pollen wall with smooth surface.

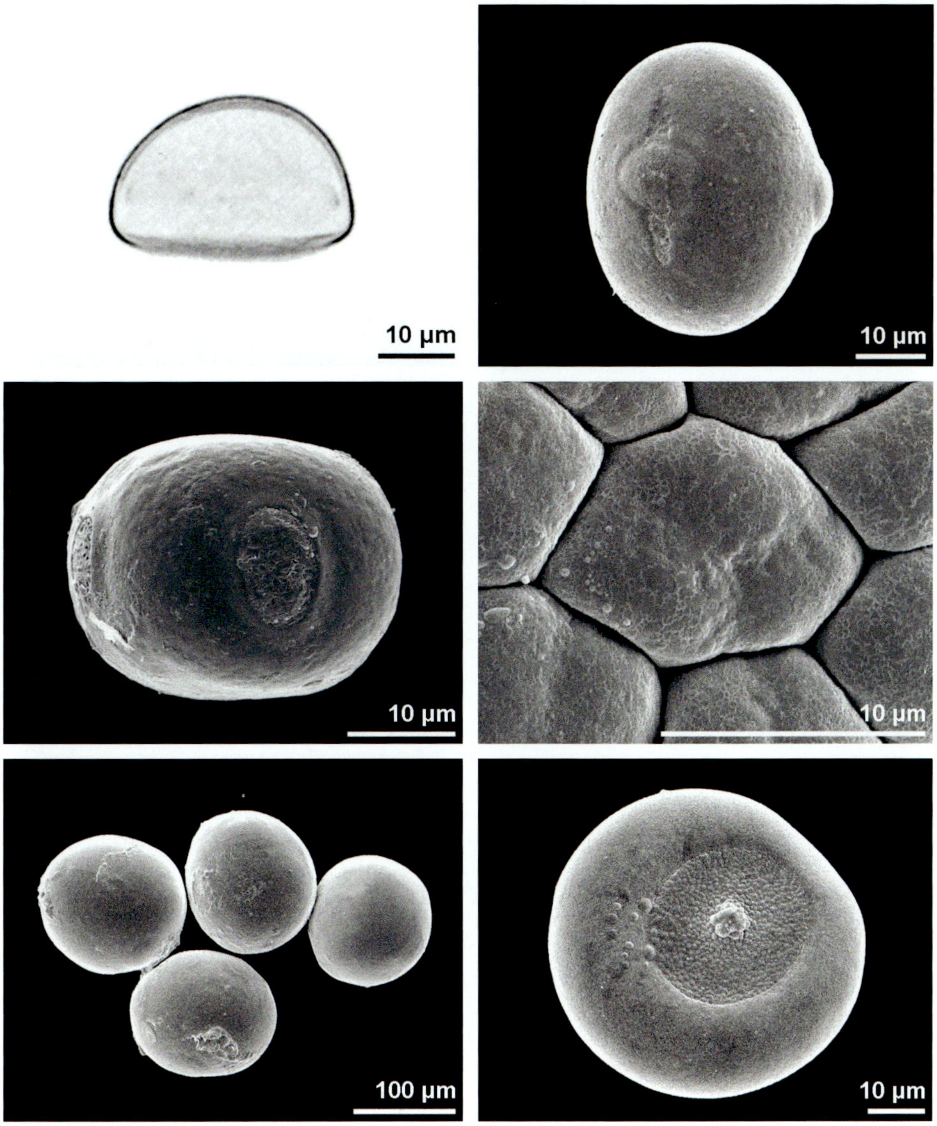

Allium ursinum
Alliaceae
equatorial view

Lathyrus niger
Fabaceae
equatorial view

Trigonia nivea
Trigoniaceae
equatorial view

Maxillaria densa
Orchidaceae
tetrad, part of massula

Costus speciosus
Zingiberaceae
pantoporate

Whitfieldia lateralis
Acanthaceae
equatorial view

LM SEM TEM mo ana fnc

perforate: pollen wall with holes less than 1 µm in diameter.

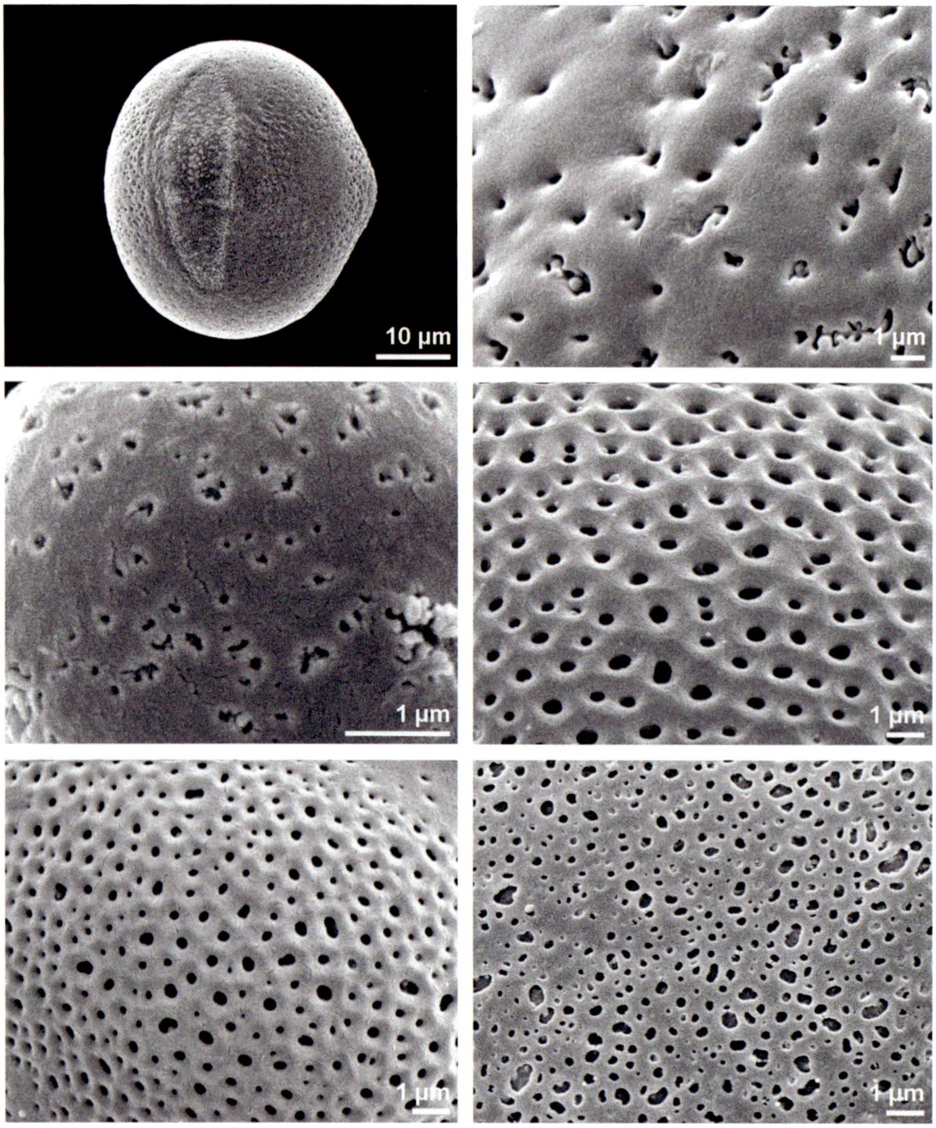

Colutea arborescens
Fabaceae
equatorial view

Gonatopus angustus
Araceae

Myosotis arvensis
Boraginaceae

Euphorbia palustris
Euphorbiaceae

Lysimachia nemorum
Primulaceae

Cucumis sativa
Cucurbitaceae

perforate: pollen wall with holes less than 1 µm in diameter.

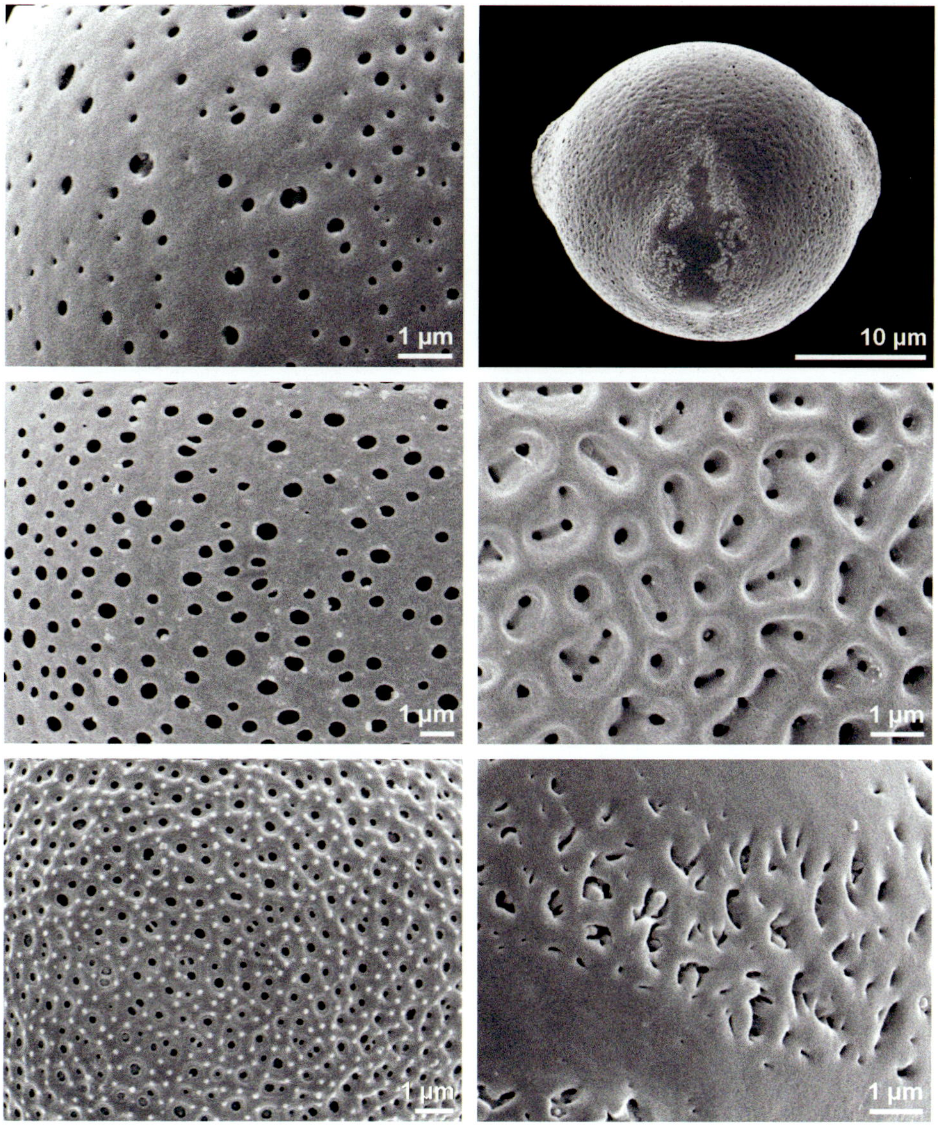

Pulmonaria officinalis
Boraginaceae

Hippocrepis emerus
Fabaceae
equatorial view

Napoleonaea imperialis
Napoleonaeaceae

Euphorbia helioscopia
Euphorbiaceae

Rumex acetosa
Polygonaceae
perforate, granulate

Trifolium montanum
Fabaceae

LM SEM TEM mo ana fnc

foveola: roundish lumen more than 1 µm in diameter; distance between two adjacent lumina larger than their diameter.

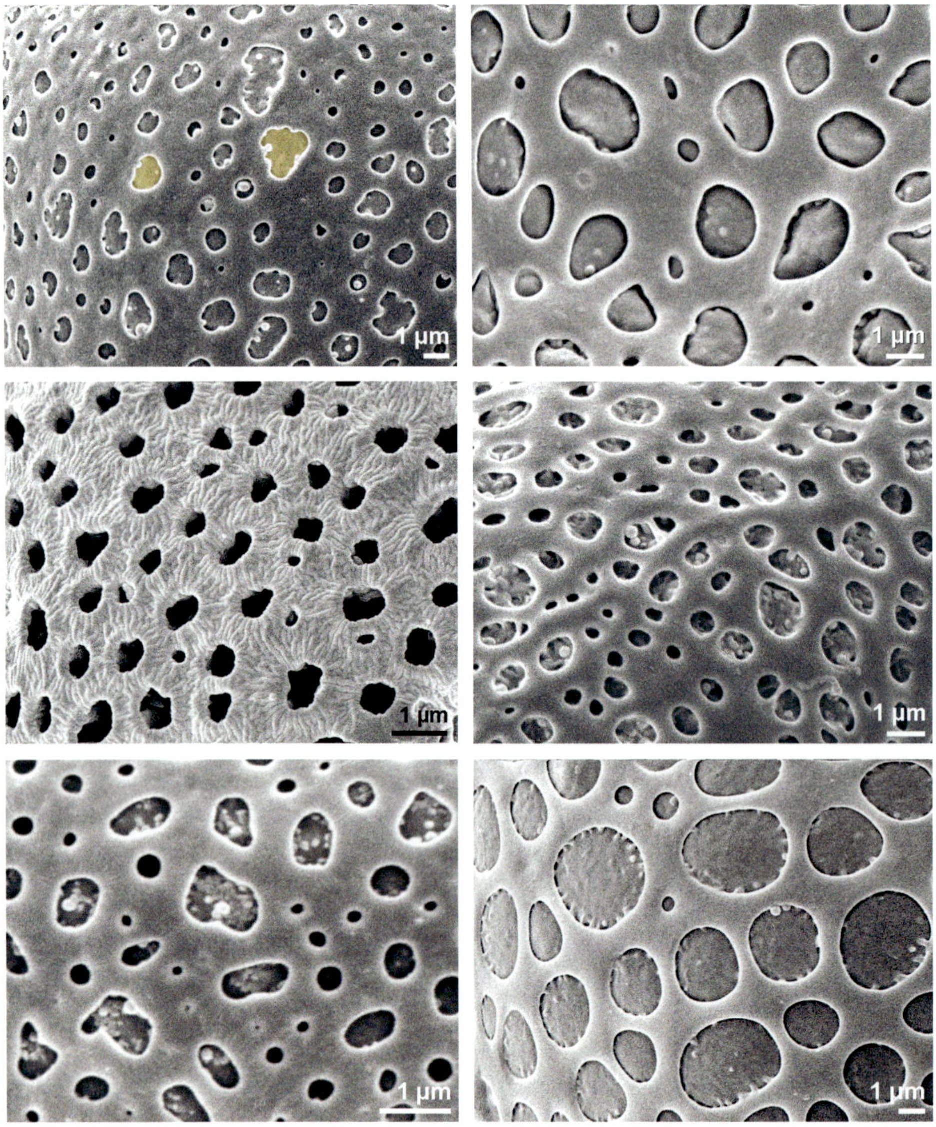

Streptocalyx poeppigii
Bromeliaceae

Canistrum camacaense
Bromeliaceae

Lavandula angustifolia
Lamiaceae

Smilacina stellata
Convallariaceae

Cyrtosperma beccarianum
Araceae

Hohenbergia stellata
Bromeliaceae
foveolate to reticulate

LM SEM TEM mo ana fnc

fossula: irregularly shaped groove in the surface of a pollen wall.

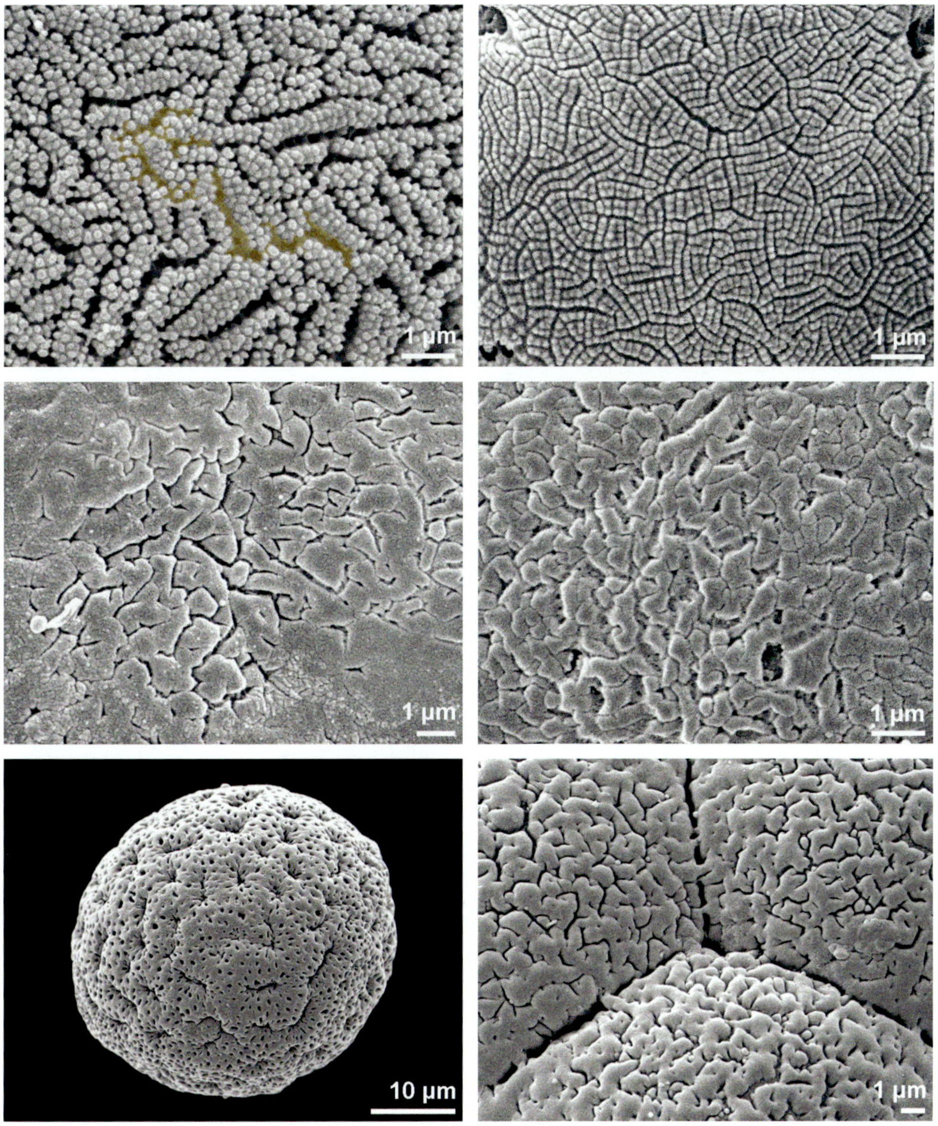

Mendoncia albida
Acanthaceae
fossulate, granulate

Rhododendron hirsutum
Ericaceae

Aristolochia manshuriensis
Aristolochiaceae
fossulate, perforate

Erica herbacea
Ericaceae

Ledum palustre
Ericaceae

Moneses uniflora
Pyrolaceae
tetrad, y-shaped fissure between the monads

LM SEM TEM mo ana fnc

lophae: a network-like pattern of ridges (= lophae) formed by the outer exine surrounding window-like spaces or depressions (= lacunae).

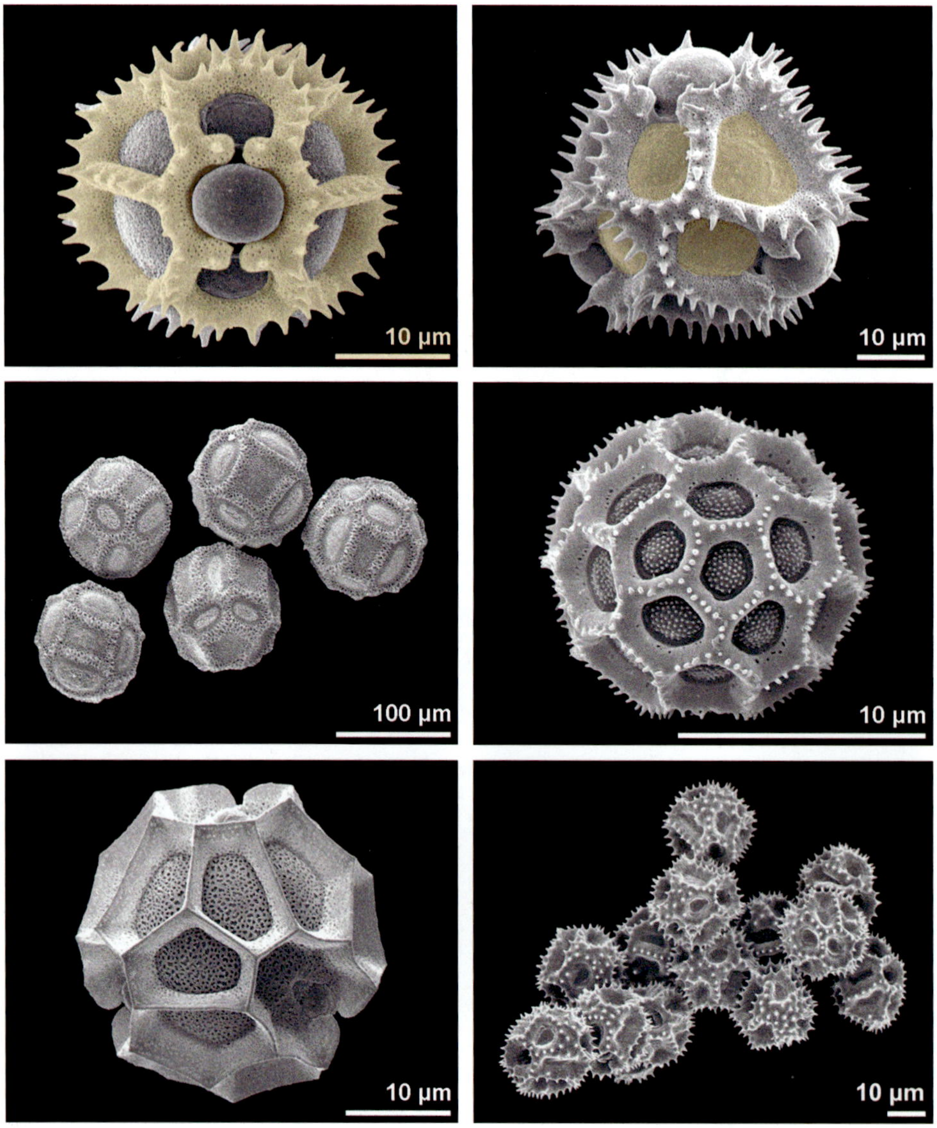

Leontodon saxatilis
Asteraceae
equatorial view, lophae colored

Cichorium intybus
Asteraceae
polar view, lacunae colored

Opuntia basilaris
Cactaceae

Pfaffia tuberosa
Amaranthaceae

Gazania sp.
Asteraceae
polar view

Hieracium hoppeanum
Asteraceae
dry pollen

LM SEM TEM mo ana fnc

lophae: a network-like pattern of ridges (= lophae) formed by the outer exine surrounding window-like spaces or depressions (= lacunae).

Taraxacum sp.
Asteraceae, fossil
equatorial view

Scorzonera cana
Asteraceae
polar view

Opuntia polyacantha
Cactaceae

Crepis biennis
Asteraceae
polar view

Tragopogon dubius
Asteraceae
equatorial view

Ipomoea caerulea
Convolvulaceae

LM SEM TEM mo ana fnc

echinus: pointed ornamentation element longer and/or wider than 1 µm.

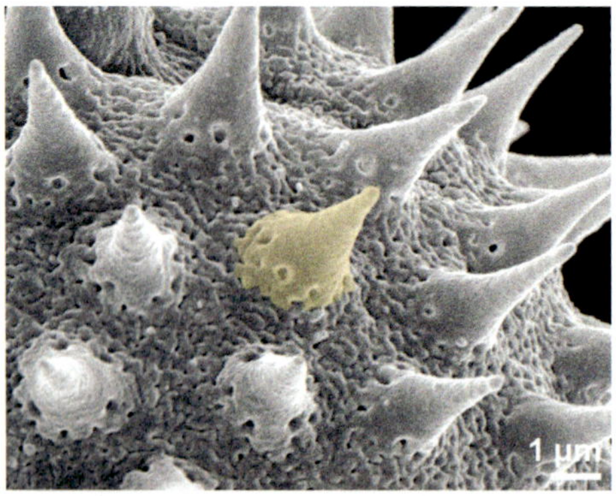

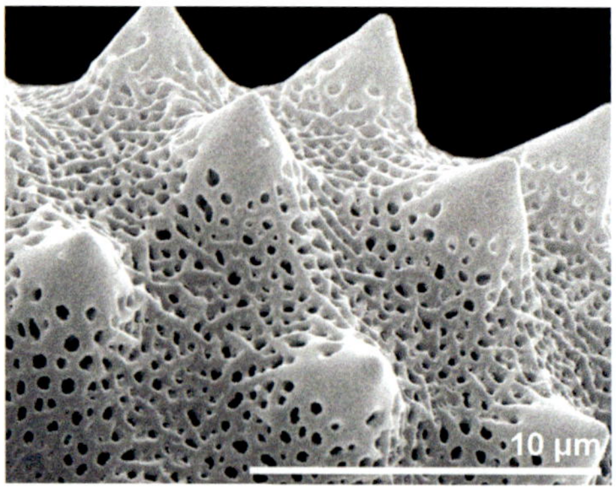

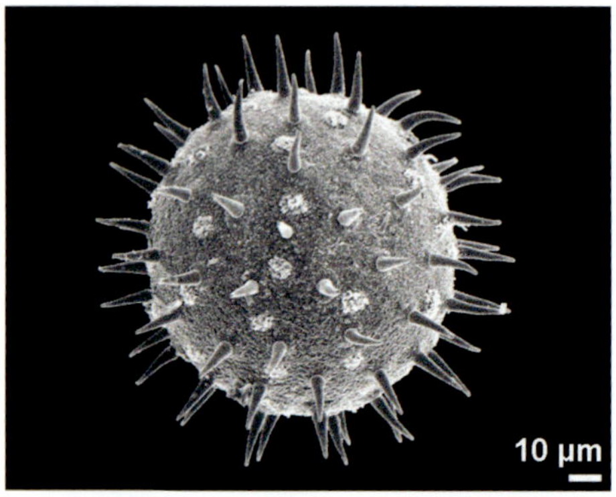

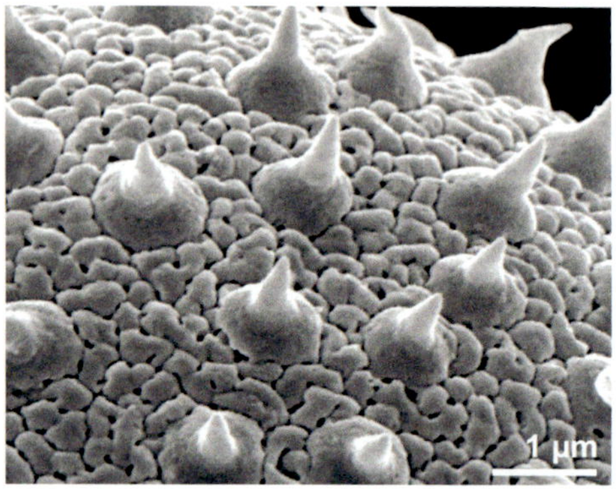

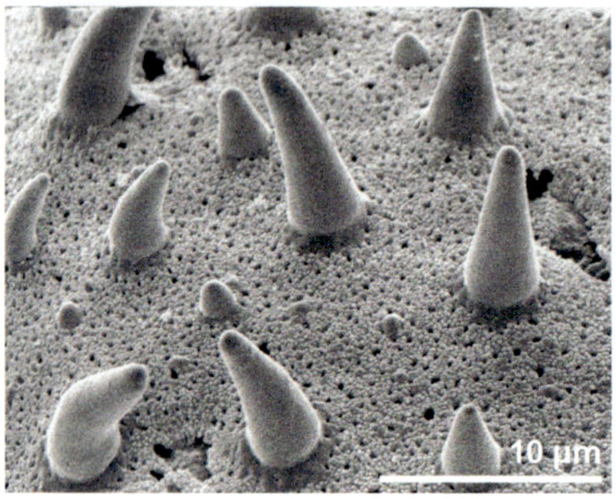

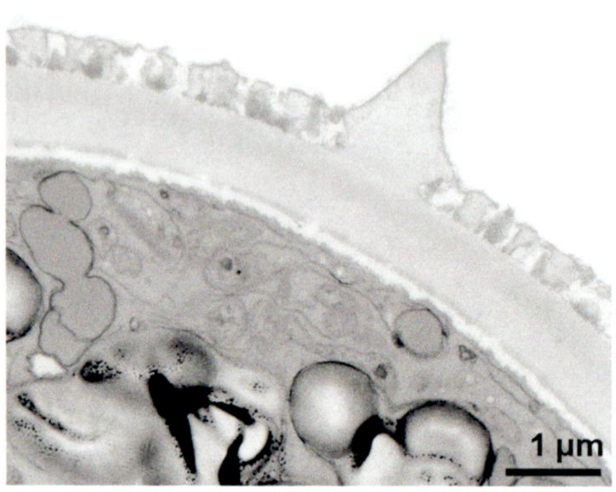

Galinsoga ciliata
Asteraceae

Hibiscus trionum
Malvaceae

Lavatera thuringiaca
Malvaceae

Carduus acanthoides
Asteraceae

Pinellia ternata
Araceae

Pinellia ternata
Araceae
PA+TCH+SP (short)

LM SEM TEM mo ana fnc

echinus: pointed ornamentation element longer and/or wider than 1 µm.

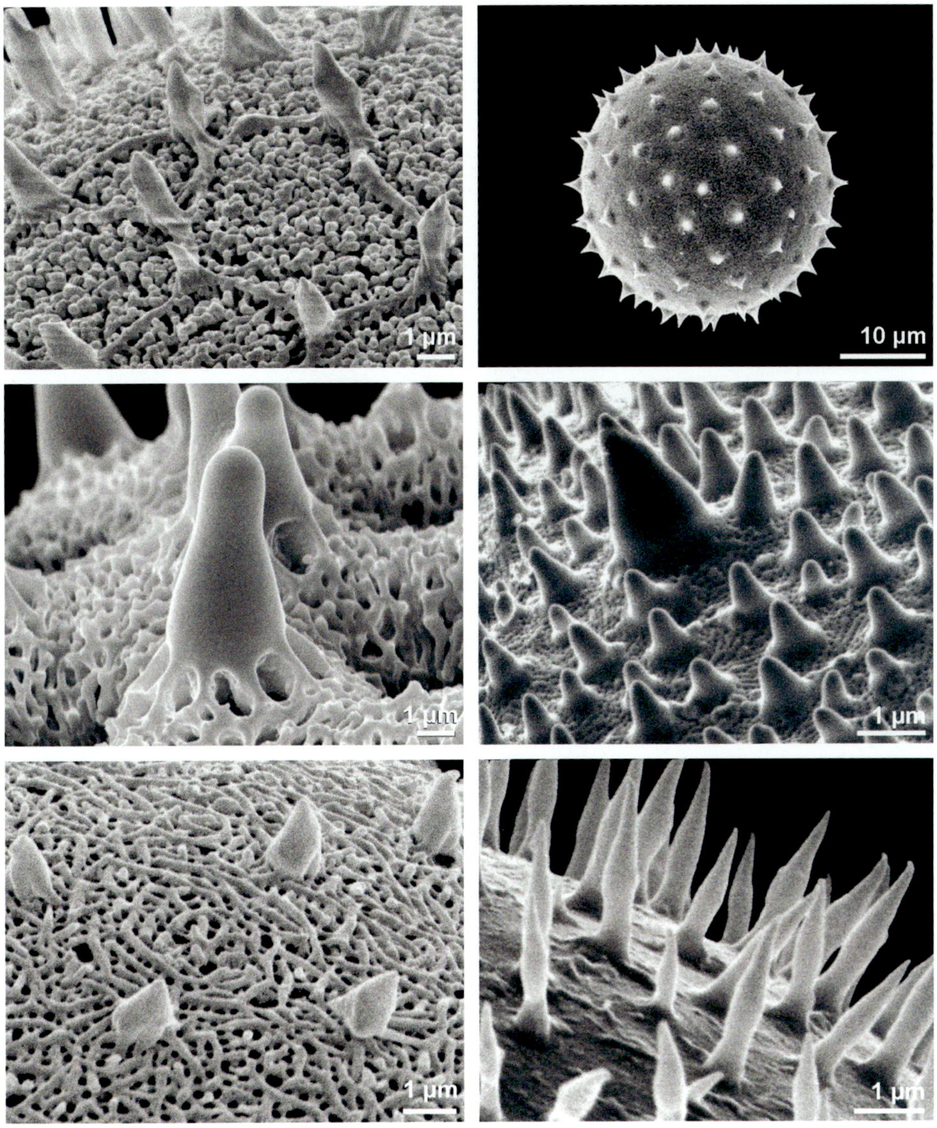

Stratiotes aloides
Hydrocharitaceae

Ipomoea batatas
Convolvulaceae

Campanula alpina
Campanulaceae

Zomicarpa riedeliana
Araceae

Knautia drymeia
Dipsacaceae

Ulearum sagittatum
Araceae

LM SEM TEM mo ana fnc

echinus: pointed ornamentation element longer and/or wider than 1 µm.

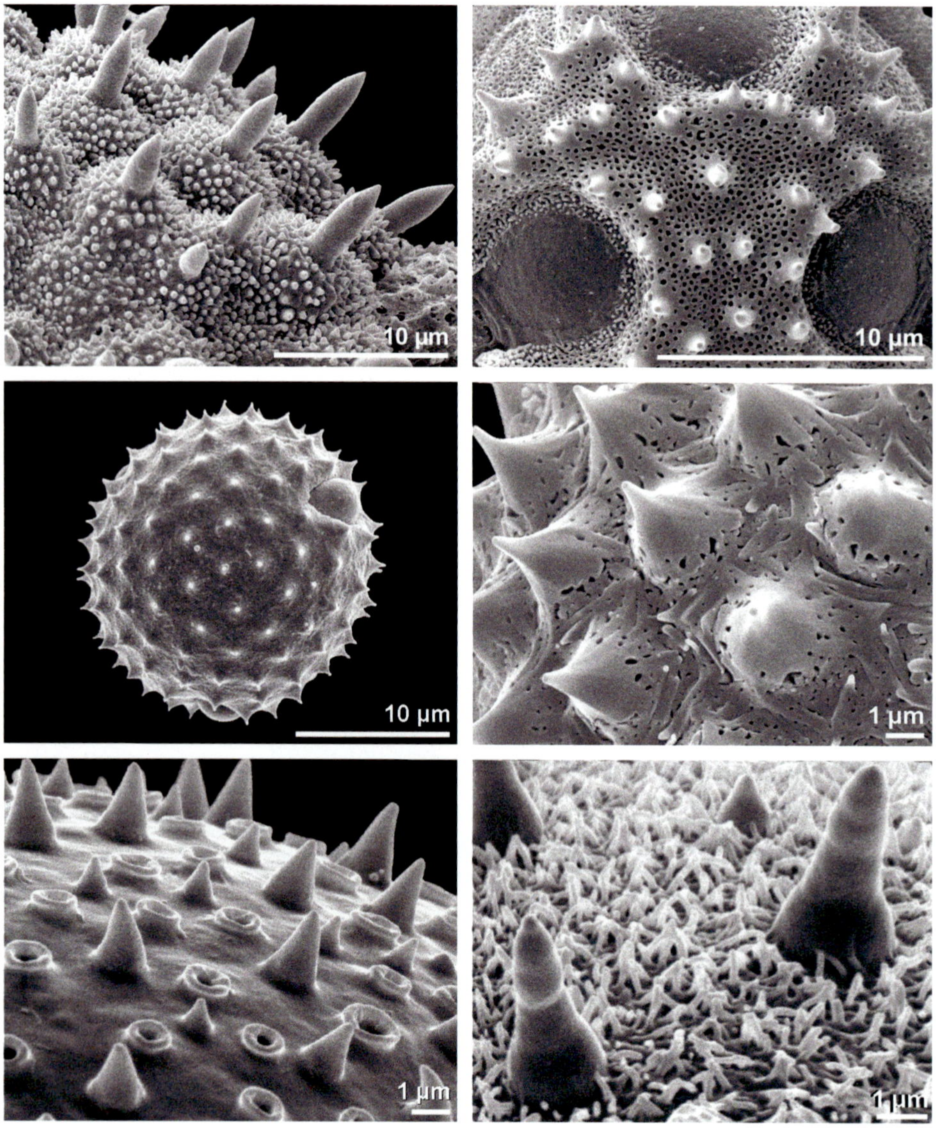

Patrinia gibbosa
Valerianaceae

Hieracium hoppeanum
Asteraceae

Ambrosia artemisiifolia
Asteraceae

Aster amellus
Asteraceae

Portulaca grandiflora
Portulacaceae

Nuphar luteum
Nymphaeaceae

LM SEM TEM mo ana fnc

echinus: pointed ornamentation element longer and/or wider than 1 µm.

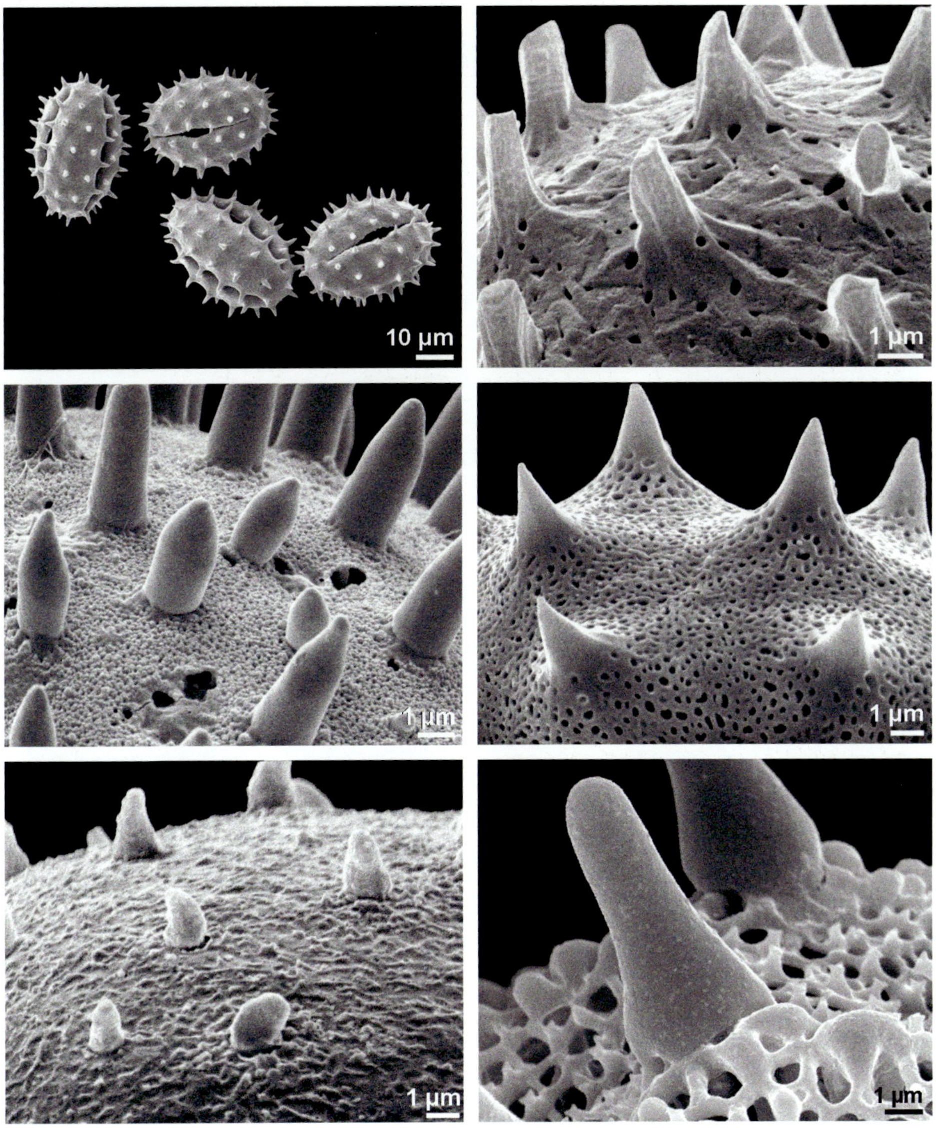

Petasites albus
Asteraceae
dry pollen

Antennaria dioica
Asteraceae

Malva neglecta
Malvaceae

Tanacetum corymbosum
Asteraceae

Lonicera fragrantissima
Caprifoliaceae

Pharbitis purpurea
Convolvulaceae

LM SEM TEM mo ana fnc

microechinate

micro-: prefix for small; features smaller as 1 µm; see "Alphabetic Glossary".

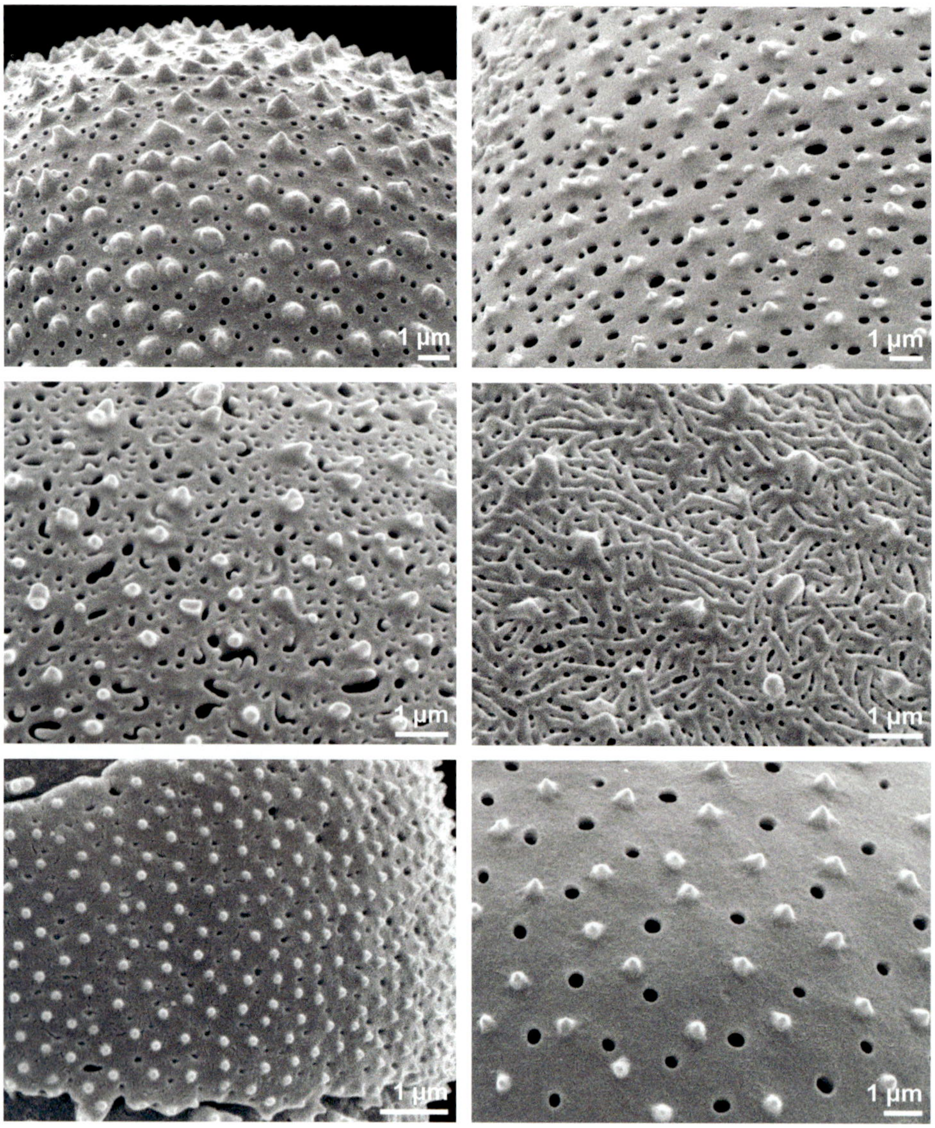

Claytonia perfoliata
Portulacaceae
microechinate, perforate

Babiana velutina
Iridaceae
microechinate, perforate

Pulsatilla pratensis
Ranunculaceae
microechinate, perforate

Platycodon grandiflorum
Campanulaceae
microechinate, striato-microreticulate

Galium lucidum
Rubiaceae
microechinate, perforate

Petrorhagia prolifera
Caryophyllaceae
microechinate, perforate

LM SEM TEM mo ana fnc

verruca: wart-like element more than 1 µm broad, broader than high.

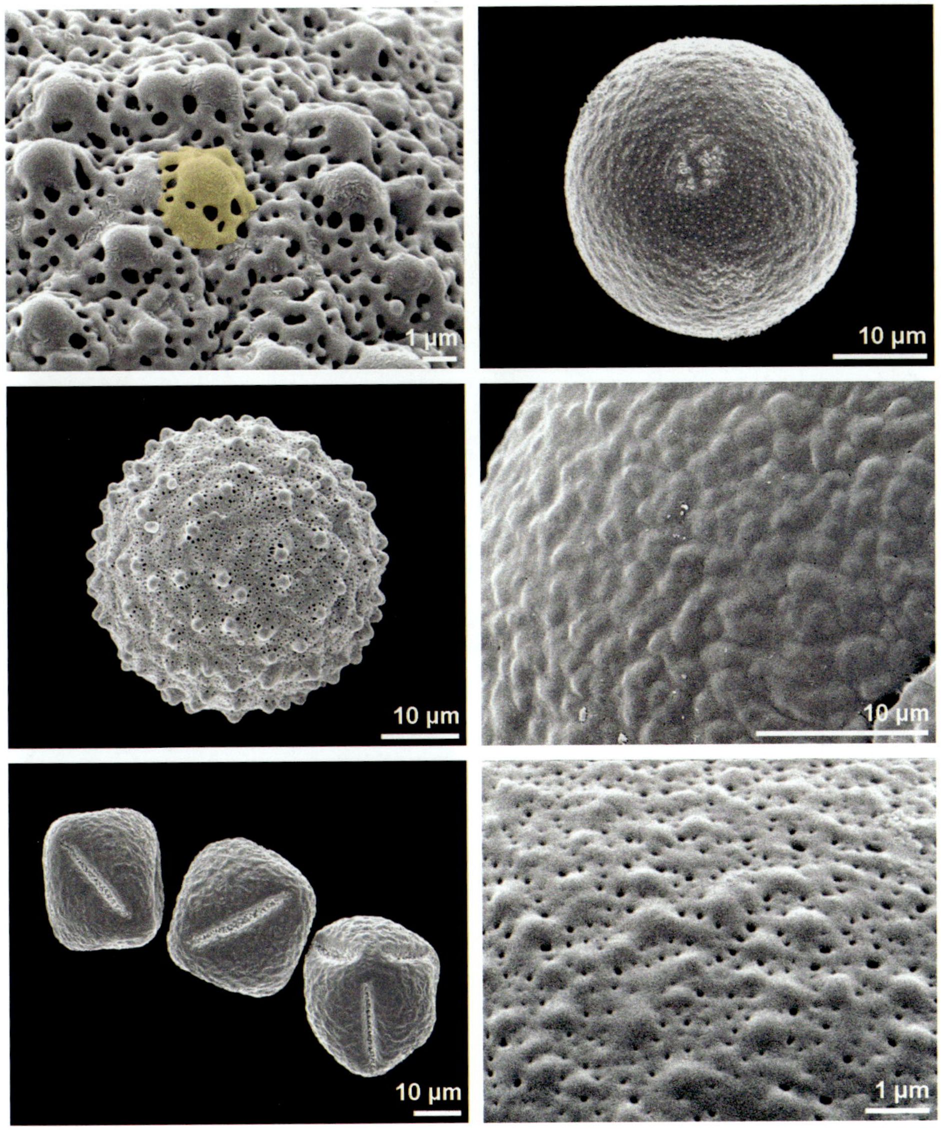

Aristolochia salvadorensis
Aristolochiaceae
verrucate, perforate

Aristolochia tricaudata
Aristolochiaceae
verrucate, perforate

Corydalis cava
Fumariaceae
dry pollen

Plantago media
Plantaginaceae
verrucate, microechinate

Calliandra emarginata
Mimosaceae

Teucrium chamaedrys
Lamiaceae
microverrucate, perforate

LM SEM TEM mo ana fnc

baculum: rod-like, free standing element, more than 1 µm in height and never pointed.

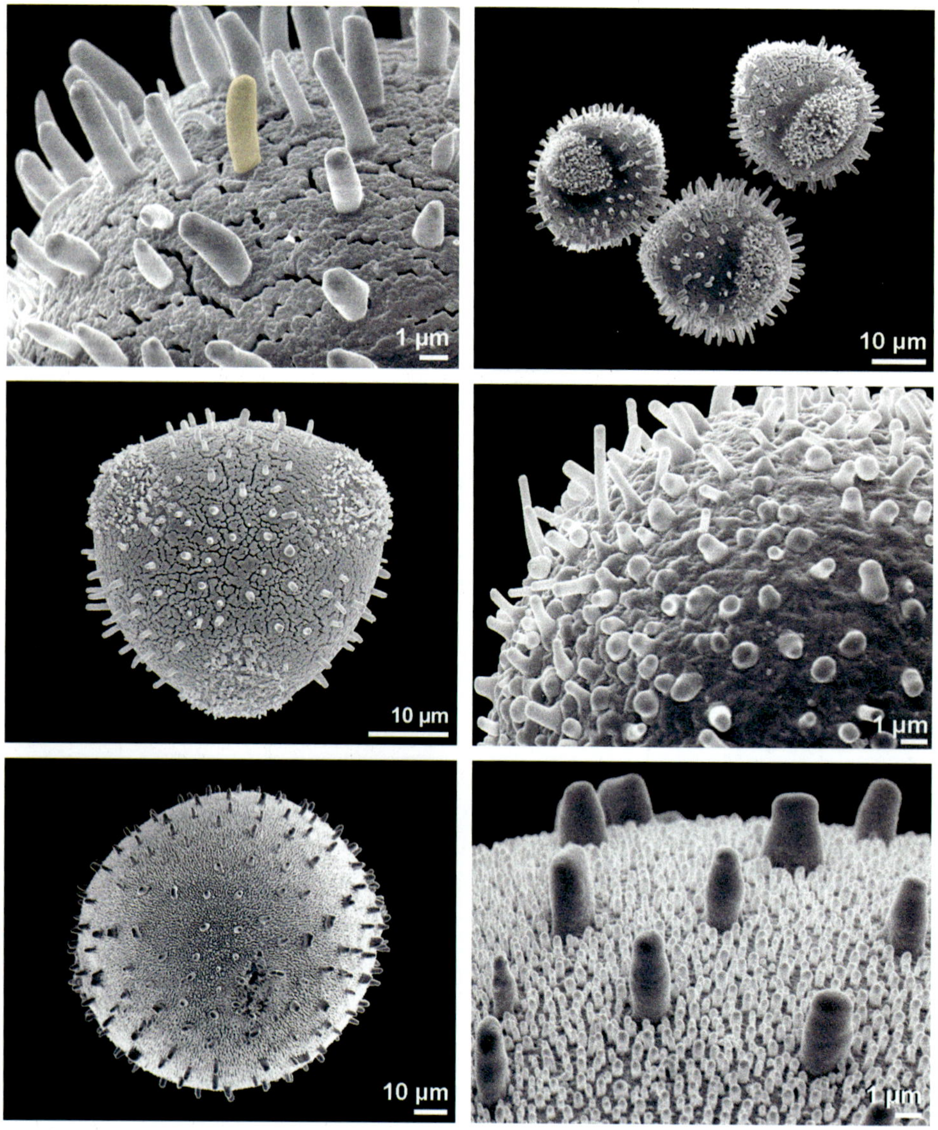

Viscum laxum
Viscaceae

Viscum laxum
Viscaceae

Viscum album
Viscaceae
polar view

Nymphaea alba
Nymphaeaceae

Erythrochiton brasiliensis
Rutaceae
equatorial view

Erythrochiton brasiliensis
Rutaceae
large bacula and smaller bacula or clavae

LM SEM TEM mo ana fnc

clava: club-shaped element, higher than 1 µm.

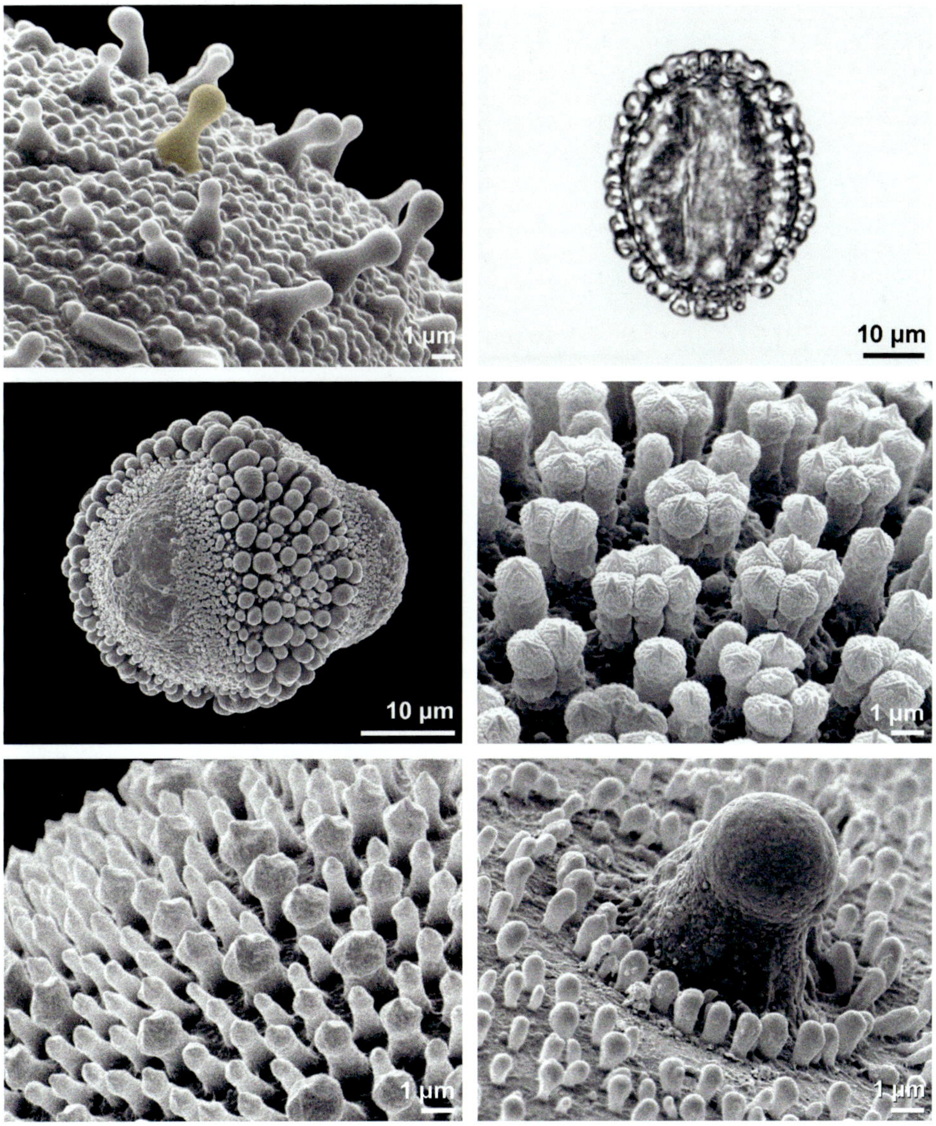

Iris alata
Iridaceae

Ilex sp.
Aquifoliaceae, fossil
equatorial view

Ilex aquifolium
Aquifoliaceae
equatorial view

Plumbago auriculata
Plumbaginaceae

Linum flavum
Linaceae
heterostylous, long-styled morph

Aratitiyopea lopezii
Xyridaceae
clavae of two different sizes

LM SEM TEM mo ana fnc

clava: club-shaped element, higher than 1 µm.

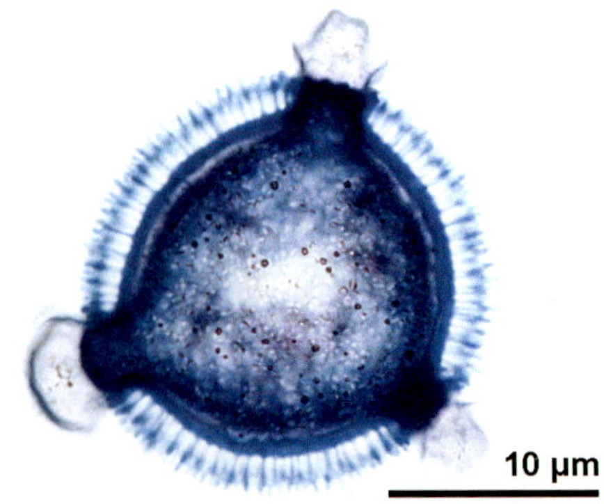

Geranium robertianum
Geraniaceae
toluidine blue

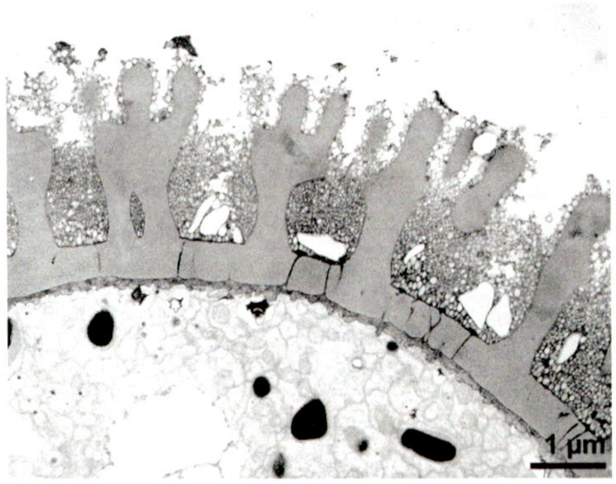

Geranium robertianum
Geraniaceae
PA+TCH+SP (short)

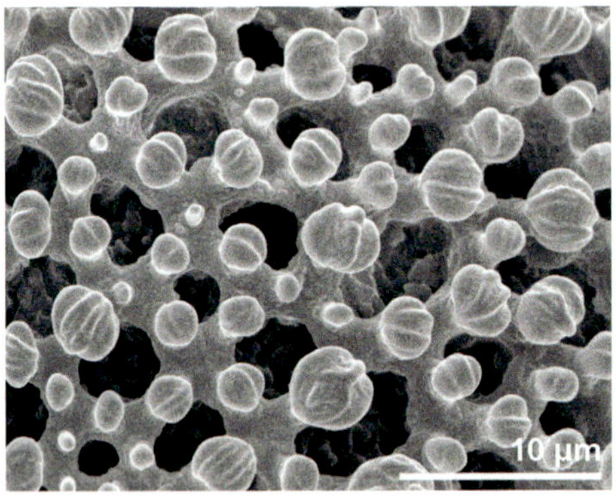

Geranium pratense
Geraniaceae
reticulum cristatum with clavae

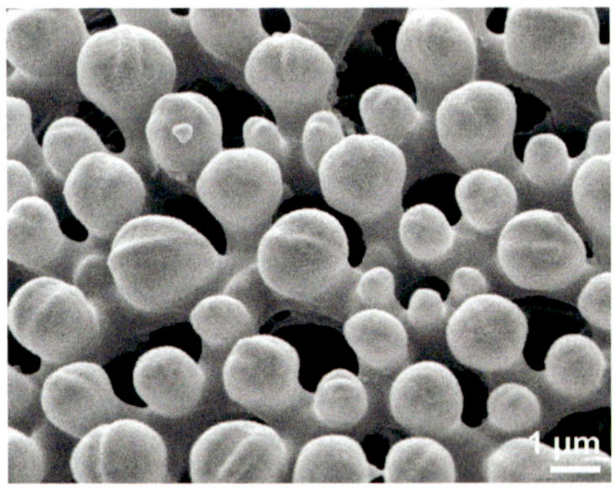

Geranium sibiricum
Geraniaceae
reticulum cristatum with clavae

LM SEM TEM mo ana fnc

free-standing columellae: columellae in the infratectal layer not covered by a tectum in semitectate pollen grains.

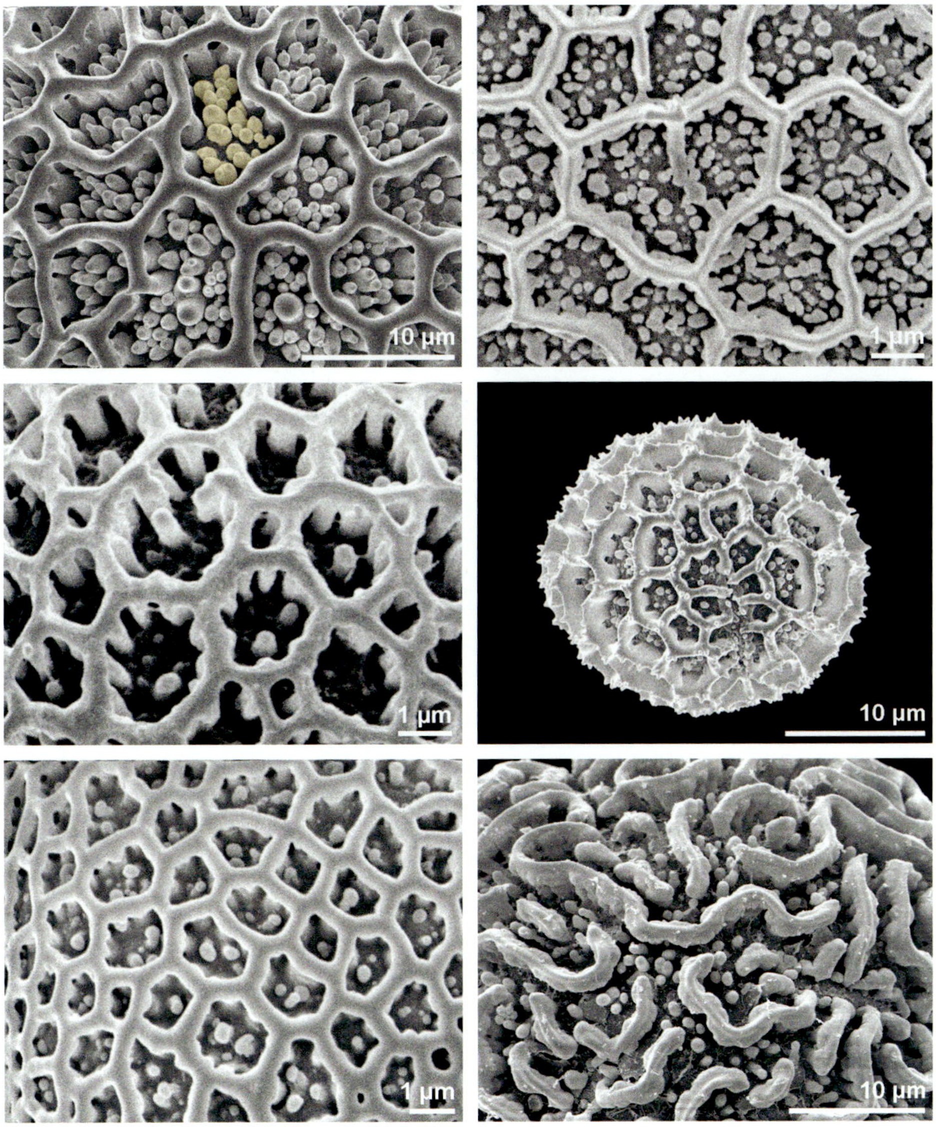

Dipteracanthus devosianus
Acanthaceae

Erophila verna
Brassicaceae

Viburnum opulus
Caprifoliaceae

Impatiens parviflora
Balsaminaceae

Bougainvillea sp.
Nyctaginaceae

Thladiantha hookeri
Cucurbitaceae

LM SEM TEM mo ana fnc

gemma: globular exine element more than 1 µm in diameter.

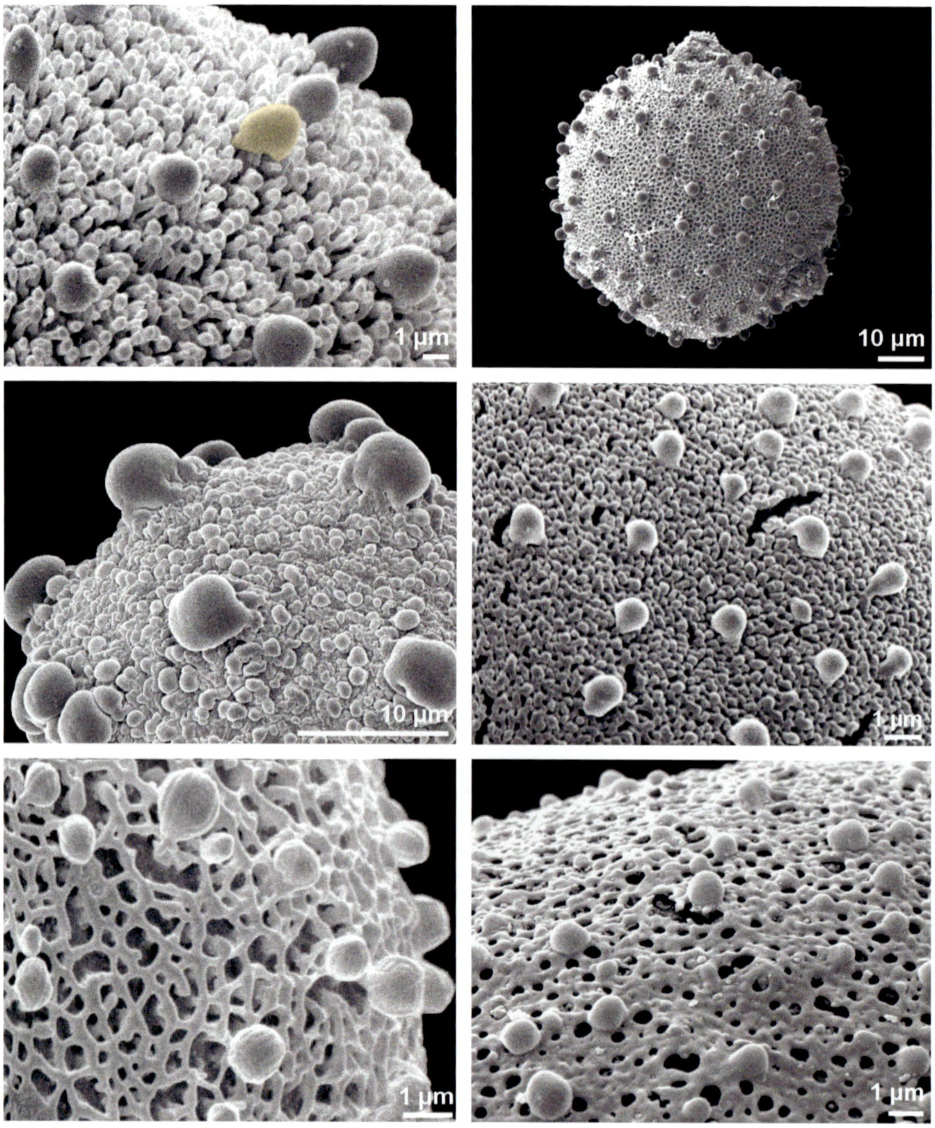

Dionaea muscipula
Droseraceae
gemmate, clavate

Stenandrium dulce
Acanthaceae
large and small gemmae and granules

Fatsia japonica
Araliaceae
gemmate, reticulate

Cephalopentandra ecirrhosa
Cucurbitaceae
gemmate, reticulate, polar view

Asarum europaeum
Aristolochiaceae
gemmate, microgemmate

Hakea kippistiana
Proteaceae
gemmate, microreticulate

LM SEM TEM mo ana fnc

areola: small, mostly convex exine island.

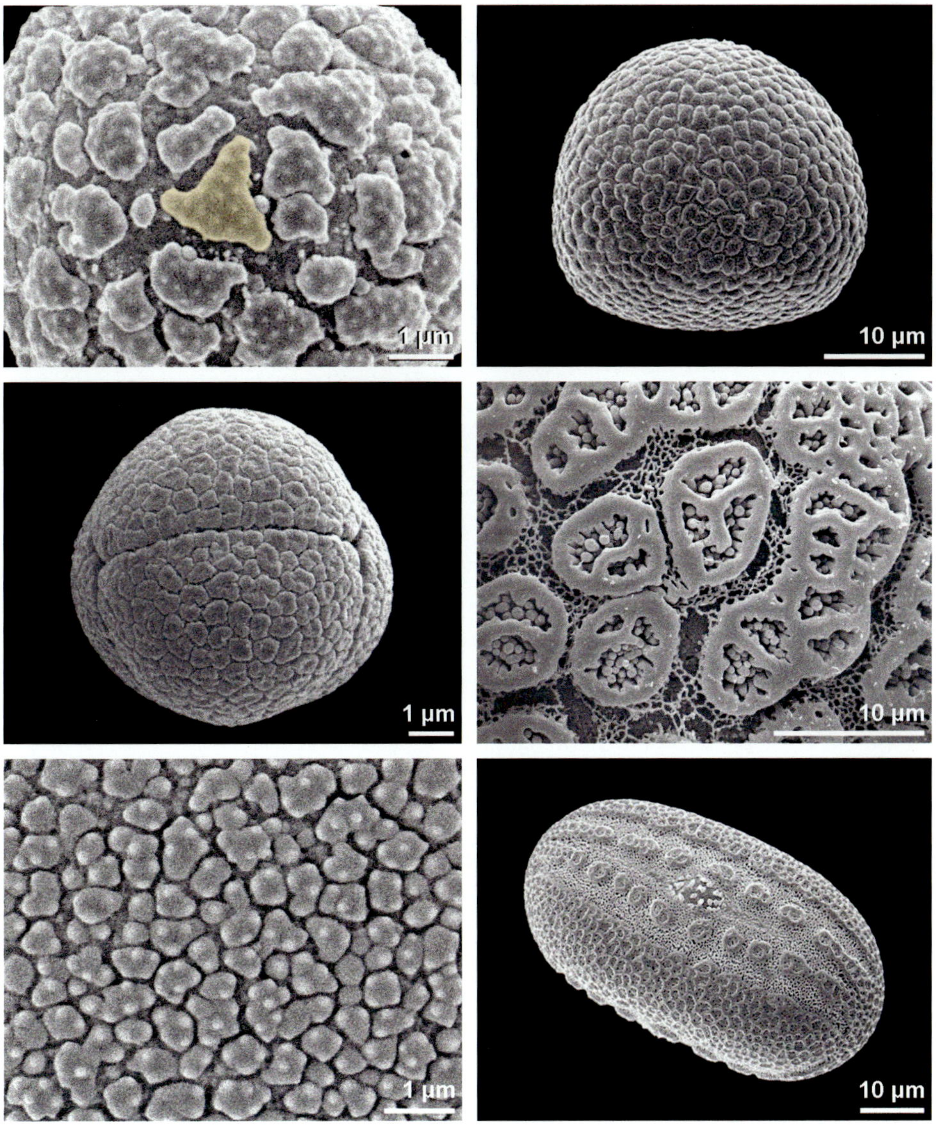

Peperomia rubella
Piperaceae

Mimosa pudica
Mimosaceae
tetrad

Cynodon dactylon
Poaceae

Dracunculus vulgaris
Araceae

Poikilacanthus macranthus
Acanthaceae
areolae reticulate

Beloperone guttata
Acanthaceae
areolae in apertural area

LM SEM TEM mo ana fnc

clypeate: pollen wall where the exine is subdivided into shields.

Ibicella lutea
Martyniaceae
inaperturate

Phyllanthus sp.
Euphorbiaceae
pantoporate

Banisteria argentea
Malpighiaceae
pantocolporate

Corydalis lutea
Fumariaceae
pantocolpate, syncolpate

Iris bucharica
Iridaceae
inaperturate

Catalpa bignonioides
Bignoniaceae
tetrads, inaperturate, dry pollen

LM SEM TEM mo ana fnc

pollen wall: layer(s) enclosing the cytoplasm of a pollen grain.

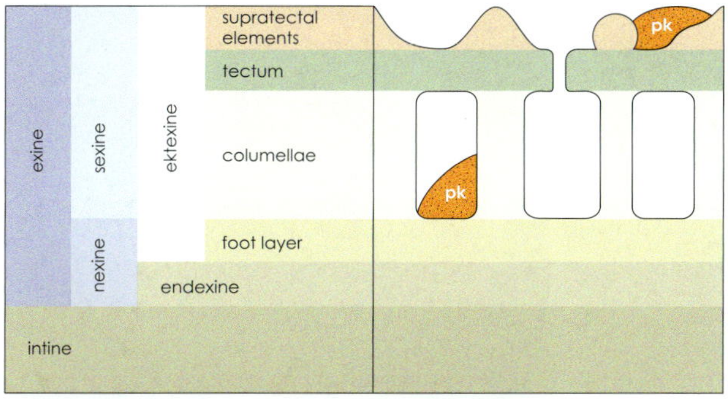

structure of the pollen wall
schematic drawing
pollenkitt (pk)

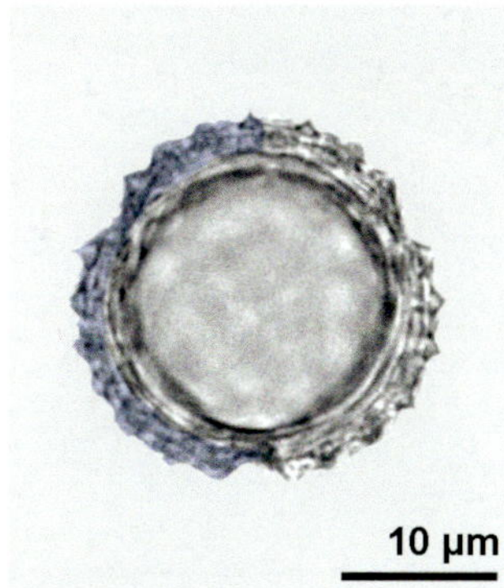

Ambrosia artemisiifolia
Asteraceae
acetolyzed

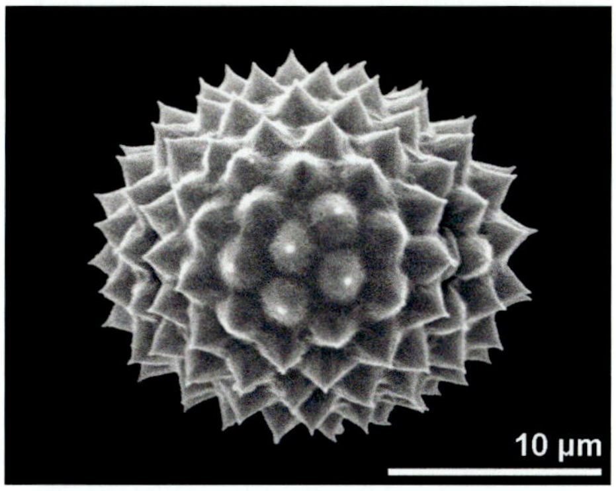

Ambrosia artemisiifolia
Asteraceae

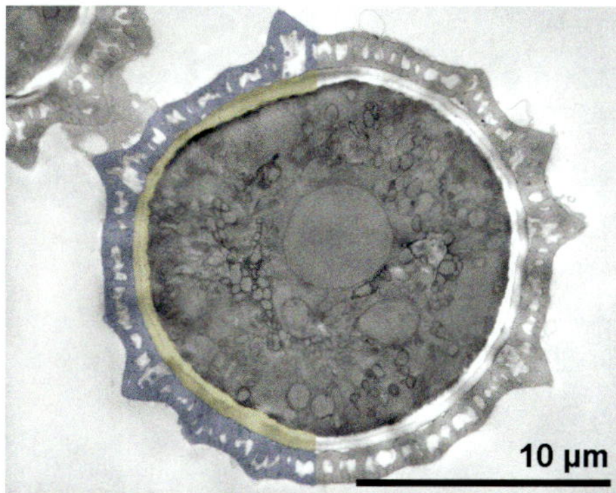

Ambrosia artemisiifolia
Asteraceae
PA+TCH+SP (short)

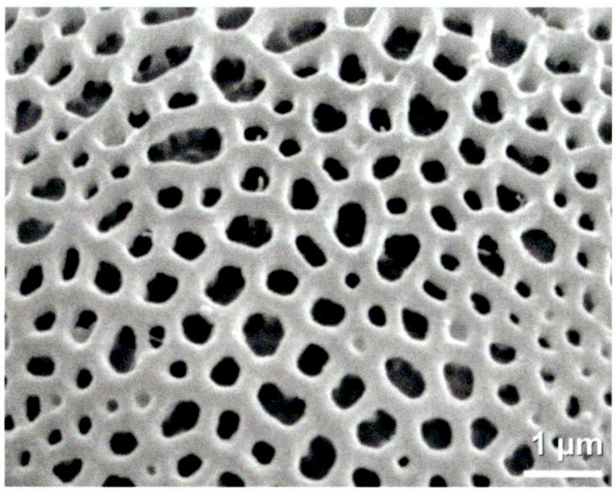

Scrophularia nodosa
Scrophulariaceae

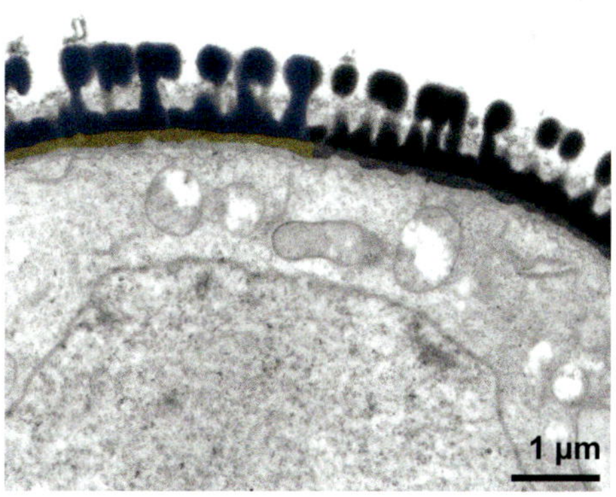

Scrophularia nodosa
Scrophulariaceae
$KMnO_4$

LM SEM TEM mo ana fnc

infratectum: layer between tectum and foot layer or endexine (if foot layer is missing); infratectum can be **alveolate**, columellate, granular or absent.

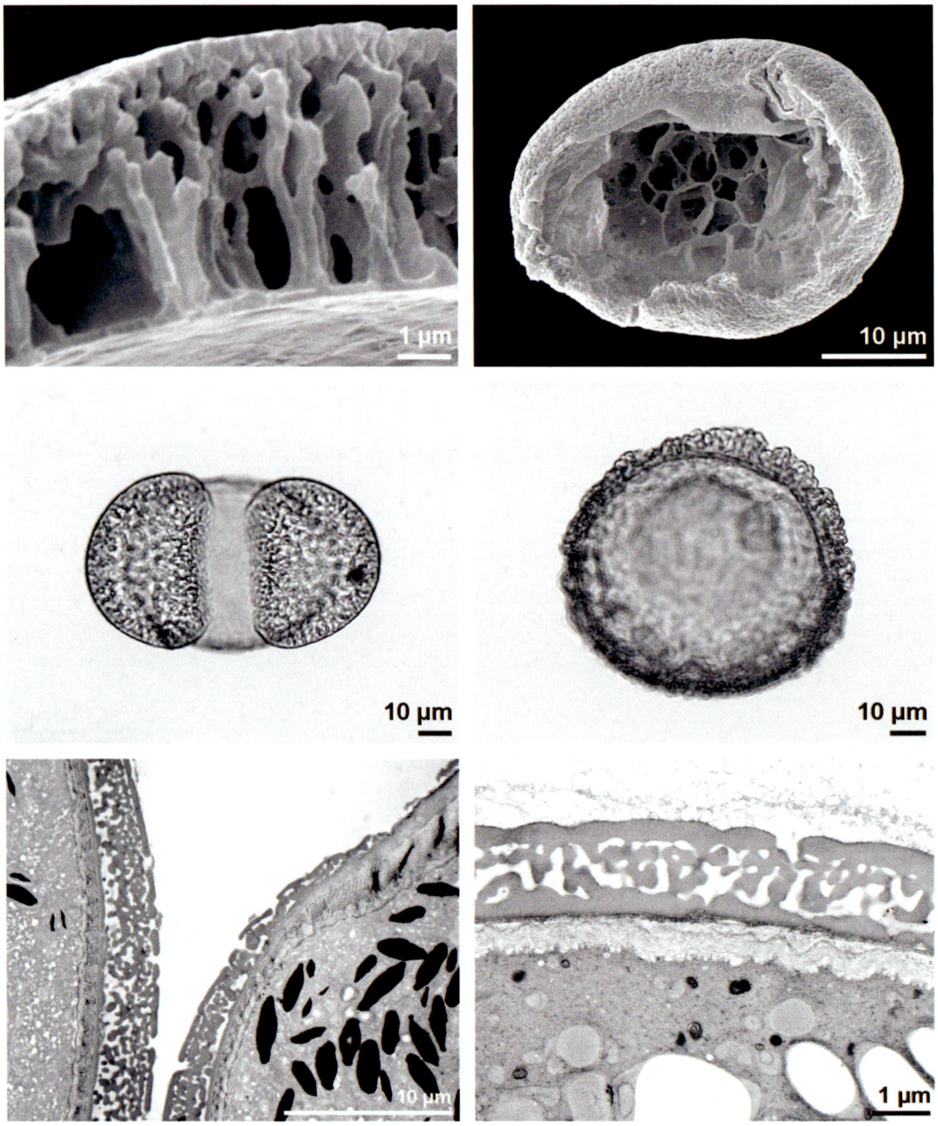

Abies sp.
Pinaceae, fossil
fractured pollen wall, proximal area (cappa)

Pinus sp.
Pinaceae, fossil
alveolae inside detached saccus

Pinus sp.
Pinaceae, fossil

Tsuga sp.
Pinaceae, fossil

Gonatopus angustus
Araceae
PA+TCH+SP

Zamioculcas zamiifolia
Araceae
U+Pb

infratectum: layer between tectum and foot layer or endexine (if foot layer is missing); infratectum can be alveolate, **columellate**, granular or absent.

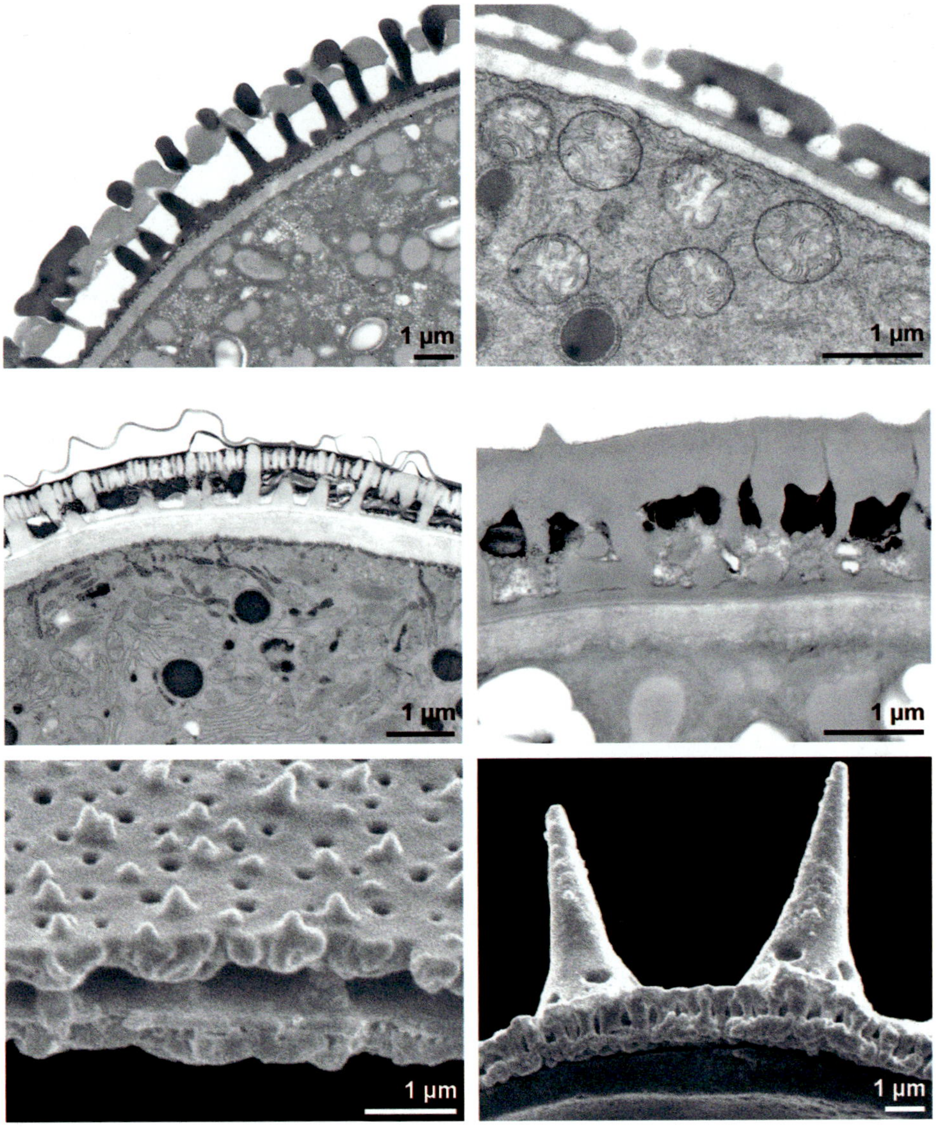

Mentha aquatica
Lamiaceae
$KMnO_4$

Microrrhinum minus
Scrophulariaceae
PA+TCH+SP (short)

Melampyrum pratense
Scrophulariaceae
PA+TCH+SP (short)

Bassia scoparia
Chenopodiaceae
U+Pb

Gladiolus illyricus
Iridaceae
fractured pollen wall

indet.
Asteraceae
acetolyzed honey sample
fractured pollen wall

LM SEM TEM mo ana fnc

infratectum: layer between tectum and foot layer or endexine (if foot layer is missing); infratectum can be alveolate, columellate, **granular** or absent.

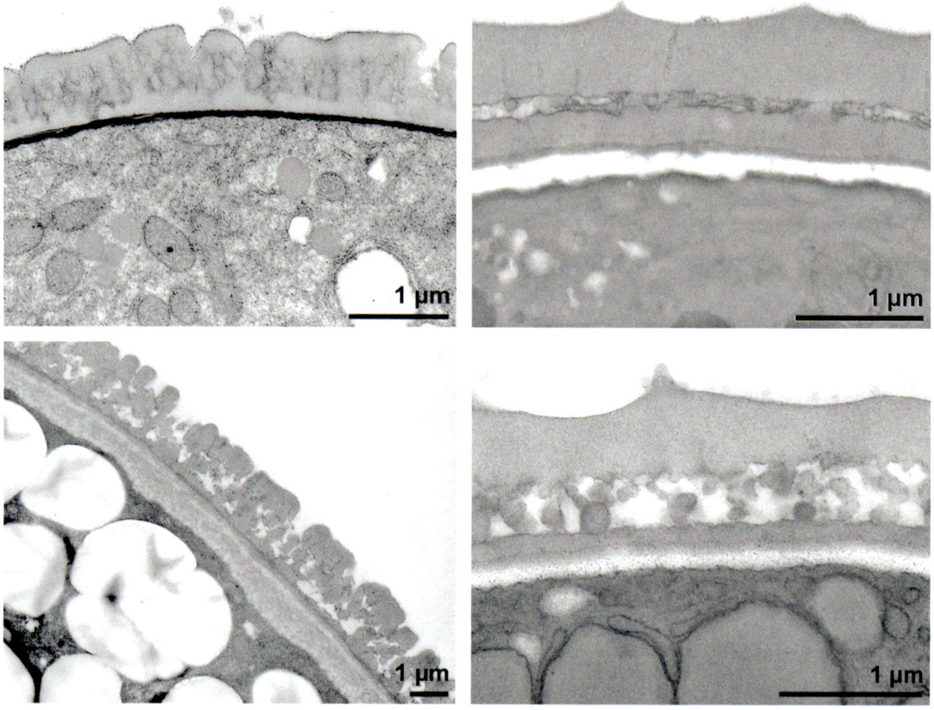

infratectum: layer between tectum and foot layer or endexine (if foot layer is missing); infratectum can be alveolate, columellate, granular or **absent**.

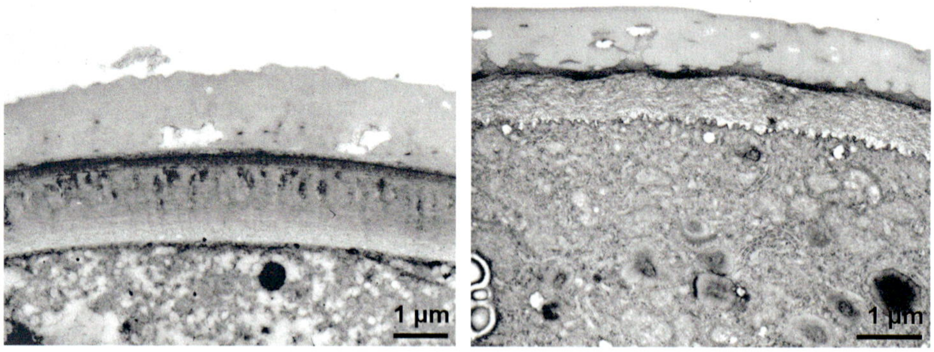

Amydrium medium
Araceae
U+Pb

Corylus colurna
Betulaceae
PA+TCH+SP (short)

Viola tricolor
Violaceae
U+Pb

Juglans regia
Juglandaceae
PA+TCH+SP (short)

Dieffenbachia humilis
Araceae
U+Pb

Mahonia aquifolium
Berberidaceae
U+Pb

LM SEM TEM mo ana fnc

internal tectum: a ± continuous layer between foot layer and tectum, separated from them by columellae.

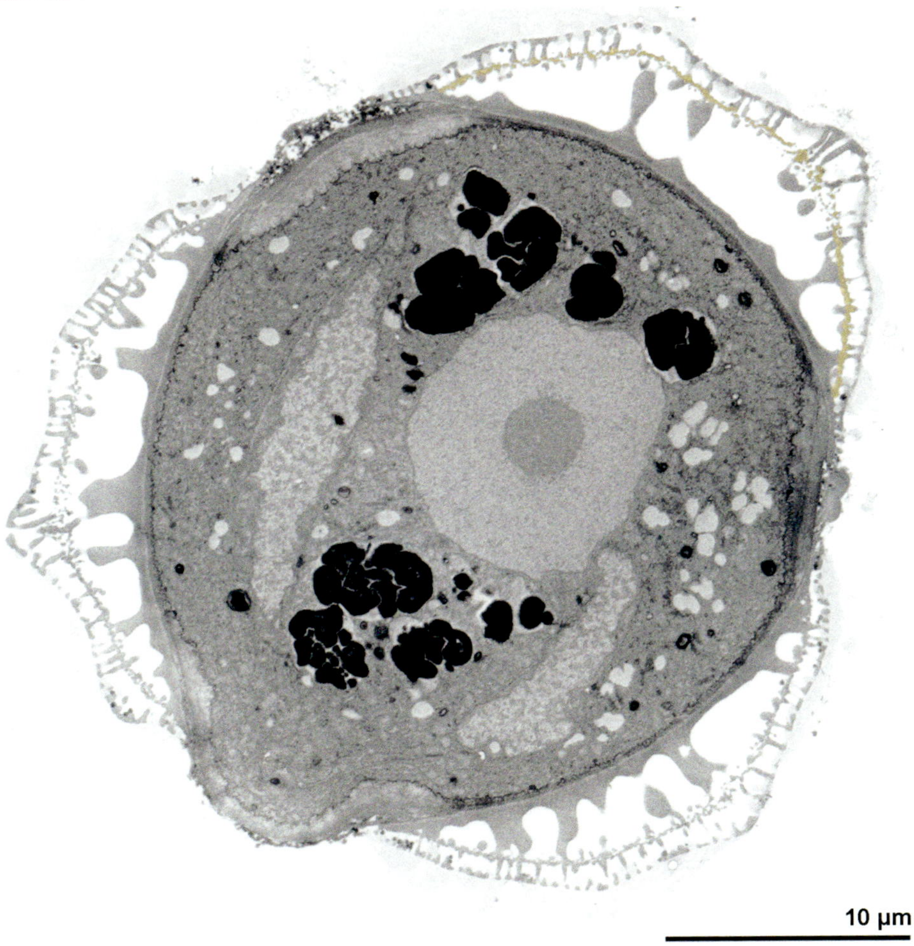

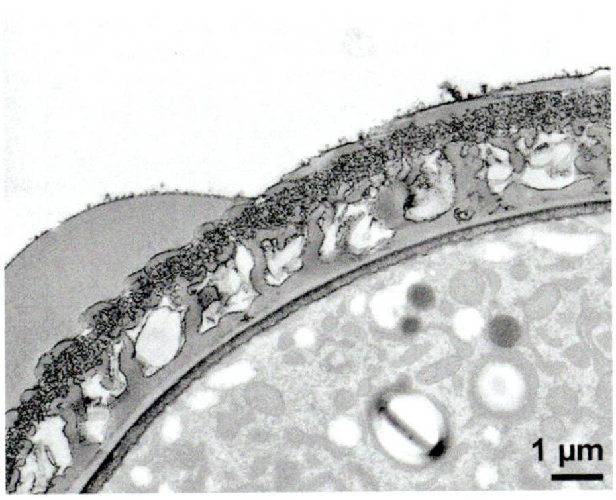

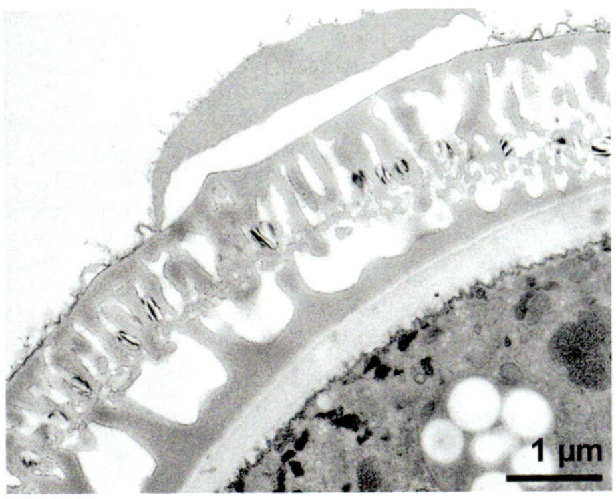

Argyranthemum sp.
Asteraceae
U+Pb

Agrimonia eupatoria
Rosaceae
PA+TCH+SP (short)

Centaurea cyanus
Asteraceae
PA+TCH+SP (short)

LM SEM TEM mo ana fnc

foot layer: inner layer of the ektexine; foot layer can be **continuous**, **discontinuous**, perforated or **absent**.

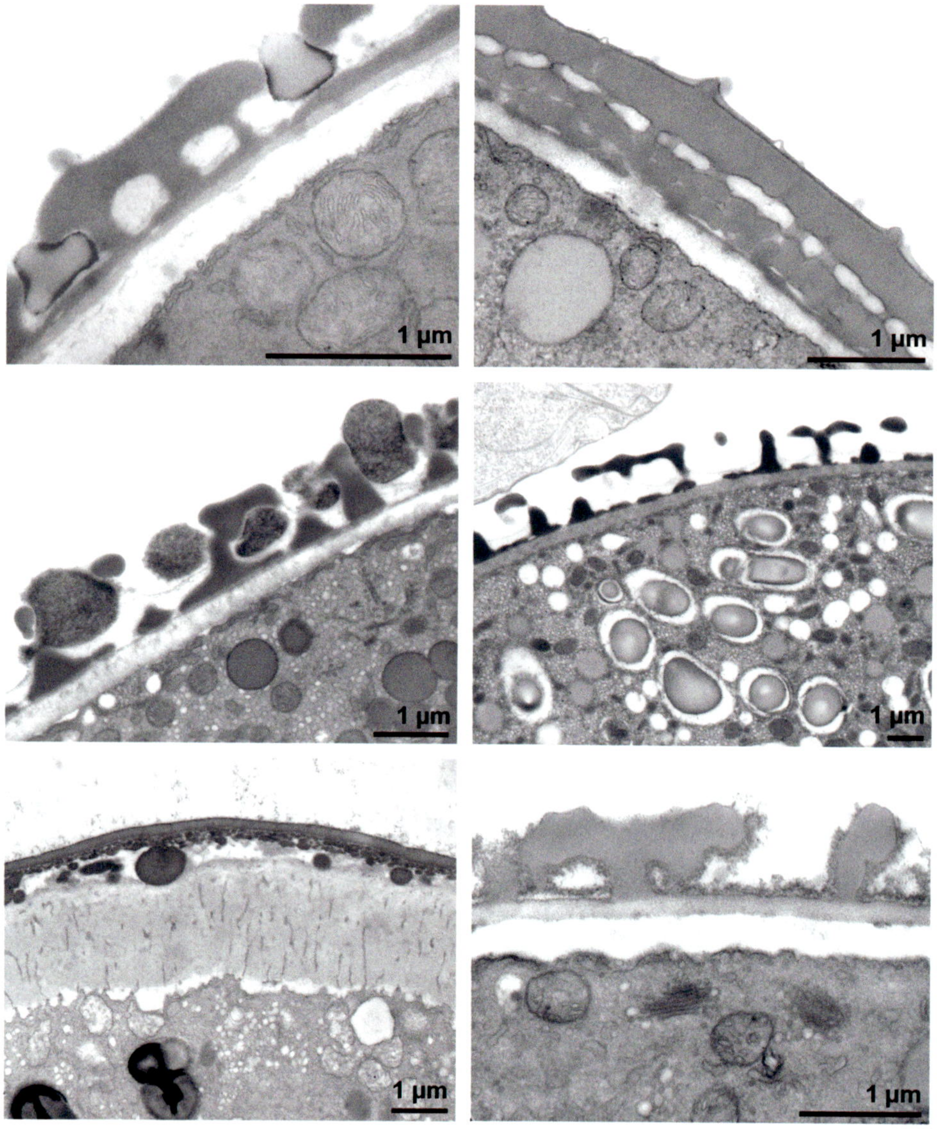

Microrrhinum minus
Scrophulariaceae
PA+TCH+SP (short)
continuous

Clinopodium vulgare
Lamiaceae
PA+TCH+SP (short)
discontinuous

Pachypodium succulentum
Apocynaceae
PA+TCH+SP (short)
absent

Plantago maritima
Plantaginaceae
TCH+SP
continuous

Acinos alpinus
Lamiaceae
$KMnO_4$
discontinuous

Fraxinus excelsior
Oleaceae
PA+TCH+SP (short)
absent

LM SEM TEM mo ana fnc

endexine: distinct exine layer between ektexine and intine; endexine (E) can be **compact**, spongy or lamellar as well as **continuous**, discontinuous, absent or in aperture only.

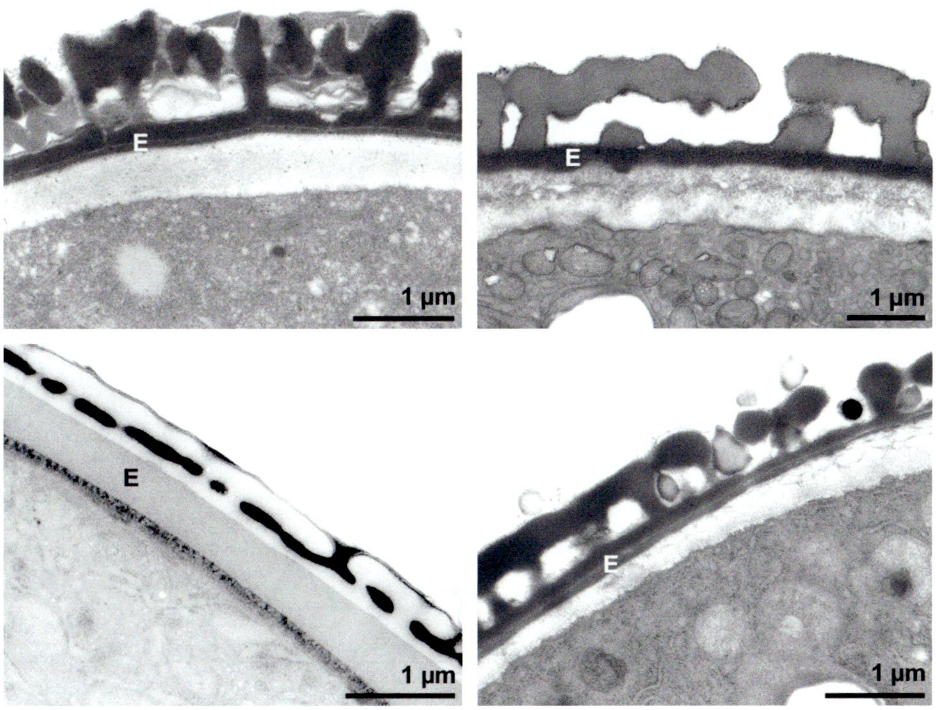

Odontites luteus
Scrophulariaceae
$KMnO_4$
compact, continuous

Ailanthus altissima
Simaroubaceae
TCH+SP
compact, continuous

Pulmonaria officinalis
Boraginaceae
PA+TCH+SP
compact, continuous

Microrrhinum minus
Scrophulariaceae
U+Pb
compact, continuous

LM SEM TEM mo ana fnc

endexine: distinct exine layer between ektexine and intine; endexine (E) can be compact, **spongy** or **lamellar** as well as **continuous**, discontinuous, absent or in aperture only.

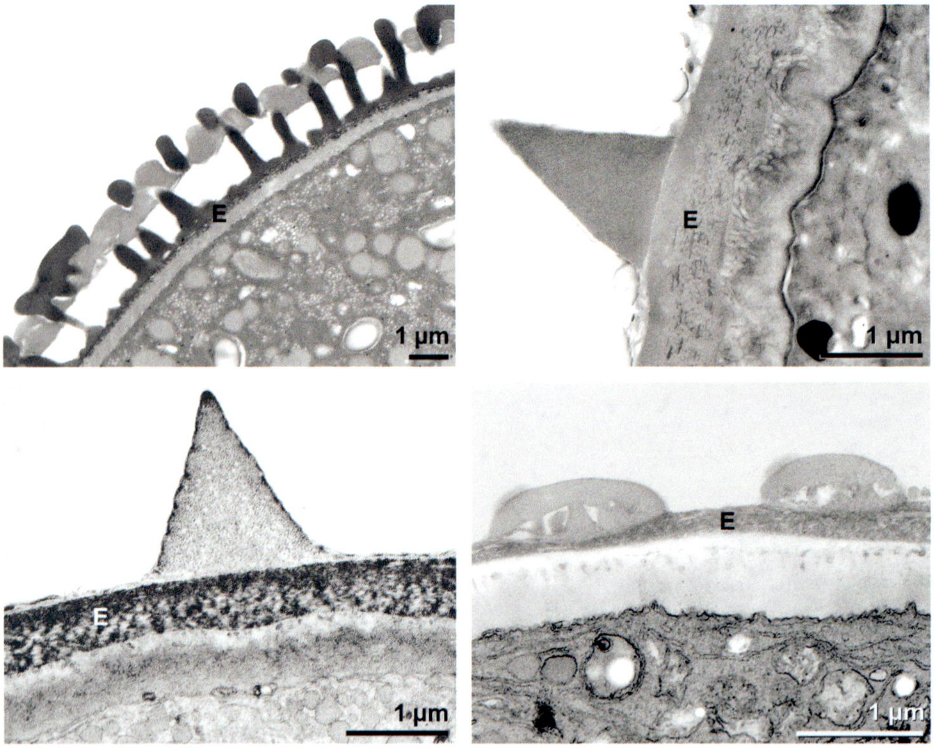

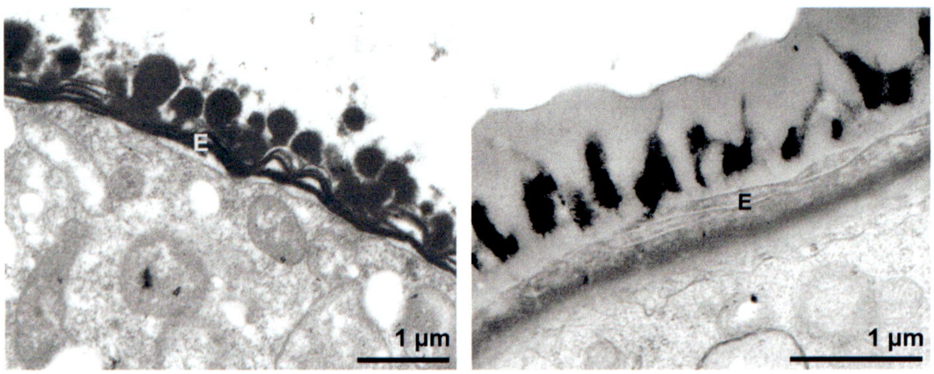

Mentha aquatica
Lamiaceae
$KMnO_4$
spongy, continuous

Arophyton buchetii
Araceae
U+Pb
spongy, continuous

Pistia stratiotes
Araceae
PA+TCH+SP
spongy, continuous

Spathiphyllum blandum
Araceae
TCH+SP
spongy, continuous

Orobanche hederae
Orobanchaceae
$KMnO_4$
lamellar, continuous

Thalictrum flavum
Ranunculaceae
PA+TCH+SP (short)
lamellar, continuous

LM SEM TEM mo ana fnc

endexine: distinct exine layer between ektexine and intine; endexine (E) can be **compact**, spongy or lamellar as well as continuous, **discontinuous**, absent or in aperture only

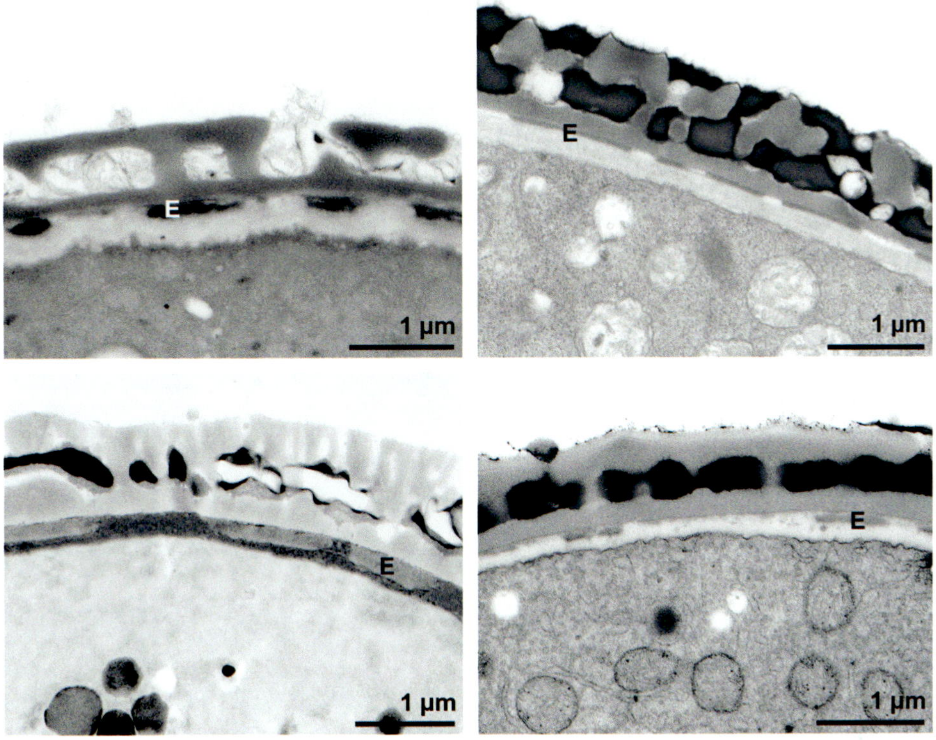

Glaucium flavum
Papaveraceae
U+Pb
compact, discontinuous

Odontites vulgaris
Scrophulariaceae
PA+TCH+SP (short)
compact, discontinuous

Ranunculus trichophyllos
Ranunculaceae
PA+TCH+SP
compact, discontinuous

Delphinium elatum
Ranunculaceae
PA+TCH+SP (short)
compact, discontinuous

LM SEM TEM mo ana fnc

endexine: distinct exine layer between ektexine and intine; endexine (E) can be compact, spongy or lamellar as well as continuous, discontinuous, **absent** or **in aperture only**.

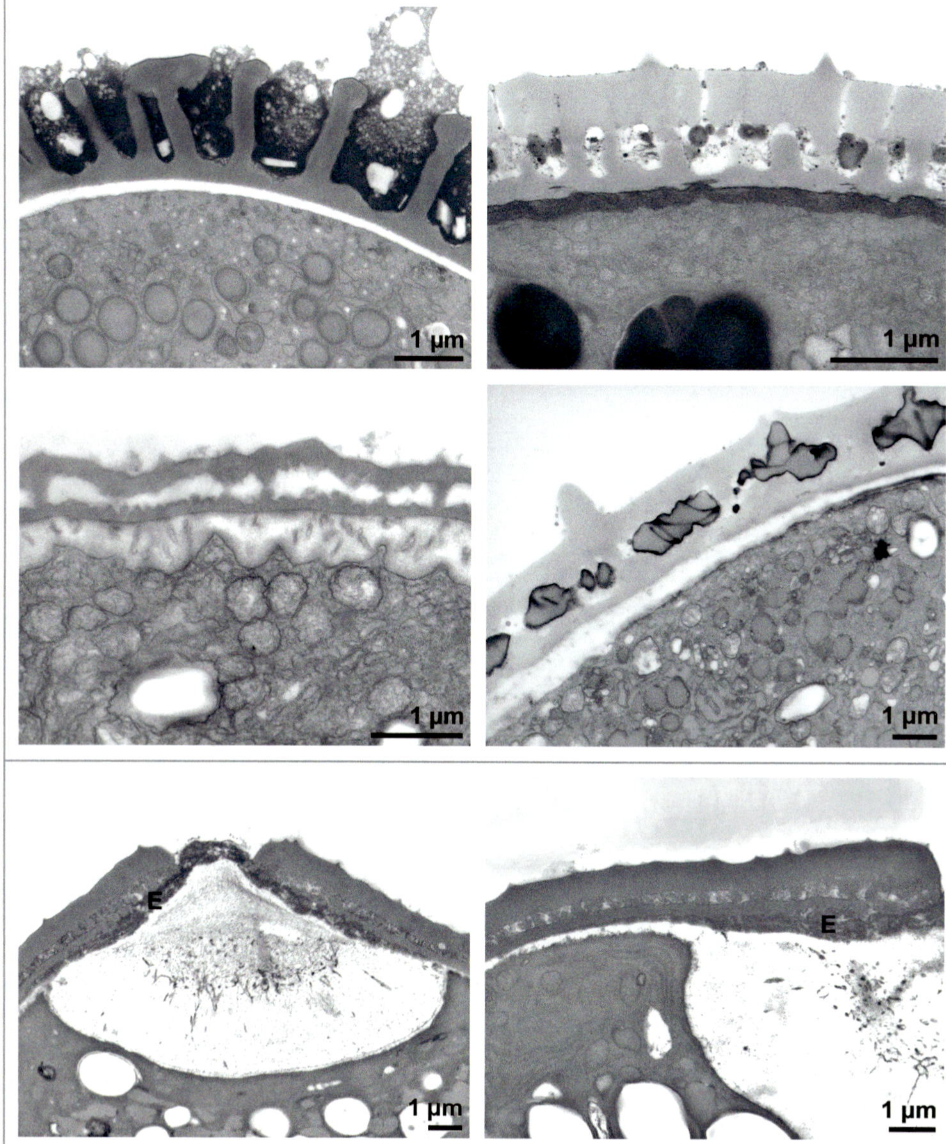

Brassica napus
Brassicaceae
PA+TCH+SP (short)
absent

Chenopodium album
Chenopodiaceae
PA+TCH+SP
absent

Trisetum flavescens
Poaceae
PA+TCH+SP (short)
absent

Cereus sp.
Cactaceae
PA+TCH+SP (short)
absent

Corylus avellana
Betulaceae
U+Pb
in aperture only

Corylus avellana
Betulaceae
TCH+SP
in aperture only

LM SEM TEM mo ana fnc

intine: part of the pollen wall next to the cytoplasm, mainly consisting of polysaccharides.

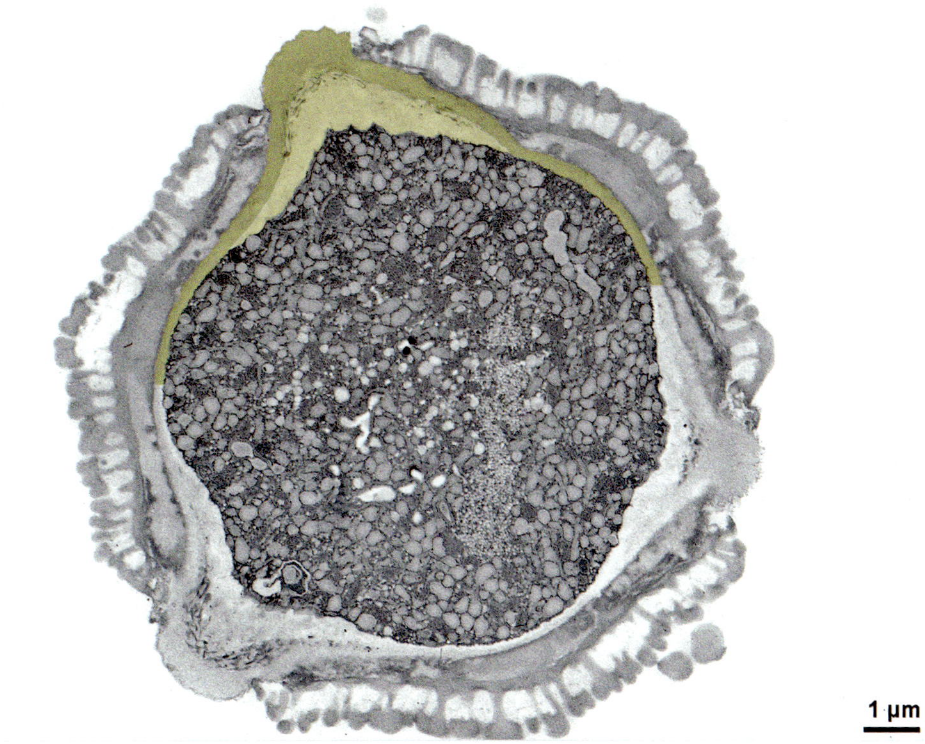

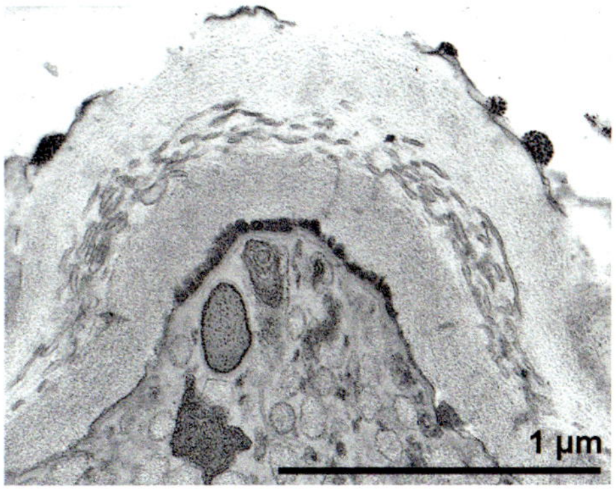

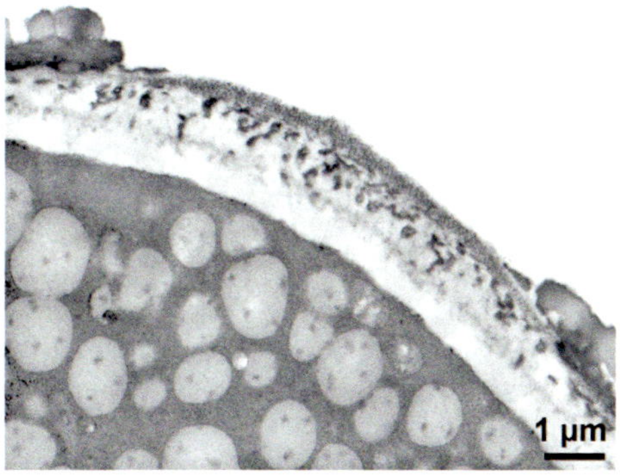

Apium nodiflorum
Apiaceae
U+Pb
ektintine (electron dense)
endintine (electron transparent)

Apium nodiflorum
Apiaceae
PA+TCH+SP

Quercus robur
Fagaceae
TCH+SP

LM SEM TEM mo ana fnc

semitectum: discontinuous tectum, covering less than 50 % of pollen grain surface.

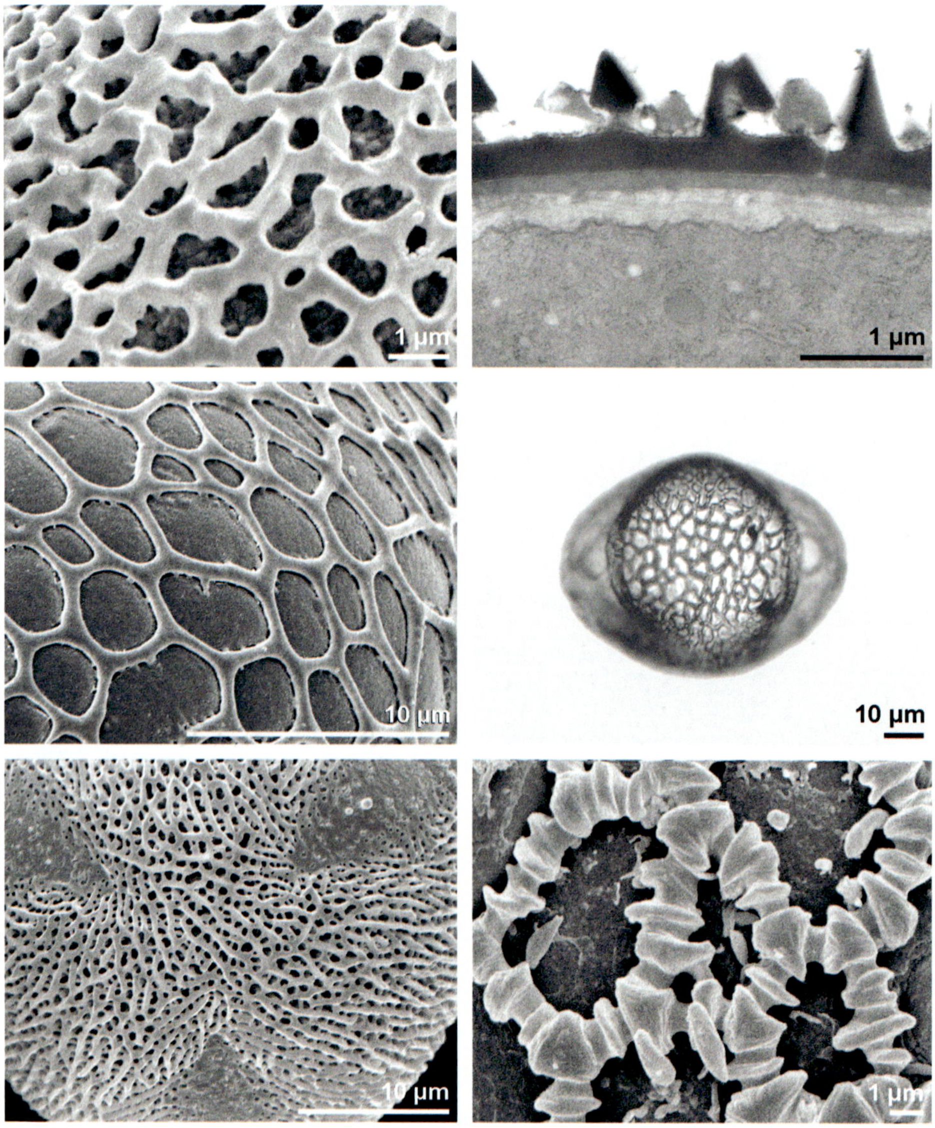

Salix fragilis
Salicaceae
reticulate

Salix fragilis
Salicaceae
U+Pb

Quesnelia lateralis
Bromeliaceae
reticulate

Alangium sp.
Cornaceae, fossil
reticulate, equatorial view

Lomatogonium carinthiacum
Gentianaceae
striato-microreticulate, polar area

Pachysandra terminalis
Buxaceae
reticulum cristatum

LM SEM TEM mo ana fnc

atectate: pollen grain lacking a tectum.

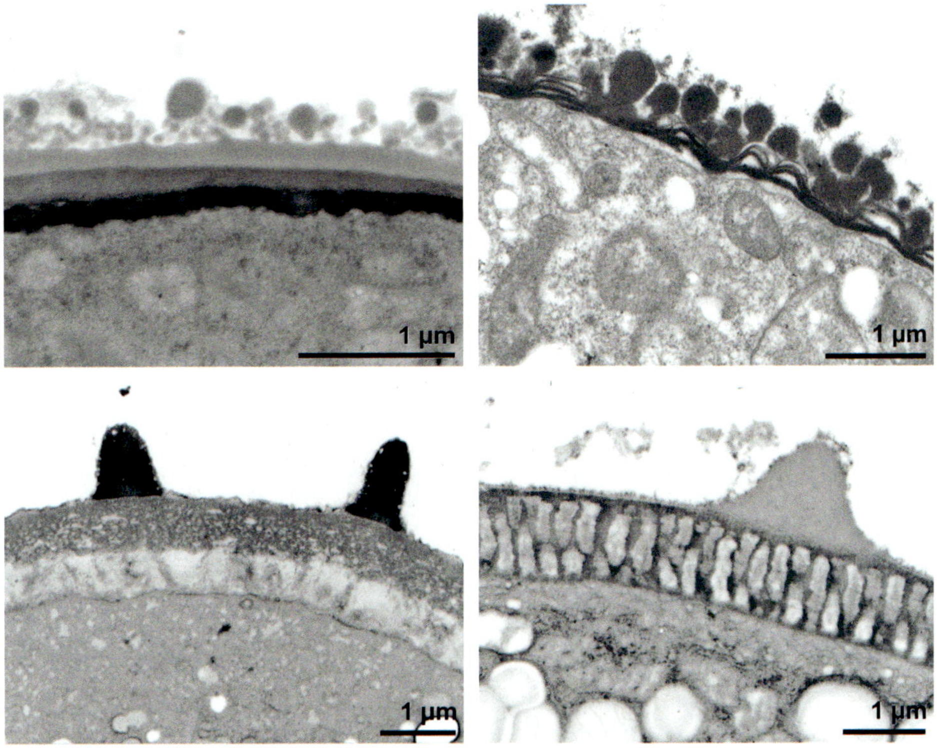

Rhaphidophora africana
Araceae
U+Pb

Sauromatum venosum
Araceae
PA+TCH+SP

Orobanche hederae
Orobanchaceae
$KMnO_4$

Globba schomburgkii
Zingiberaceae
U+Pb

LM SEM TEM mo ana fnc

primexine: polysaccharidic layer formed during early developmental stage wherein the later exine structures are preformed.

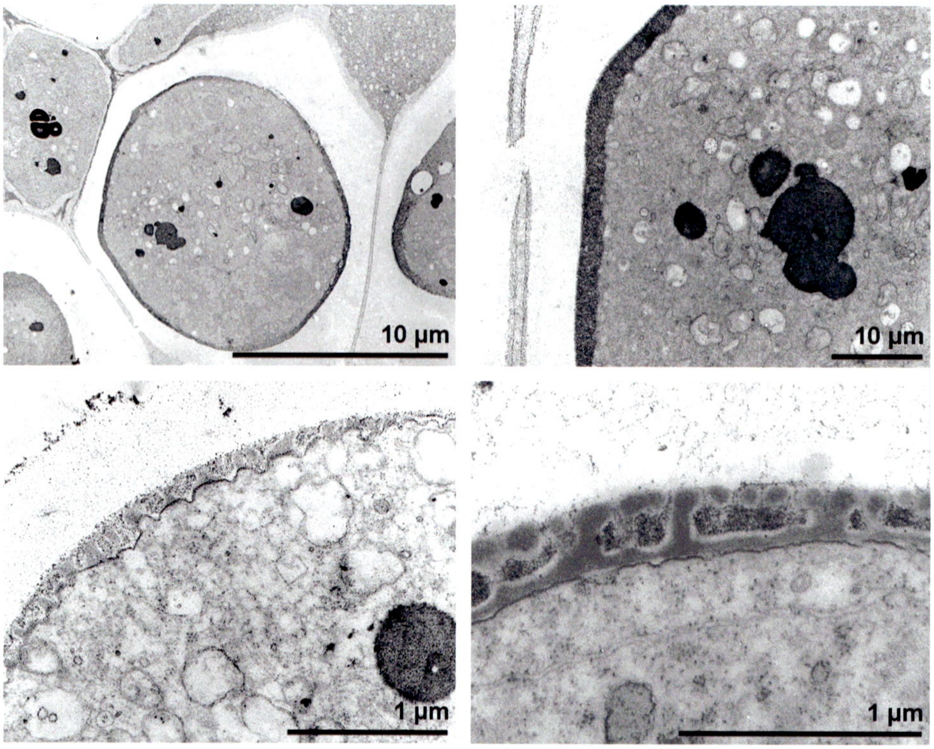

Smyrnium perfoliatum
Apiaceae
U+Pb
pollen mother cell, primexine within callose wall

Smyrnium perfoliatum
Apiaceae
U+Pb
pollen mother cell, primexine within callose wall

Smyrnium perfoliatum
Apiaceae
U+Pb
tetrad stage, first exine elements visible within primexine

Smyrnium perfoliatum
Apiaceae
U+Pb
free microspore stage, ektexine developed

vegetative nucleus (*)

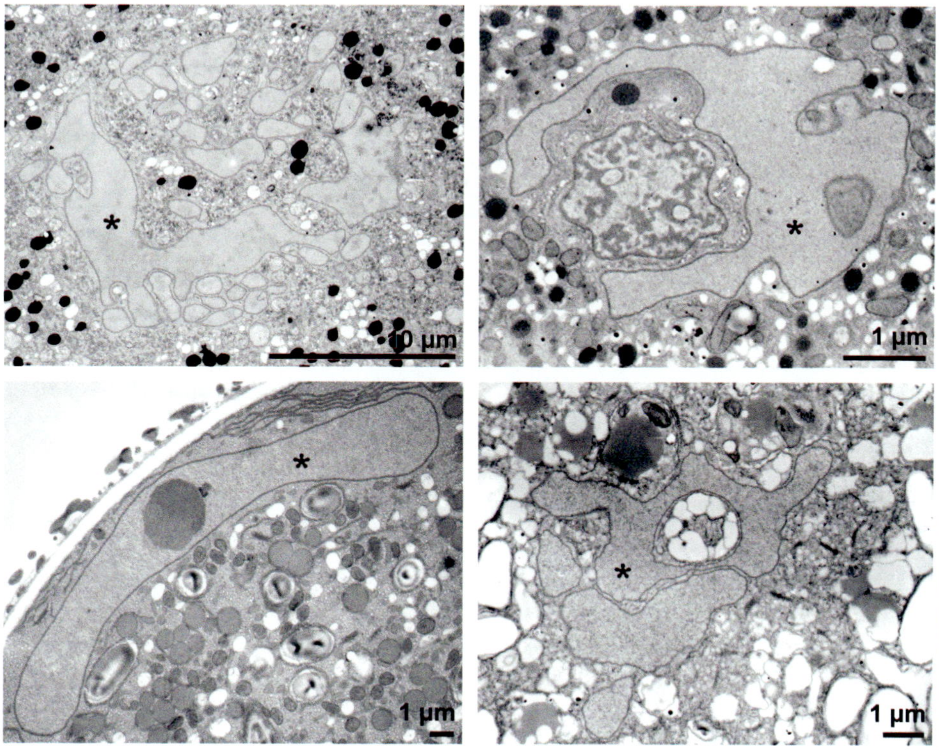

Iris pumila
Iridaceae
PA+TCH+SP (short)
complex vegetative nucleus in mature pollen grain

Consolida regalis
Ranunculaceae
PA+TCH+SP (short)
vegetative nucleus enclosing generative cell

Acinos alpinus
Lamiaceae
PA+TCH+SP (short)

Zantedeschia aethiopica
Araceae
U+Pb
complex vegetative nucleus in mature pollen grain

LM SEM TEM mo ana fnc

generative cell (*)

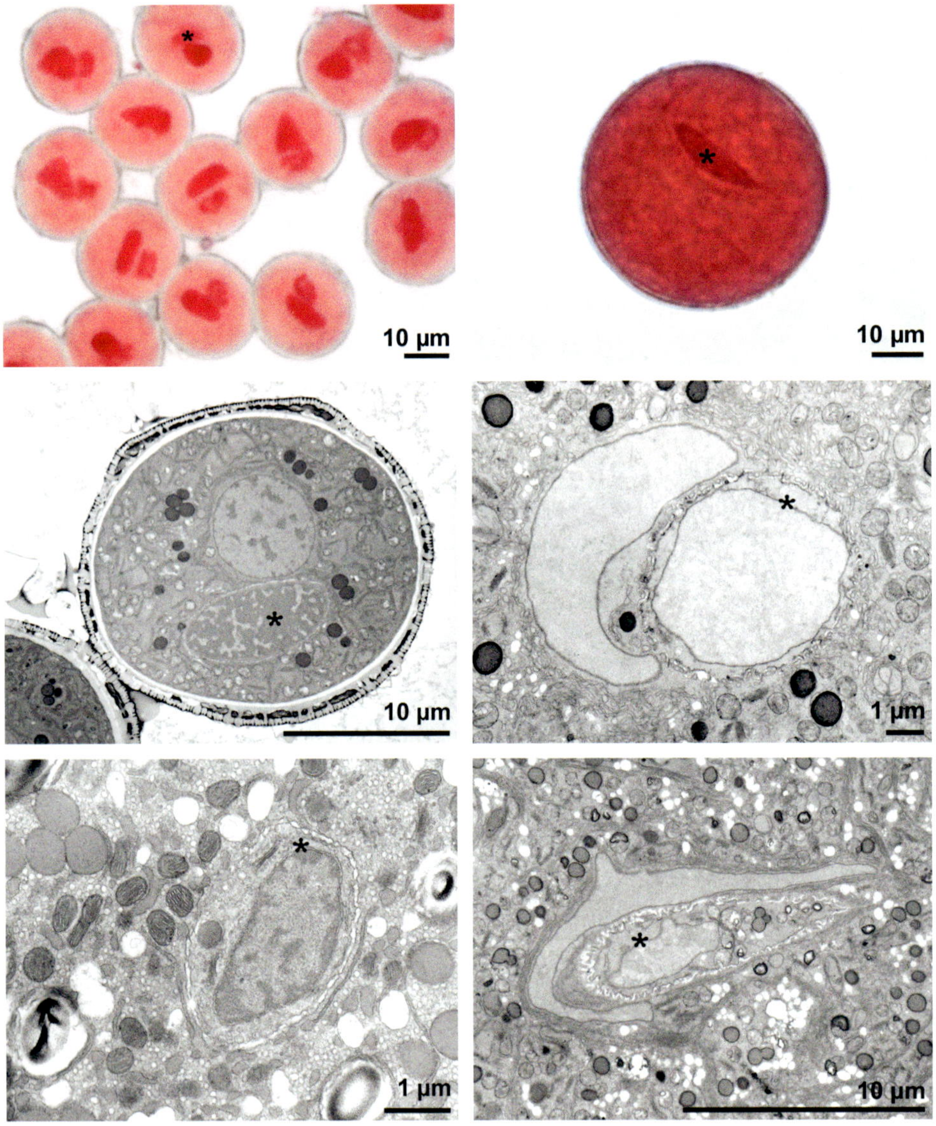

Melampyrum nemorosum
Scrophulariaceae
acetocarmine

Betonica officinalis
Lamiaceae
acetocarmine

Melampyrum nemorosum
Scrophulariaceae
PA+TCH+SP (short)

Betonica officinalis
Lamiaceae
PA+TCH+SP (short)

Acinos alpinus
Lamiaceae
PA+TCH+SP (short)

Ajuga reptans
Lamiaceae
PA+TCH+SP (short)

sperm cell: male gamete.

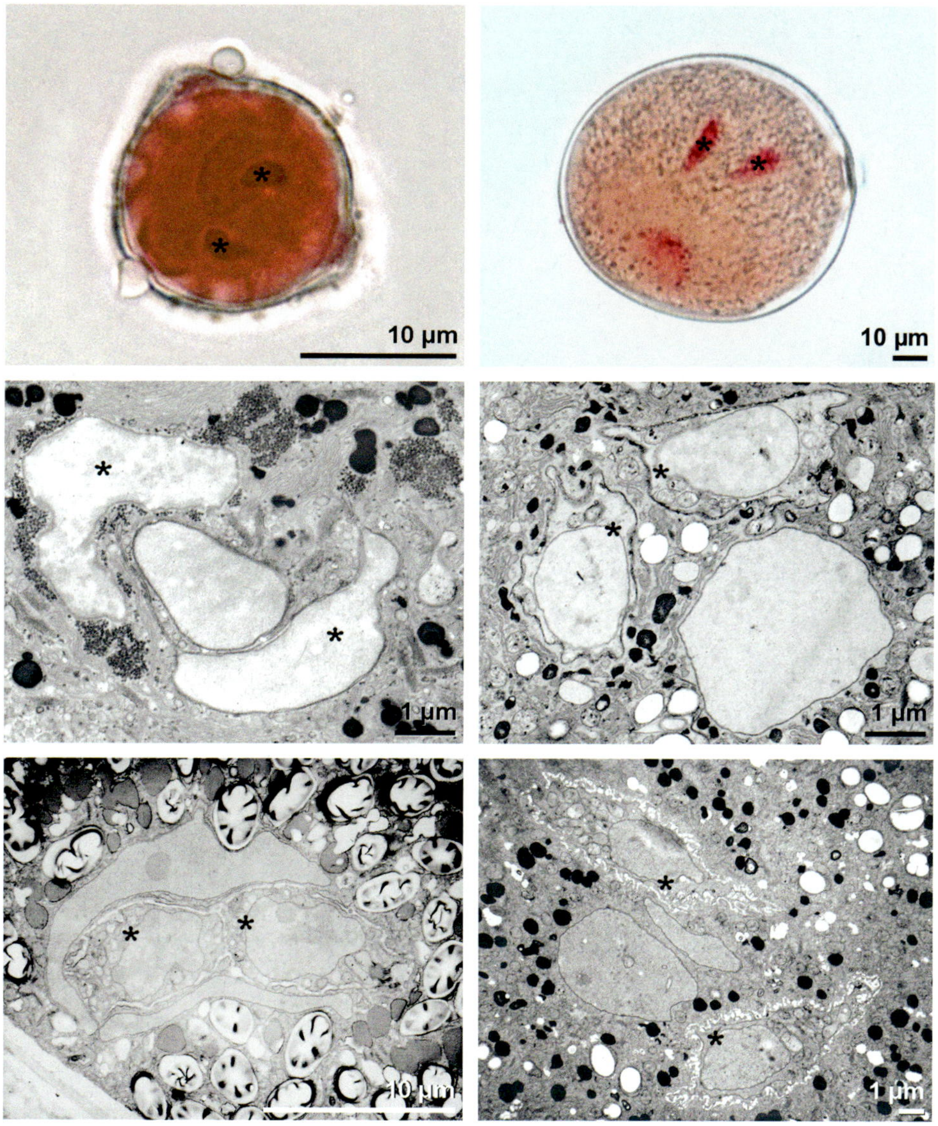

Smyrnium perfoliatum
Apiaceae
acetocarmine

Triticum aestivum
Poaceae
acetocarmine

Smyrnium perfoliatum
Apiaceae
PA+TCH+SP
sperm cells in Apiaceae extremely poor in organelles

Galium mollugo
Rubiaceae
PA+TCH+SP

Zantedeschia aethiopica
Araceae
PA+TCH+SP (short)
sperm cells still in contact with each other; enclosed by the vegetative nucleus

Jasminum nudiflorum
Oleaceae
TCH+SP

tapetum: specialized layer of cells lining the locule and participating in the nourishment of pollen grains, pollen wall formation and synthesis of pollen coatings.

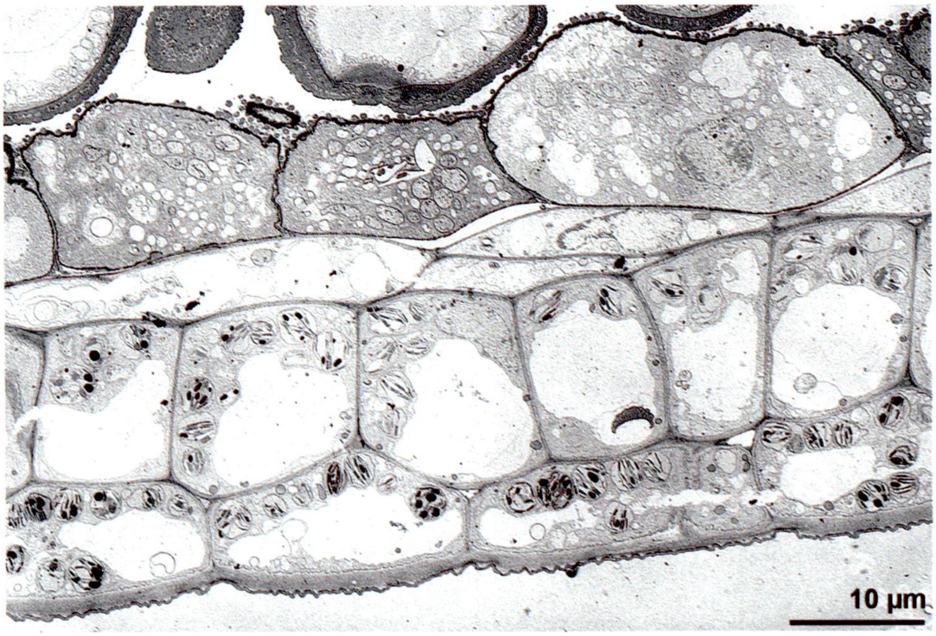

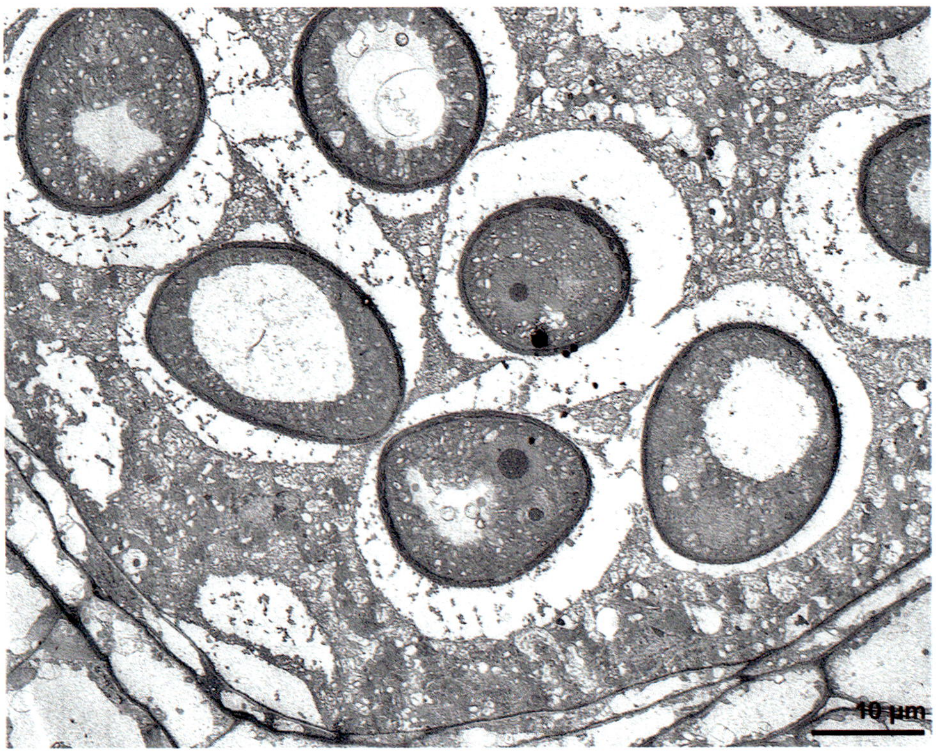

Hacquetia epipactis
Apiaceae
PA+TCH+SP
secretory tapetum in young anther

Zantedeschia aethiopica
Araceae
U+Pb
amoeboid tapetum

pollenkitt: pollen coating consisting of sticky substances, mainly lipids.

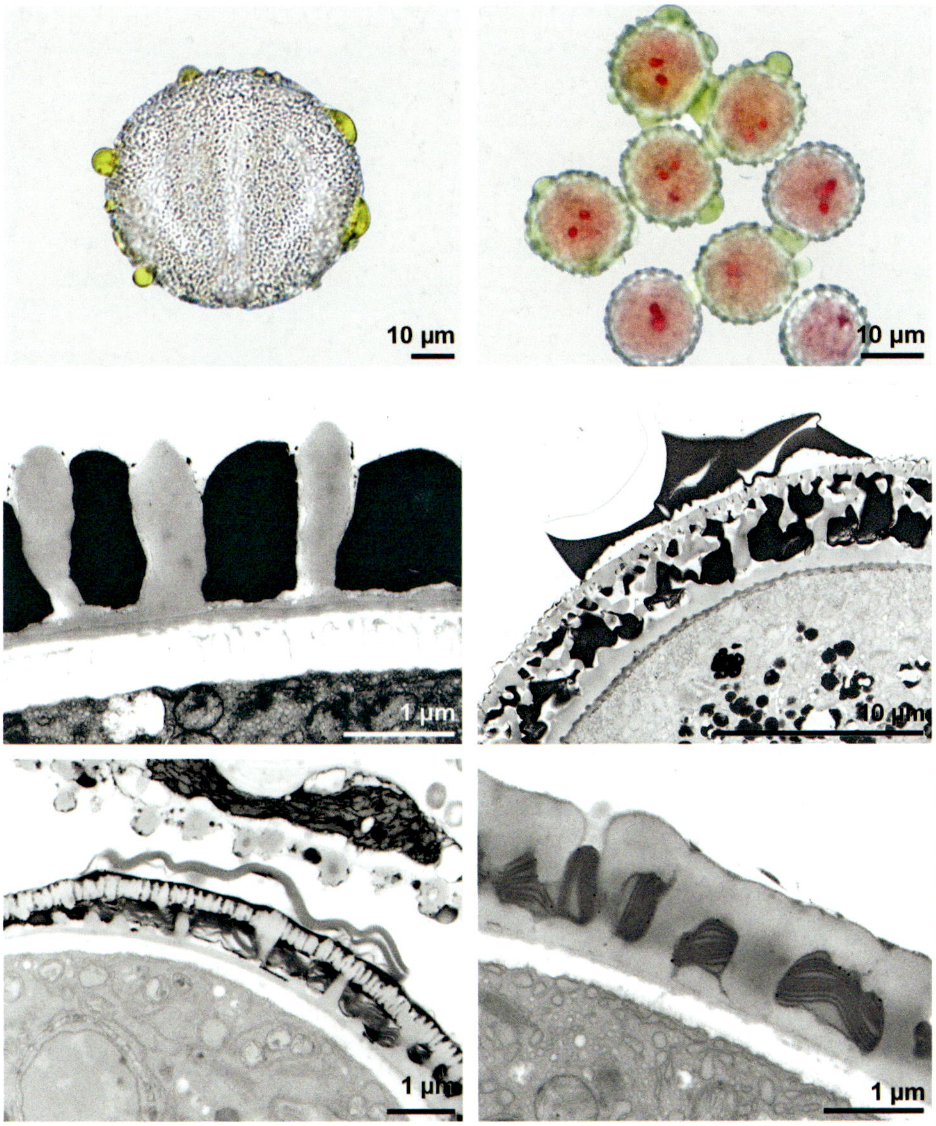

Salvia nemorosa
Lamiaceae
unstained

Ambrosia artemisiifolia
Asteraceae
acetocarmine

Jasminum nudiflorum
Oleaceae
TCH+SP

Nigella arvensis
Ranunculaceae
PA+TCH+SP

Melampyrum nemorosum
Scrophulariaceae
PA+TCH+SP (short)

Betonica officinalis
Lamiaceae
PA+TCH+SP (short)

LM SEM TEM mo ana fnc

primexine matrix: pollen coating consisting of primexine remnants in mature pollen grains.

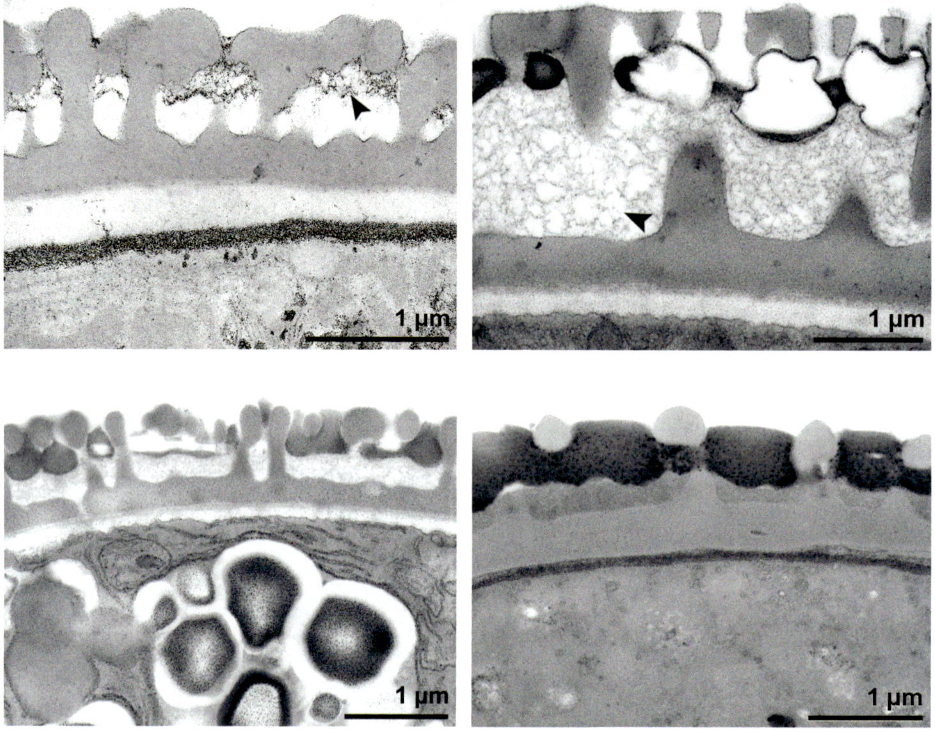

tryphine: pollen coating consisting mainly of lipids mixed with membrane remnants.

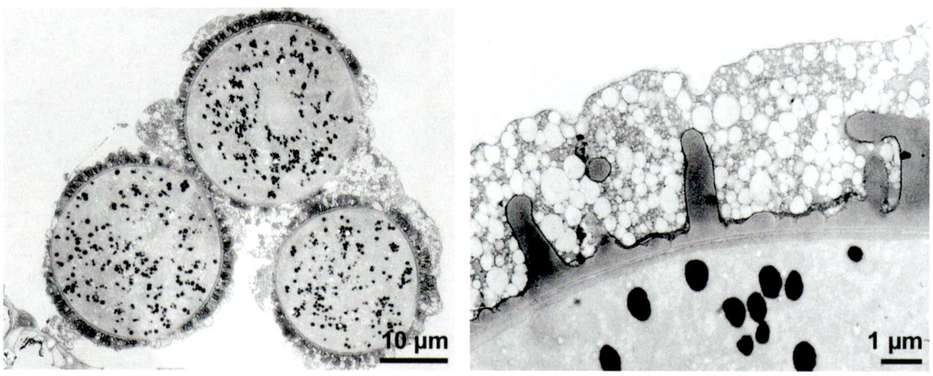

Apium nodiflorum
Apiaceae
without osmium, PA+TCH+SP

Convolvulus tricolor
Convolvulaceae
PA+TCH+SP (short)

Pseudolysimachion barrelieri
Scrophulariaceae
PA+TCH+SP (short)

Sambucus nigra
Sambucaceae
PA+TCH+SP

Brassica nigra
Brassicaceae
PA+TCH+SP (short)

Sinapis alba
Brassicaceae
PA+TCH+SP (short)

LM SEM TEM mo ana fnc

viscin thread: acetolysis resistant thread arising from the exine.

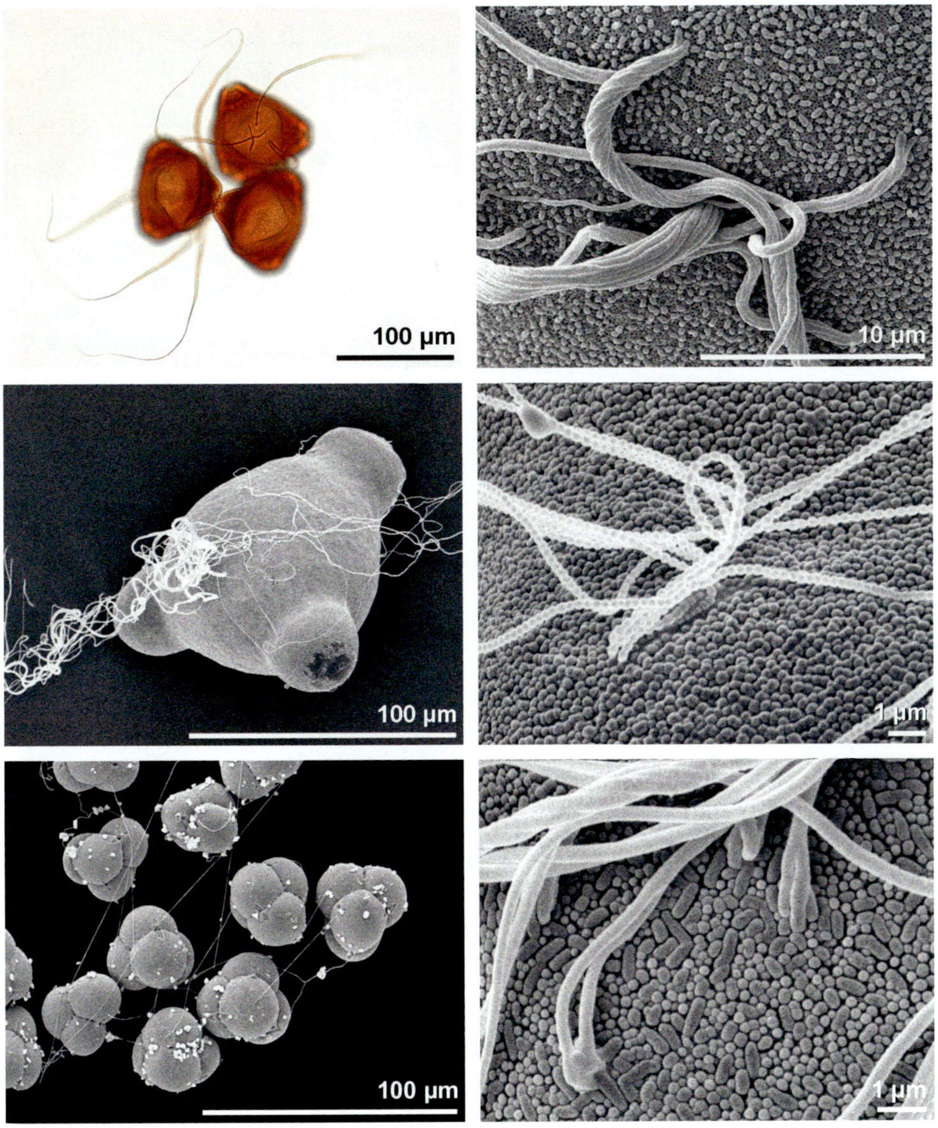

Epilobium angustifolium
Onagraceae
acetolyzed

Epilobium fleischeri
Onagraceae

Oenothera biennis
Onagraceae
oblique view

Oenothera biennis
Onagraceae

Kalmia latifolia
Ericaceae
tetrads

Godetia purpurea
Onagraceae

LM SEM TEM mo ana fnc

viscin thread: acetolysis resistant thread arising from the exine.

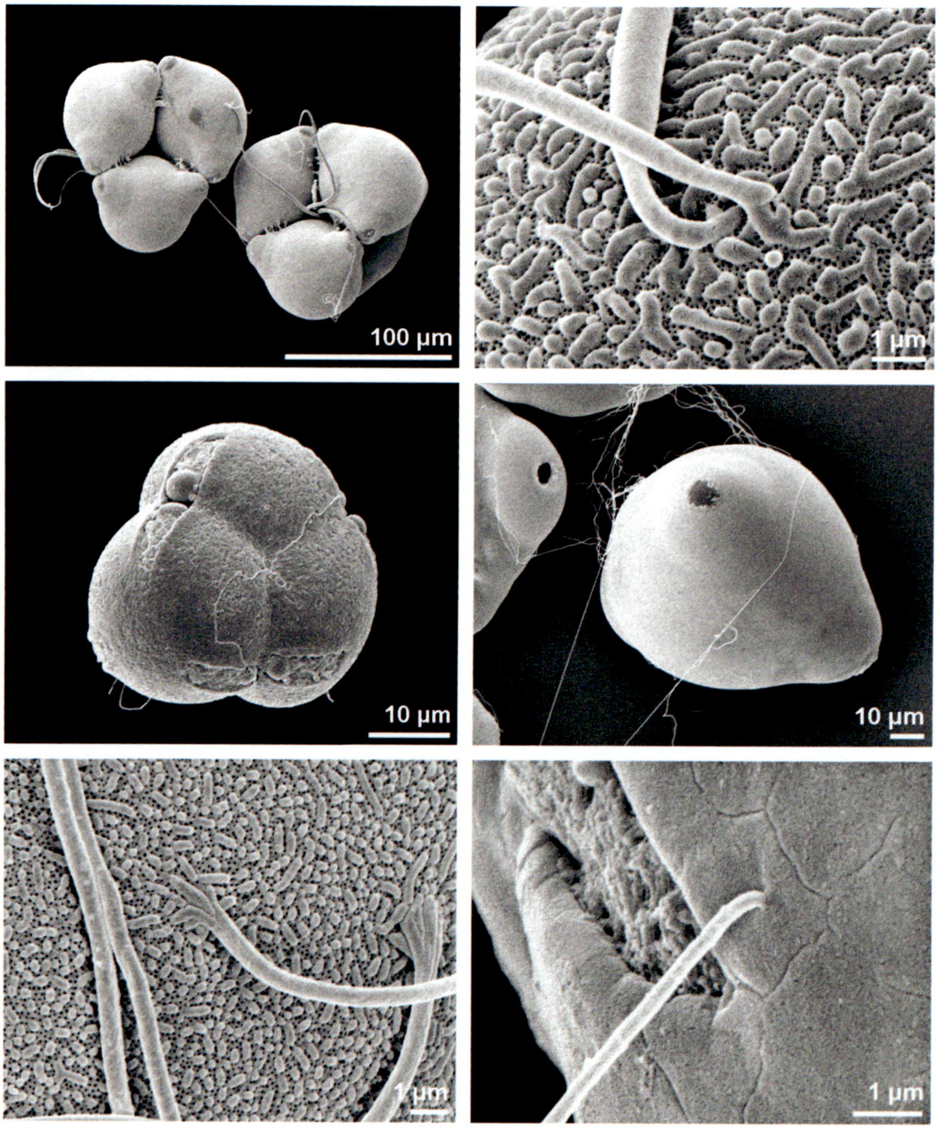

Epilobium parviflorum
Onagraceae
tetrads

Circaea lutetiana
Onagraceae

Ledum palustre
Ericaceae
tetrad

Clarkia pulchella
Onagraceae
equatorial view

Epilobium dodonaei
Onagraceae

Rhododendron hirsutum
Ericaceae

Ubisch body: sporopolleninous element produced by the tapetum.

Comment:
the "Ubisch body" is named after Gerta von Ubisch, who described these bodies for the first time.

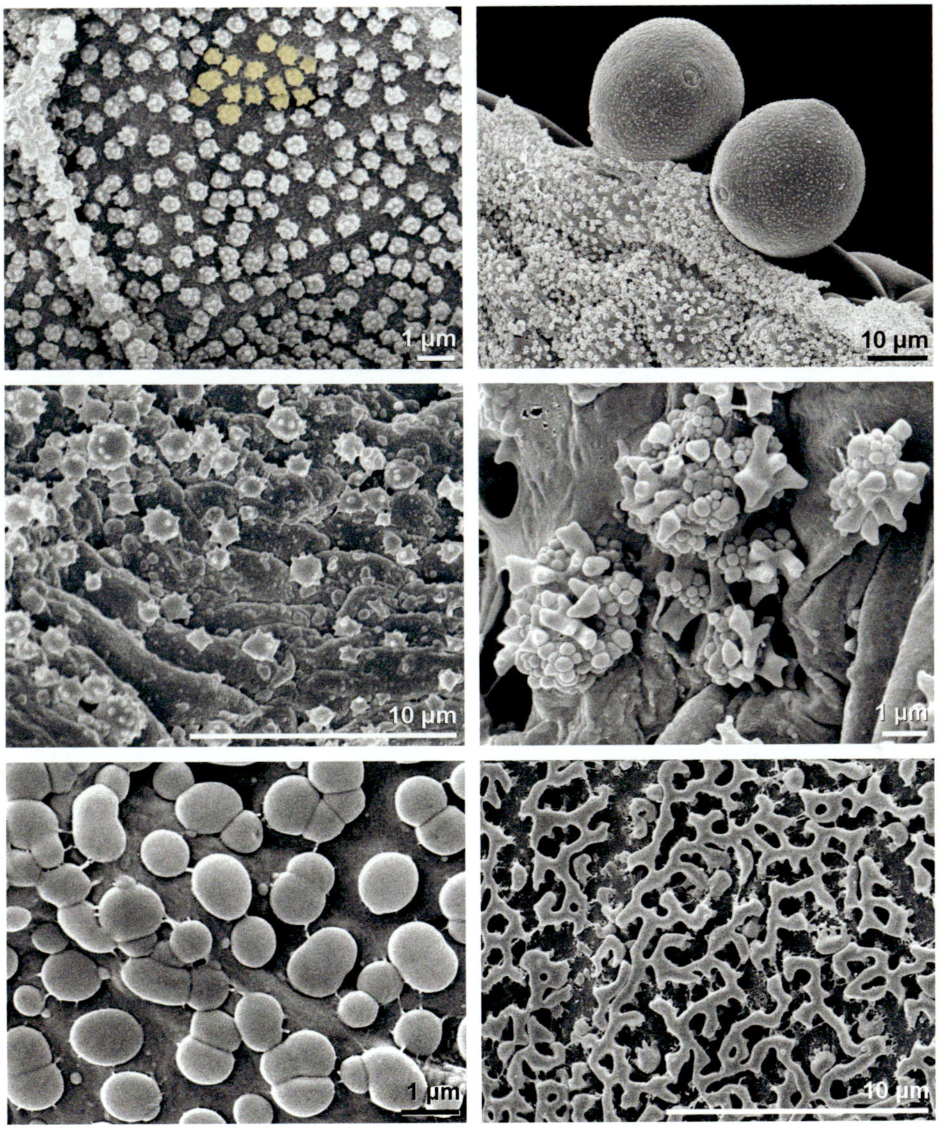

Trisetum flavescens
Poaceae

Corylus avellana
Betulaceae
two pollen grains attached to locular wall

Atriplex sagittata
Chenopodiaceae

Stellaria graminea
Caryophyllaceae

Acacia binerva
Mimosaceae

Beloperone guttata
Acanthaceae

LM SEM TEM mo ana fnc

Ubisch body: sporopolleninous element produced by the tapetum.

Comment:
the "Ubisch body" is named after Gerta von Ubisch, who described these bodies for the first time.

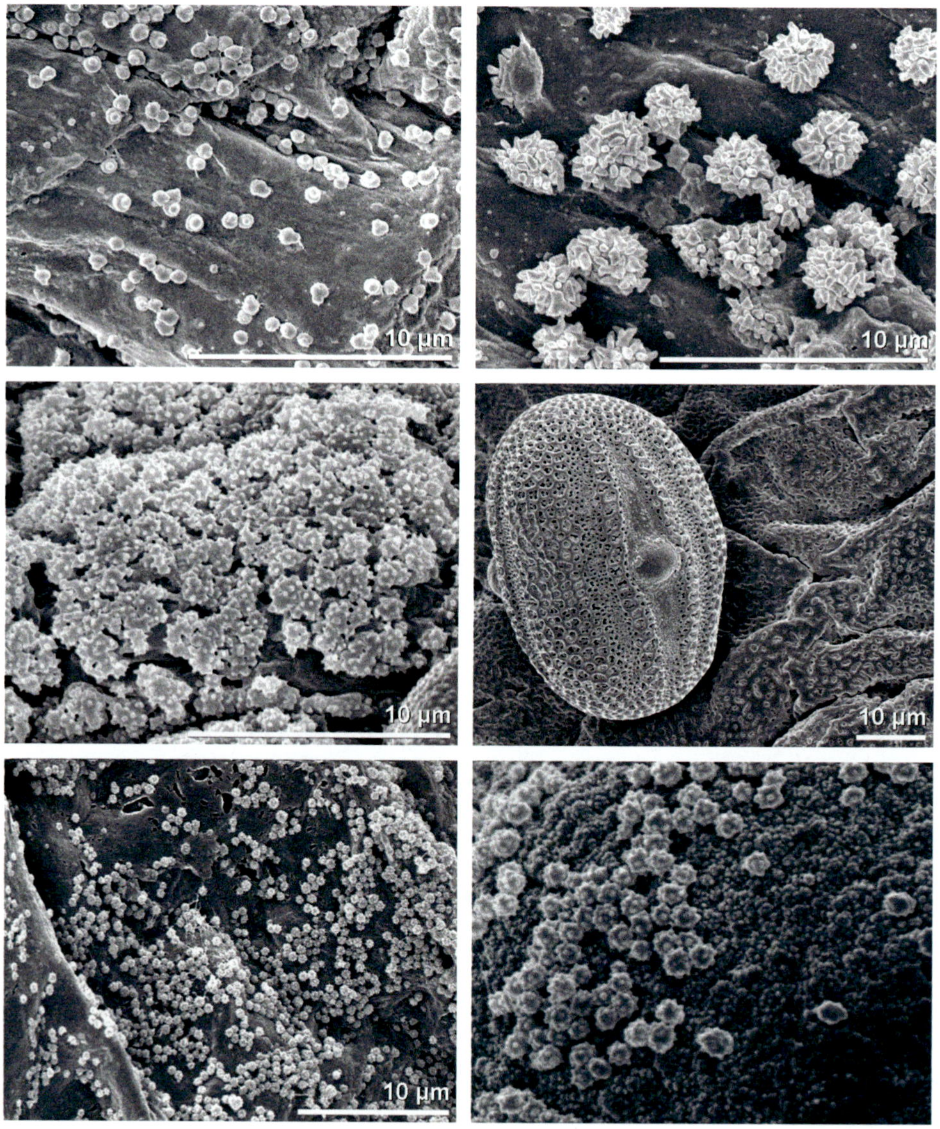

Quercus robur
Fagaceae

Gladiolus illyricus
Iridaceae

Cyperus longus
Cyperaceae

Ruspolia seticalyx
Acanthaceae
pollen grain attached to reticulate locular wall

Chamaecyparis lawsoniana
Cupressaceae
Ubisch bodies on locular wall

Chamaecyparis lawsoniana
Cupressaceae
Ubisch bodies attached on pollen surface

Ubisch body: sporopolleninous element produced by the tapetum.

Comment:
the "Ubisch body" is named after Gerta von Ubisch, who described these bodies for the first time.

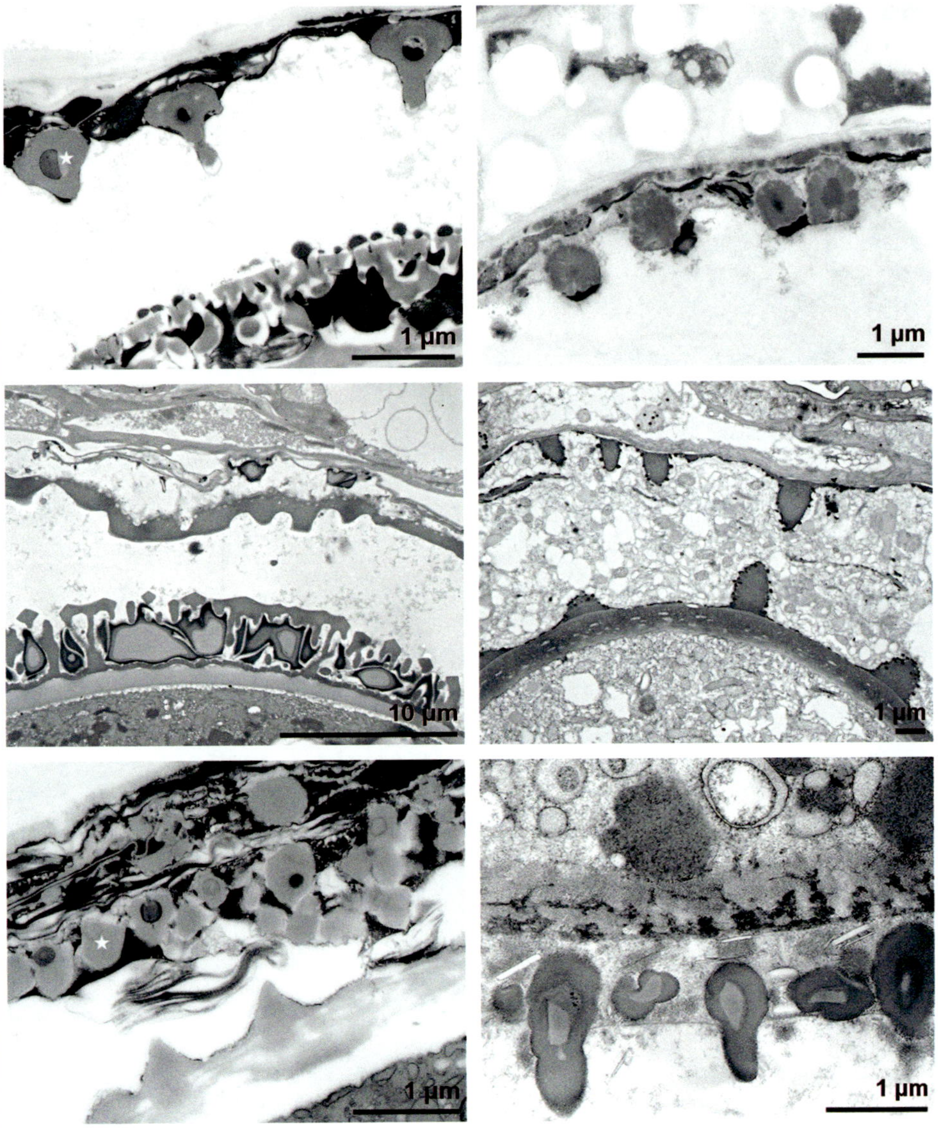

Nigella arvensis
Ranunculaceae
PA+TCH+SP (short)

Ruspolia seticalyx
Acanthaceae
U+Pb

Ranunculus ficaria
Ranunculaceae
PA+TCH+SP (short)

Odontites luteus
Scrophulariaceae
U+Pb

Sauromatum venosum
Araceae
U+Pb

Tilia platyphyllos
Tiliaceae
U+Pb

a- | acalymmate | acetolysis | actuopalynology | aeropalynolog
turate | aperture | aperture membrane | apocolpium | apoporiu
baculate | baculum | bi- | biporate | bireticulate | bisaccate | bi
| brevicolporus | brevicolpus | bridge | brochus | calymmate | c
| clavate | clypeate | colpate | colporate | colporoidate | colp
compact | compound aperture | copropalynology | corpus | cos
di- | diaperturate | dicolpate | dicolporate | diploxylon-pollentype
| echinolophate | echinus | ektexine | ektintine | ekto- | ektoap
| endoplica | equator | equatorial | equatorial diameter | equato
| Fischer's rule | foot layer | forensic palynology | fossula | fossul
| Garside's rule | gemma | gemmate | generative cell | geniculu
harmomegathy | hetero- | heteroaperturate | heterobrochate | h
| impression mark | in- | inaperturate | infoldings | infra- | infrate
internal tectum | interporium | interstitium | intine | intra- | irregula
| lamellar | leptoma | LO-analysis | lobate | lolongate | LO-patte
lynology | melittopalynology | meridian | meridional | meso- | m
monad | mono- | monoaperturate | monocolpate | monolete |
Normapolles | oblate | oblique view | omniaperturate | oncus |
| outline in equatorial view | outline in polar view | P/E-ratio | pa
panto- | pantoaperturate | pantocolpate | pantoporate | papilla
| pilate | pilum | planaperturate | plicae | plicate | pluricolume
pollen | pollen analysis | pollen class | pollen coatings | pollen
wall | pollenkitt | pollinarium | pollination | pollinium | poly- | po
| polyplicate | pontoperculate | pontoperculum | porate | poroi
prae(pre)-pollen | primexine | primexine matrix | prolate | proxim
drangular | reticulate | reticulum | reticulum cristatum | retipilat
| scabrate | sculpture | semi- | semitectate | semitectum | sexi
spiraperturate | spongy | spore | sporoderm | sporopollenin | ste
nocolporate | stephanoporate | stephano- | striae | striate | stri
| syn- | synaperturate | syncolpate | syncolporate | tapetum |
| tetrad mark | tetrad planar | tetrad stage | tetrad tetrahedral
mosulcus | tricolpate | tricolporate | trilete | triporate | tryphine
nucleus | verruca | verrucate | vesiculate | vestibulum | viscin

ALPHABETIC GLOSSARY

alveolate | angulaperturate | annulate | annulus | aper-
arcuate | arcus | areola | areolate | atectate | atrium |
te | boat-shaped | brevi- | brevicolpate | brevicolporate
a | cappula | caput | cavea | caveate | circular | clava
| colpus | colpus membrane | columella | columellate |
costate | croton pattern | cryopalynology | cup-shaped |
porate | dispersal unit | distal | disulcate | dyad | echinate
e | elliptic | endexine | endintine | endo- | endoaperture
iew | eu- | eurypalynous | eutectate | exine | fenestrate
foveola | foveolate | free-standing columellae | frustrate
ranular | granulate | granulum | haploxylon-pollentype |
polar | hexa- | homo- | homobrochate | iatropalynology
| intectate | inter- | interapertural area | intercolpium |
o- | isodiametric | isopolar | lacuna | laesura | lalongate
ophae | lophate | lumen | margo | massula | melissopa-
colpium | micro- | microspore | microspore mother cell |
porate | monosaccate | monosulcate | muri | nexine |
culate | operculum | orbicule | ornamentation | outline
palynology | palynogram | palynology | palynomorph |
dium | penta- | perforate | peri- | pharmacopalynology
| polar area | polar axis | polar view | polarity | pole |
| pollen mother cell | pollen tube | pollen type | pollen
| polychotomosulcate | polychotomosulcus | polygonal
oroidate | pororate | porus | porus membrane | prae- |
seudocolpus | pseudomonad | psilate | punctate | qua-
g-like aperture | rugulae | rugulate | saccate | saccus
hape | size | sperm cell | spheroidal | spine | spinose |
ynous | stephanoaperturate | stephanocolpate | stepha-
ticulate | structure | sub- | sulcate | sulcus | symmetry
e | tectum | tenuitas | tetra- | tetrad | tetrad decussate
triangular | triaperturate | trichotomosulcate | trichoto-
ch body | ulcerate | ulcus | vegetative cell | vegetative
d | zona-aperturate | zono-aperturate | Zwischenkörper

ALPHABETIC GLOSSARY

A strict rationalization of terms on the basis of practical criteria has been attempted. For consistency, phrases are standardized as far as possible; for example, features of ornamentation are stereotypically defined as "**pollen wall with** ....", and pollen wall features (or pollen shape and size) as "**pollen grain with** ....".

Three categories of terms are used: **important terms** are printed in **bold** and are usually illustrated; terms of **minor importance** are printed in **regular** script, usually without illustrations; terms printed in **italics** are **not recommended** and often provided with an explanatory comment.

Numbers in square brackets referring to important literature (see "Bibliography").

a-
prefix meaning absent.
acalymmate 47
feature describing a dispersal unit of two or more monads enclosed by an exine, which is discontinuous at the junctions between the monads, and is absent from the internal walls.
Antonym: calymmate
acetolysis 7, 19-20, 25, 32, 51
widely used technique for preparing pollen and spore exines especially for light microscopy. [23]
actuopalynology
the study of pollen grains and spores of extant plants.
aeropalynology 12
the study of palynomorphs found in the atmosphere. [24]
alveolate 23, **200-202**
infratectum with compartments of irregular size and shape. [122]
angulaperturate **104**
pollen grain with an angular outline where the apertures are situated at the angles. [21]
Antonym: planaperturate
annulate **143-144**
pollen grain with an annulus or annuli.
annulus (lat., pl. annuli) 21, **143-144**
ring-like thickening of the pollen wall surrounding a porus or ulcus. [4]
Comment: "anulus" is an orthographical variant of "annulus".
aperturate 41, 49
pollen grain with one or more apertures. [20]
Antonym: inaperturate.
aperture 15-17, 19-20, 23, 25, **101-154**
region of the pollen wall which differs significantly morphologically and/or anatomically from the rest of the pollen wall, presumed to function usually as germination site and to play a role in harmomegathy. [20]
aperture membrane 19, **145-148**
exine layer covering an aperture; aperture membrane can be smooth or ornamented. [21]
Comment: the terms "smooth" and "ornamented" should be used when the feature is remarkably expressed.
apocolpium, see **polar area**
Comment: "polar area" is the more general term independent of the aperture type
apoporium, see **polar area**
Comment: "polar area" is the more general term independent of the aperture type.
arcuate **100**
pollen grain with curved wall thickenings interconnecting apertures. [18]
arcus (lat., pl. arcus) 21, **100**
a curved wall thickening interconnecting apertures. [18]
areola (lat., pl. areolae) 23, 47, **197**
small, mostly convex exine island.
areolate 32-33, 46-47, **197**
pollen wall with areolae.
atectate **211**
pollen grain lacking a tectum. [128]
Antonym: tectate
atrium (lat., pl. atria)
space between diverging exine layers within the aperture. [115]
baculate 8, 30, **192**
pollen wall with bacula. [64]
baculum (lat., pl. bacula) **192**
rod-like, free standing element, more than 1 µm in height and never pointed. [83]
bi-
prefix for two.
biporate, see **diporate**
Comment: "diporate" is the more common term.
bireticulate 29, **160-161**
special type of reticulate ornamentation, where the brochi of the large-meshed reticulum are filled by a small-meshed reticulum.
bisaccate 8, 49, **59, 97-99**
pollen grain with two sacci. [84]
Comment: nomen conservandum
bisulcate, see **disulcate** [20]
Comment: "disulcate" is the more common term.
boat-shaped 7, 24-25, 44-45, **92-93**
characteristic shape of sulcate pollen grains caused by an infolding as a consequence of harmomegathy; see "Pollen Morphology".

brevi-
prefix meaning short.

brevicolpate **113**
pollen grain with brevicolpi. [21]

brevicolporate **113**
pollen grain with brevicolpori.

brevicolporus (lat., pl. brevicolpori) **113**
short colpus in a compound aperture.

brevicolpus (lat., pl. brevicolpi) **113**
short colpus. [21]

bridge 43, **153**
exine connection between the margins of a colpus in the equatorial region. [30]
Comment: the term is often used in a more general context, e.g., for exine connections within tetrads.

brochus (lat., pl. brochi) **162-163**
mesh of a reticulum consisting of one lumen and the adjoining half of the muri. [21]

calymmate 47
feature describing a dispersal unit of two or more monads enclosed by a continuous ektexine.
Antonym: acalymmate

cappa (lat., pl. cappae) 22-23
the thick-walled proximal side of the corpus of a saccate pollen grain. [22]

cappula, see **leptoma** [22]
Comment: may be confused with "cappa" which points to the proximal side, while "cappula" refers to distal.

caput (lat., pl. capita)
apex of a clava. [21]

cavea (lat., pl. caveae)
infratectal cavity in the interapertural area. [109]

caveate
pollen wall with caveae.

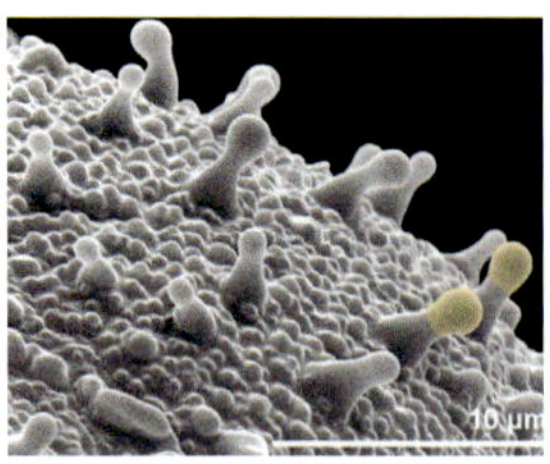

caput
Iris alata
Iridaceae
pollen surface with clavae

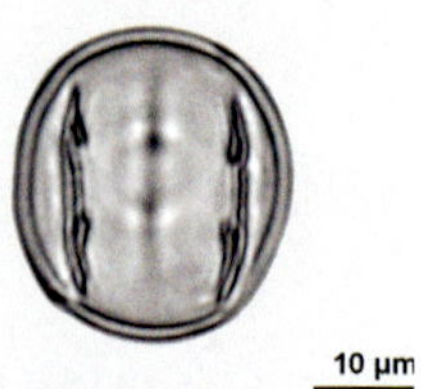

colporoidate
Eucommia sp.
Eucommiaceae, fossil
equatorial view

circular, see **outline**

clava (lat., pl. clavae) 23, 30, **193-194**
club-shaped element, higher than 1 µm. [20, 64, 83]

clavate 8, 27, 30, **193-194**
pollen wall with clavae. [20, 64]

clypeate 9, 42-43, 47, **68, 198**
pollen wall, in which the exine is subdivided into shields.

colpate 9, 19, 43, **68, 118**
pollen grain with colpi.

colporate 9, **68, 128-132**
pollen grain with colpori.

colporoidate
pollen grain with compound apertures composed of a colpus (ektoaperture) with an indistinct endoaperture. [21]
Comment: a rare character, e.g., for deciduous Quercus pollen.

colporus (lat., pl. colpori) 17, 30, 45, 48, **128-132**
compound aperture composed of a colpus (ektoaperture) combined with an endoaperture of variable size and shape.

colpus (lat., pl. colpi) 17, 41, 43, 48, **113, 118-120**
elongated aperture (length/width ratio > 2) situated at the equatorial region or regularly distributed over the pollen grain. [18]

colpus membrane
aperture membrane of a colpus; see **aperture membrane**.

columella (pl. columellae) 21, 23, 30 **200-202**
rod-like structure element, supporting a tectum. [64]

columellate 21, 23, **200-202**
infratectum with rod-like elements. [117]

compact, see **endexine**

compound aperture **113, 128-129**
aperture with two or more components that are situated in more than one wall layer, e.g., colporus. [24]

copropalynology
the study of palynomorphs in coprolites or faeces. [24]

corpus (lat., pl. corpora) 22-23
body of a saccate pollen grain. [22]

costa (lat., pl. costae) ________ 21
thickening of the nexine/endexine bordering an endoaperture. [64]

costate
pollen grain with costae.

croton pattern ________ 8, **166-167**
special type of reticulum cristatum formed by regularly arranged sculpture elements on muri.

cryopalynology ________ 12
the study of palynomorphs found in ice.

cup-shaped ________ 7, 25, 45, **91**
characteristic shape of pollen grains caused by infoldings as a consequence of harmomegathy; see "Pollen Morphology".

di-
prefix meaning two.

diaperturate
pollen grain with two apertures.

dicolpate ________ **118**
pollen grains with two colpi. [107]

dicolporate
pollen grain with two colpori. [107]

diploxylon-pollen-type ________ 22-23
bisaccate pollen grain with balloon-like sacci.

diporate ________ **121-122, 145**
pollen grains with two pori.

dispersal unit ________ 15, 47, **59-67**
unit in which pollen is shed (monad, dyad, tetrad, polyad, massula, pollinium, pollinarium).

distal ________ 15, 18-19, 23, 40-41, 44
pollen features that face or are directed outwards in the tetrad. [65]
Antonym: proximal

disulcate ________ **138**
pollen grain with two sulci. [20, 107]

dyad ________ **60, 69**
dispersal unit of two pollen grains.

echinate ________ 8, 27, 32, 55, **186-189**
pollen wall with echini. [133]

echinolophate
lophate pollen grains with echinate ridges. [133]

echinus (lat., pl. echini) ________ 23, 28, 32, **186-189**
pointed ornamentation element longer and/or wider than 1 µm. [133]

ektexine ________ 20-21, 23, 25, 36, 47, 52
the outer layer of the exine. [18]

ektintine ________ **209**
the outer layer of a two-layered intine which is adjacent to the exine. [72]

ekto-
prefix meaning outer.

ektoaperture ________ **128-129**
outer part of a compound aperture. [121]

elliptic, see **outline**

endexine ________ 20-21, 23, 25, 51-52, **205-208**
distinct exine layer between ektexine and intine; endexine can be compact, spongy or lamellar as well as continuous, discontinuous, absent or in aperture only. [83, 107]

endintine ________ **209**
inner layer of a two-layered intine which is adjacent to the cytoplasm. [72]

endo-
prefix meaning inner.

endoaperture ________ 21, 48, 51, 55, **128-129**
inner part of a compound aperture. [121]

endoplica
fold of the inner exine layer. [115]

equator ________ 15-19
imaginary line around a pollen grain at the distance half-way between the (proximal and distal) poles. [134]

equatorial ________ 15-19
preposition indicating a direction on the pollen surface; see **equator**.

equatorial diameter ________ 16
diameter of a pollen grain or spore in the equatorial plane. [18]

equatorial view ________ 15-19
the view of a pollen grain or spore perpendicular to the polar axis. [18]

eu-
prefix meaning true.

eurypalynous
plant taxa characterized by a significant variation in pollen (or spore) morphology. [21]
Antonym: stenopalynous

eutectate
pollen grain with a continuous tectum.

exine ________ 12-13, 19-21, 23, 25, **205-208, 210**
outer layer of the pollen wall which is usually resistant to acetolysis. [36]

fenestrate, see **lophate**
Comment: as there is no corresponding substantive to "fenestrate", we prefer the terms "lophate" and "lophae".

Fischer's law/rule ___ 12, 16
see "Pollen Morphology". [21]

foot layer ___ 21, 23, **200-204**
inner layer of the ektexine; foot layer can be continuous, discontinuous, perforated or absent. [29]

forensic palynology ___ 12
the study of palynomorphs found in forensic samples.

fossula (lat., pl. fossulae) ___ 23, **183**
irregularly shaped groove in the surface of a pollen wall. [30]

fossulate ___ 32, **183**
pollen wall with fossulae. [30]

foveola (lat., pl. foveolae) ___ 23, **182**
roundish lumen more than 1 µm in diameter; distance between two adjacent lumina larger than their diameter. [21]

foveolate ___ 27, 29, **182**
pollen wall with foveolae. [21]

free-standing columellae ___ 23, 30
columellae in the infratectal layer not covered by a tectum in semitectate pollen grains.

frustrate ___ 49
special mental condition of palynologists discussing terminology of pollen and spores; see "Controversial and Fuzzy Terms". [58]

Garside's law/rule ___ 16
see "Pollen Morphology".

gemma (lat., pl. gemmae) ___ 23, 48, **196**
globular exine element more than 1 µm in diameter. [64]

gemmate ___ 8, 27, **196**
pollen wall with gemmae. [64]

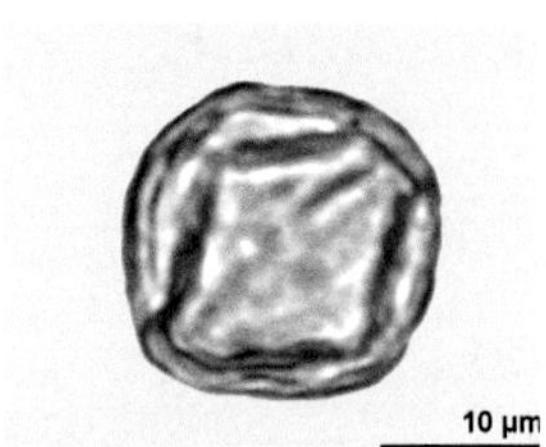

geniculum
Quercus sp.
Fagaceae, fossil
equatorial view

generative cell ___ 36, **213-214**
see "Pollen Development".

geniculum (lat., pl. genicula)
bulge of a colpus in the equatorial region of a pollen grain. [83]

granular ___ 8, 21, 23, 46, **200-202**
infratectum composed of granula, cluster of granula or elements of different size and shape (never solid and rod-like). [124]
Comment: not to be confused with "granulate", which is a type of ornamentation.

granulate ___ 8, 32
pollen wall with granula. [20]
Comment: not to be confused with "granular", which is a feature of the pollen wall structure.

granulum (lat., pl. granula) ___ 8, 23, **177**
structure- or sculpture element of different size and shape; smaller than 1 µm. [20, 124]

haploxylon-pollen-type ___ 22-23
bisaccate pollen grain with hemispherical sacci.

harmomegathy ___ 23-25, **91-93**
mechanism permitting changes in shape and size of the pollen grain (by varying the hydration status). [134]

hetero-
prefix meaning different.

heteroaperturate ___ 17, 45, 48, **116-117**
pollen grain with two different types of apertures; only one type presumed to function a germination site.
Comment: the term "heterocolpate" is commonly used for pollen grains with alternating colpi and colpori; but "heterocolpate" [64] means two different types of colpi; therefore we prefer the more general term "heteroaperturate".

heterobrochate ___ 8, **163**
reticulate pollen wall with lumina of different sizes. [21]
Comment: the term should be used when the feature is remarkably expressed
Antonym: homobrochate

heteropolar ___ 15, 41, **95-96**
pollen grain with different proximal and distal faces. [21]
Antonym: isopolar

hexa-
prefix meaning six.

homo-
prefix meaning equal.

homobrochate ____ **162**
reticulate pollen wall with lumina of uniform size. [21]
Comment: the term should be used when the feature is remarkably expressed
Antonym: heterobrochate

iatropalynology ____ 12
the study of palynomorphs causing human allergies.

impression mark ____ 39
a linear or Y-shaped mark on the proximal polar area of a pollen grain retained from the tetrad stage.

in-
prefix meaning absent.

inaperturate ____ 17, **103**
pollen grain without distinct aperture(s). [64, 113]
Antonym: aperturate.

infoldings ____ **88-93**
a consequence of harmomegathy (aperture sunken, interapertural area sunken, irregularly infolded).

infra-
prefix meaning beneath.

infratectum ____ 21, 23, **200-202**
layer between tectum and foot layer or endexine (if foot layer is missing); infratectum can be alveolate, columellate, granular or absent. [1]

intectate, see **atectate**
Comment: "atectate" is the more common term

inter-
prefix for in between.

interapertural area ____ 25, 29
region between apertures.

intercolpium, see **interapertural area**
Comment: "interapertural area" is the more general term independently from the aperture type.

internal tectum ____ 21, **203**
a ± continuous layer between foot layer and tectum, separated from them by columellae. [109]

interporium, see **interapertural area**
Comment: "interapertural area" is the more general term independently from the aperture type

interstitium, see **infratectum**
Comment: outdated term

intine
part of the pollen wall next to the cytoplasm, mainly consisting of polysaccharides. [36]

intra-
prefix for within.

irregular, see **outline**

iso-
prefix meaning identical.

isodiametric , see **shape**

isopolar ____ 15, **94**
pollen grain with identical proximal and distal faces. [20]
Antonym: heteropolar.

lacuna (lat., pl. lacunae) ____ **184-185**
depressed area surrounded by ridges (lophae) in lophate pollen grains. [133]

laesura (lat., pl. laesurae) ____ 19
a single arm of a tetrad mark; abbreviation is "-lete" (see also: "tetrad mark", "monolete", "trilete").

lalongate
endoaperture elongated equatorially.

lamellar, see **endexine**

leptoma (gr., pl. leptomata) ____ 17, 20, 22-23
thinning of the pollen wall at the distal pole (of a pollen grain) in conifers, presumed to function as germination area; special case of tenuitas. [22, 27]

LO-analysis ____ 12
light microscopical method for analysing pollen surfaces; see "Palynology".

lobate, see **outline**

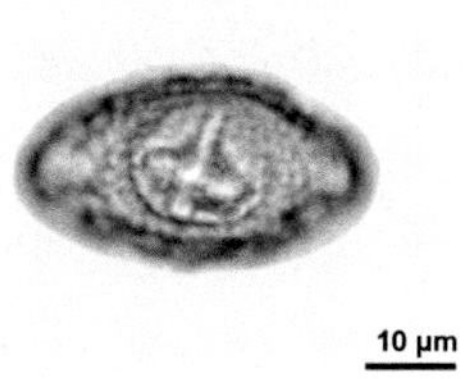

lalongate
Symplocos sp.
Symplocaceae, fossil
equatorial view

lolongate
endoaperture elongated meridionally.

LO-pattern
ornamentation identified by LO-analysis; see "Palynology".

lophae (lat., sing. lopha) ___ **184-185**
a network-like pattern of ridges (=lophae) formed by the outer exine surrounding window-like spaces or depressions (=lacunae).

lophate ___ **184-185**
pollen wall with lophae. [133]

lumen (lat., pl. lumina) ___ **182**
general term for the space enclosed by e.g., muri. [83]

margo (lat., pl. margines) ___ **142**
exine area surrounding an aperture and differentiated in ornamentation. [64]

massula (lat., pl. massulae) ___ 15, 42, **66, 179**
dispersal unit of more than four pollen grains and fewer than the locular content. [83]

melissopalynology ___ 12
the study of palynomorphs found in honey. [21]

melittopalynology, see **melissopalynology**
Comment: the term melittopalynology is the Greek variant of the Latin "melissopalynology".

meridian
imaginary line on the pollen surface connecting proximal and distal poles. [64]

meridional
preposition indicating a direction on the pollen surface; see meridian. [64]

meso-
prefix meaning middle.

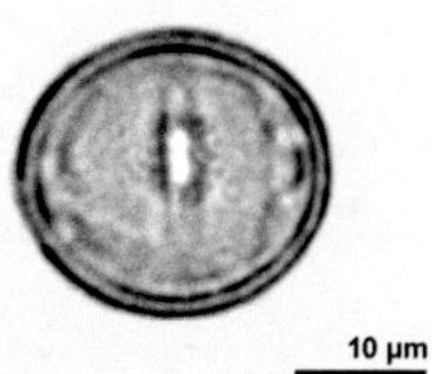

lolongate
Rumex sp.
Polygonaceae, fossil
equatorial view

mesocolpium, see **interapertural area**
Comment: "interapertural area" is the more general term independent of the aperture type.

micro-
prefix for small; features smaller as 1 µm: -baculate, -clavate, -echinate, -gemmate, -pilate, -rugulate, -reticulate, -verrucate; not used in combination with striate, foveolate, perforate.

microspore ___ 15-16, 19, 35-36
see "Pollen Development".

microspore mother cell, see **pollen mother cell** and "Pollen Development".

monad ___ **59**
dispersal unit consisting of a single pollen grain. [107]

mono-
prefix meaning one.

monoaperturate
pollen grain with a single aperture. [80]

monocolpate, see **sulcate** [134]
Comment: superfluous term; as far as known there is no example of a pollen grain with a single colpus (situated equatorially); in all pollen grains with a single elongated aperture the latter is situated distally (sulcus).

monolete, see **laesura** and **tetrad mark**

monoporate, see **ulcerate**
Comment: superfluous term; as far as known there is no example of a pollen grain with a single porus (situated equatorially); in all pollen grains with a single porus the latter is situated distally (ulcus).

monosaccate ___ **97**
pollen grain with a single saccus. [84]

monosulcate, see **sulcate** [20]
Comment: superfluous term, because "sulcate" implies a single elongated aperture (sulcus).

muri (lat., sing. murus) ___ 48, **155-167**
exine elements forming the meshes in a reticulum. [18]

nexine ___ 21
term used for light microscopy, describing the inner, unstructured layer/part of the exine. [21, 30]

Normapolles ______ 33
group of Cretaceous and Lower Paleogene pollen, usually triaperturate, with a complex pore apparatus.

oblate ______ 16, 24, **78-79**
pollen grain with a polar axis shorter than the equatorial diameter. [21]
Antonym: prolate

oblique view
view of a pollen grain neither in polar nor in equatorial view.

omniaperturate, see **inaperturate**
Comment: the term refers to the functional aspect only, therefore we prefer "inaperturate".

oncus (lat., pl. onci)
lens-shaped body located in the apertural region. [62]

operculate ______ **149-152**
aperture with an operculum. [89]

operculum (lat., pl. opercula) ______ 19, 44, **149-152**
coherent exine structure covering an aperture. [89]

orbicule, see **Ubisch body** [25]
Comment: "orbicule" implies a globular element, a too restrictive term; we recommend "Ubisch body" because they are polymorphic.

ornamentation ______ 23, **155-198**
general term, applied in palynology to surface features. [89]

outline ______ 24, **80-93**
general term used to describe the contour of pollen grains in polar and/or equatorial view (can be circular, elliptic, triangular, quadrangular, polygonal, irregular, lobate). [70]

outline in equatorial view
outline of a pollen grain formed by two opposite meridians.

outline in polar view
outline of a pollen grain formed by the equator.

P/E-ratio ______ 16
ratio of the length of the polar axis to the equatorial diameter.

palaeopalynology
the study of fossil palynomorphs.

palynogram ______ 15
diagram summarising the main morphological features of a palynomorph. [21]

palynology ______ 11-13
the study of palynomorphs.

palynomorph ______ 11, 15
general term for all entities found in palynological preparations.

panto-
prefix for global.

pantoaperturate ______ 19, **110-112**
pollen grain with apertures distributed more or less regularly over the whole surface.

pantocolpate, see **pantoaperturate**

pantoporate, see **pantoaperturate**

papilla (lat., pl. papillae) ______ 20, **154**
small protuberance typical for Taxodioideae-pollen (Cupressaceae). [134]

pedium, see **foot layer**
Comment: outdated term

penta-
prefix meaning five.

perforate ______ 27-29, **180-181**
pollen wall with holes less than 1 µm in diameter. [64, 83]

peri-, see **panto-**

pharmacopalynology ______ 12
the study of palynomorphs in drugs.

pilate, see **clavate** [20]
Comment: see pilum

pilum (lat., pl. pila), see **clava** [83]
Comment: the term "pilum" does not refer to the palynological feature; "pilum" means "dart" or "javelin".

planaperturate ______ 42, **105**
pollen grain with an angular outline, where the apertures are situated in the middle of the sides. [21]
Antonym: angulaperturate

plicae (lat., sing. plica) ______ 23, **168**
cirumferential, parallel ridge-like folds. [21]

plicate ______ 30, **70, 168**
pollen wall with plicae. [21]

pluricolumellate
reticulate pollen wall with more than one row of columellae beneath a murus.

polar area
region at and around the pole(s).

polar axis ______ 15-16
imaginary line between the proximal and the distal pole of a pollen grain. [134]

polar view
view of a pollen grain in which the polar axis is directed towards the observer. [18]

polarity ________________ 15, 17-19, 35, 49
orientation of a pollen grain in tetrad stage.

pole
outermost proximal and/or distal point of a pollen grain. [20]

pollen, see **pollen grain**

pollen analysis ________________ 11
study of assemblages of dispersed palynomorphs.
Comment: it does not mean the morphological description of a pollen grain; see palynogram.

pollen class ________________ 8-9, **68-72**
artificial grouping of pollen grains that share a single distinctive character.

pollen coatings__________23, 25, 36, **217-218**
generic term applied to organic compounds usually produced by the tapetum, located on the exine and/or in exine cavities.

pollen grain (pl. pollen grains or pollen)
the male gametophyte of seed plants; the point of origin and the carrier for the male gametes (spermatozoids or sperm cells).

pollen mother cell
see "Pollen Development".

pollen tube
see "Pollen Development".

pollen type______________ 5, 8, 22-23, 30, 46
a general term categorising pollen grains; often used in connection with a distinct taxon.

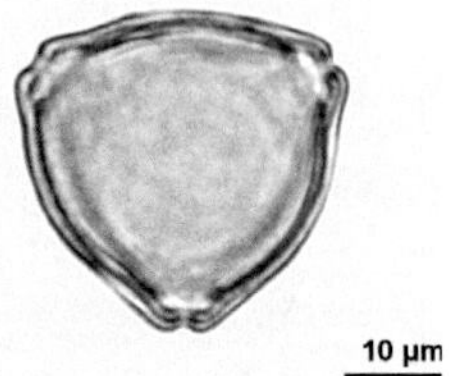

pororate
Corylus sp.
Betulaceae, fossil
polar view

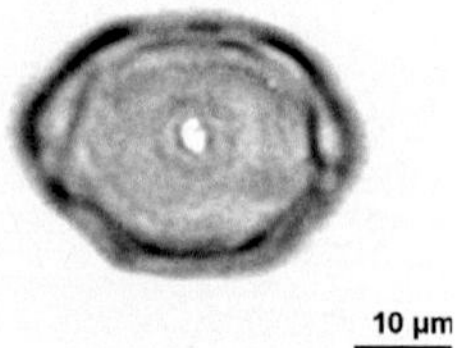

equatorial view

pollen wall ______________ 20-21, 23, **199-211**
layer(s) enclosing the cytoplasm of a pollen grain.

pollenkitt________________ 21, 23, 36, **217**
pollen coating consisting of sticky substances, mainly lipids.

pollinarium (lat., pl. pollinaria) ________15, **67**
dispersal unit of pollinium (or pollinia) and a single interconnecting sterile appendage.

pollination ________________ 12, 25
transfer of pollen from the male to the female reproductive organs in seed plants.

pollinium (lat., pl. pollinia) ___________15, **67**
dispersal unit of a more or less interconnected loculiform pollen mass. [65]

poly-
prefix for many.

polyad________________ 42, **70**
dispersal unit of more than four united pollen grains. [64]

polychotomosulcate
pollen grain with a polychotomosulcus.

polychotomosulcus
sulcus with more than three arms.

polygonal, see **outline**

polyplicate, see **plicate**
Comment: a "plicate" pollen grain has always more than one plica, therefore the term "polyplicate" is superfluous.

pontoperculate ________________ **152**
aperture with a pontoperculum.

pontoperculum (lat., pl. pontopercula) _ **152**
operculum covering a colpus, not completely isolated from the remainder of the sexine.

porate ______________17, 43-44, **70, 121-125**
pollen grain with pori. [21]

poroid ________________ 17, 44, **126-127**
circular or elliptic aperture, with indistinct margin. [20]

poroidate
pollen grain with poroid aperture(s).

pororate
pollen grain with compound apertures composed of a circular ekto- (porus) and endoaperture. [21]

porus (lat., pl. pori; engl. pore, pl. pores) __8, 17, 30, **121-125**
more or less circular aperture situated at the equator or regularly spread over the pollen grain. [84]

porus membrane
aperture membrane of a porus; see **aperture membrane**.

prae-
prefix for before.

prae(pre)-pollen ______________________ 19
microspores of certain extinct seed plants characterised by proximal and distal apertures, and presumed proximal germination.

primexine ____________________ 35, **212, 218**
polysaccharidic layer formed during early developmental stage wherein the later exine structures are preformed.

primexine matrix ____________________ **218**
pollen coating consisting of primexine remnants in mature pollen grains.

prolate ____________________ 16, 24, 55, **76-77**
pollen grain with a polar axis longer than the equatorial diameter. [18]
Antonym: oblate

proximal ________________________ 15, 18-19
pollen features that face or are directed towards the centre of the tetrad. [83]
Antonym: distal

pseudocolpus ____________________ 48, **117**
colpus in heteroaperturate pollen grains, presumably non-functional. [64]

pseudomonad ________________________ **61**
dispersal unit of a permanent tetrad with three rudimentary pollen grains. [107]

psilate ____________________ 29, 32, **178-179**
pollen wall with smooth surface. [134]

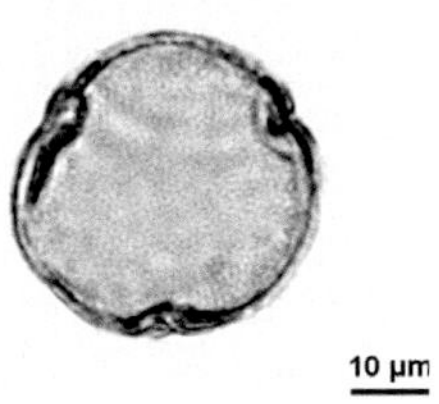

scabrate
Fagus sp.
Fagaceae, fossil
polar view

punctate, see **perforate** [83]
Comment: "punctum" [21] does not describe the three dimensional character of a perforation.

quadrangular, see **outline**

reticulate ______________ 27, 29-30, **155-167**
pollen wall with reticulum. [134]

reticulum (lat., pl. reticula) ________ **155-167**
network like pattern formed by exine elements (muri), where the lumina are wider than 1 µm. [134]

reticulum cristatum ___________ 48, **165-167**
special type of reticulum; muri with prominent sculpture elements. [84]

retipilate [21] ________________________ 48
Comment: to the best of our knowledge there is no example of a reticulum formed by rows of pila instead of muri. Earlier observations where based on light microscopy, SEM-investigations reveal that the given examples of Cuscuta and Callitriche do not fit the definition.

ring-like aperture ________ 44-45, 49, **140-141**
circumferential aperture (situated more or less equatorially or, rarely, meridionally).

rugulae (lat., sing. rugula) ___ 23, 31, **175-176**
elongated exine elements longer than 1 µm; irregularly arranged.

rugulate __________________ 8, 31, **175-176**
pollen wall with rugulae.

saccate ________________ 9, 22-23, **71, 97-99**
pollen grain with one or more air sacs.

saccus (lat., pl. sacci) __________ 22-23, **97-99**
exinous expansion forming an air sac.

scabrate ______________________________ 32
a term used for light microscopy only, describing minute sculpture elements of undefined shape and of a size close to the resolution limit of the light microscope.

sculpture ______________________________ 23
elements of ornamentation on the pollen surface.

semi-
prefix for half.

semitectate __________________________ **210**
pollen grain with a semitectum.

semitectum __________________________ **210**
discontinuous tectum, covering less than 50 % of pollen grain surface.

sexine 21, **152**
term used for light microscopy, describing the structured/sculptured outer layer of the exine.

shape 15-16, 25, **74-100**
see "Pollen Morphology".

size **74**
see "Pollen Morphology".

sperm cell 11, 36, **215**
male gamete; see "Pollen Development".

spheroidal, see **shape**

spine, see **echinus**
Comment: the terms "spine", "spinulate", "spinus", "spinous" and "spinose" are linguistically inconsequent.

spinose, see **echinate**
Comment: see spine

spiraperturate **71, 115**
pollen grain with one or more spiral aperture(s).

spongy, see **endexine**

spore
general term for a reproductive unit (sexual, asexual) of cryptogams and fungi.

sporoderm 11, 20
general term for the wall of spores and/or pollen.

sporopollenin 35
the main component of the exine, consisting of acetolysis-resistant biopolymers.

stenopalynous
plant taxa characterised by only slight variation in pollen (or spore) morphology. [21]
Antonym: eurypalynous

stephanoaperturate 19-20, 49, **106-109**
apertures situated at the equator (term usually used for more than three apertures).

stephanocolpate, see **stephanoaperturate**

stephanocolporate, see **stephanoaperturate**

stephanoporate, see **stephanoaperturate**

stephano-
prefix meaning equatorially situated.

striae (lat., sing. stria) 23, **169-174**
elongated exine elements separated by grooves predominantly parallel arranged. [64]
Comment: the term "striae" is used inconsistently in the literature. We use the term for the elevated elements and not for the grooves.

striate 30, 32, **169-172**
pollen wall with striae. [64]

striato-reticulate 29, **173-174**
ornamentation intermediate between striate and reticulate. [21]

structure 20, 23, 25, **199**
the construction of a pollen wall.

sub-
prefix for less than.

sulcate 44, **71, 135-137**
pollen grain with a sulcus. [20, 134]

sulcus (lat., pl. sulci) 17, 40-41, 49, **135-139**
elongated aperture situated distally. [20, 134]

symmetry 15, 35
see "Pollen Morphology".

syn-
prefix for together.

synaperturate 9, 40, **72, 114**
pollen grain with anastomosing apertures.

syncolpate 40, **114**
pollen grain with anastomosing colpi.

syncolporate 40, **114**
pollen grain with anastomosing colpori.

tapetum 35-36, **216**
specialized layer of cells lining the locule and participating in the nourishment of pollen grains, pollen wall formation and synthesis of pollen coatings.

tectate
pollen grain with a tectum. [30]
Antonym: atectate

tectum (lat., pl. tecta)
outer more or less continuous ektexine layer; tectum condition can be eutectate, semitectate or atectate. [30]

tenuitas (lat., pl. tenuitates) 20-21, **127**
general term for a thinning of the pollen wall. [84]

tetra-
prefix meaning four.

tetrad 15-16, 35, 39, **62-65, 72**
dispersal unit of four pollen grains (spores). [83, 128]

tetrad decussate
dispersal unit of four pollen grains arranged in two planes with two pairs at right angles.
tetrad mark ______ 18, 39
a mark on the proximal face of a spore retained from the postmeiotic stage functioning as germination area (linear = monolete, y-shaped = trilete).
tetrad planar ______ 16
dispersal unit of four pollen grains arranged in one plane; can be: tetragonal, T-shaped, linear.
tetrad stage
see "Pollen Morphology" and "Pollen Development".
tetrad tetrahedral ______ 16
dispersal unit of four pollen grains in which the centers of the grains define a tetrahedron.
tri-
prefix meaning three.
triangular, see **outline**
triaperturate
pollen grain with three apertures.
trichotomosulcate ______ 40, **139**
pollen grain with a trichotomosulcus. [21]
trichotomosulcus ______ 40, **139**
three-radiate sulcus. [21]
tricolpate
pollen grain with three colpi. [64]
tricolporate
pollen grain with three colpori. [64]
trilete ______ 19
see "laesura" and "tetrad mark".
triporate
pollen grain with three pori. [30]
tryphine ______ 23, 36, **218**
pollen coating consisting mainly of lipids mixed with membrane remnants. [24]
Ubisch body ______ 33, 36, **221-223**
sporopolleninous elements produced by the tapetum. [25]
Comment: the "Ubisch body" is named after Gerta von Ubisch, who described these bodies for the first time.
ulcerate ______ **72, 133-134**
pollen grain with an ulcus. [21]
ulcus (lat., pl. ulci) ______ 17, 20, **133-134**
more or less circular aperture situated distally. [21]
vegetative cell
see "Pollen Development".
vegetative nucleus ______ **213**
see "Pollen Development".
verruca (lat., pl. verrucae) ______ 23, **191**
wart-like element more than 1 µm, broader than high. [64]
verrucate ______ 27-28, 31-33, **191**
pollen wall with verrucae. [64]
vesiculate, see **saccate**
Comment: "saccate" is the more general term.
vestibulum (lat., pl. vestibula), see **atrium**
Comment: "atrium" is the more common term.
viscin thread ______ **219-220**
acetolysis resistant thread arising from the exine. [65]
zona-aperturate, see **ring-like aperture**
Comment: source of constant confusion.
zono-aperturate, see **stephanoaperturate**
Comment: source of constant confusion.
Zwischenkörper, see **oncus**
Comment: "oncus" is the more common term.

NOTES

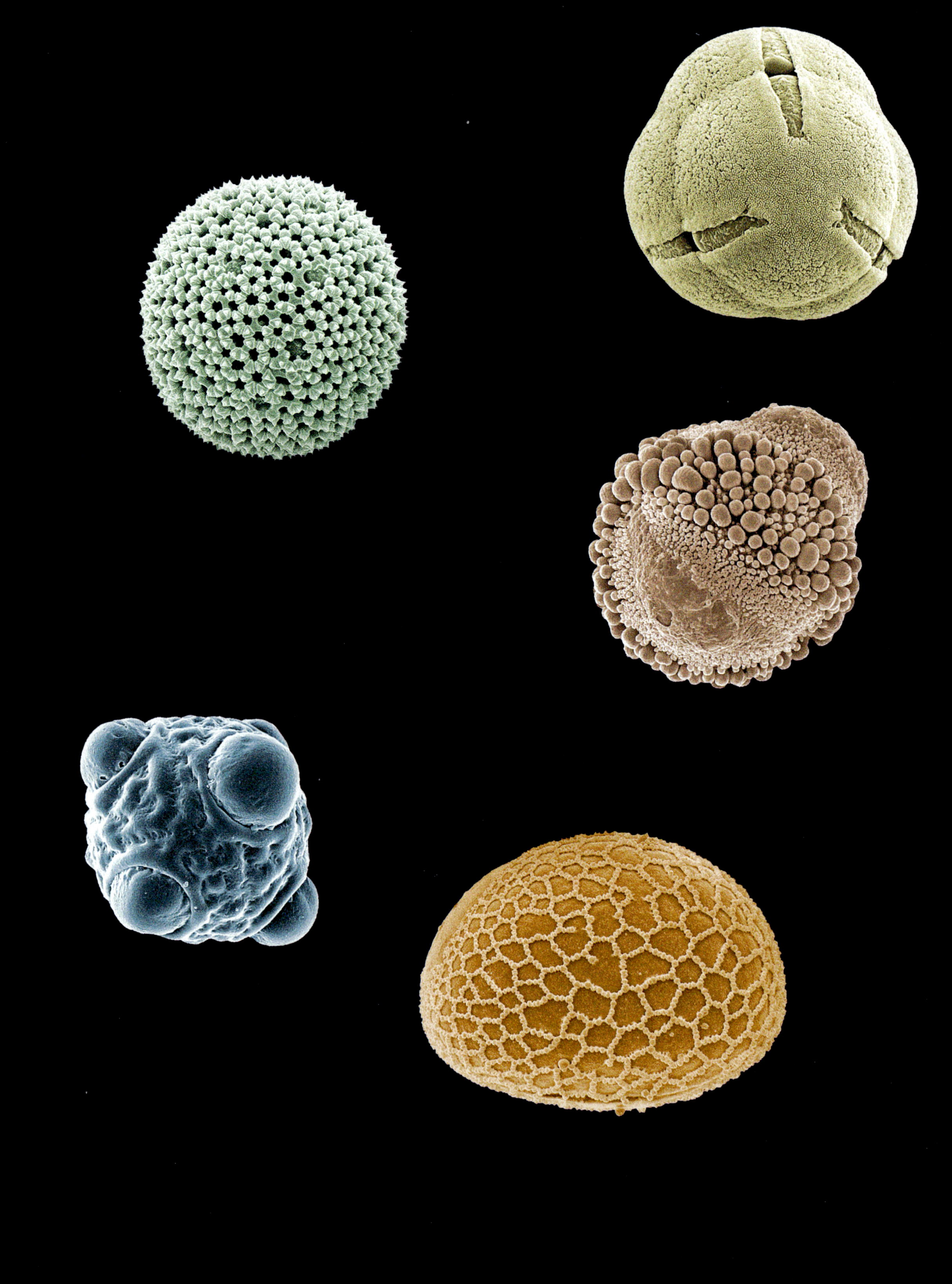

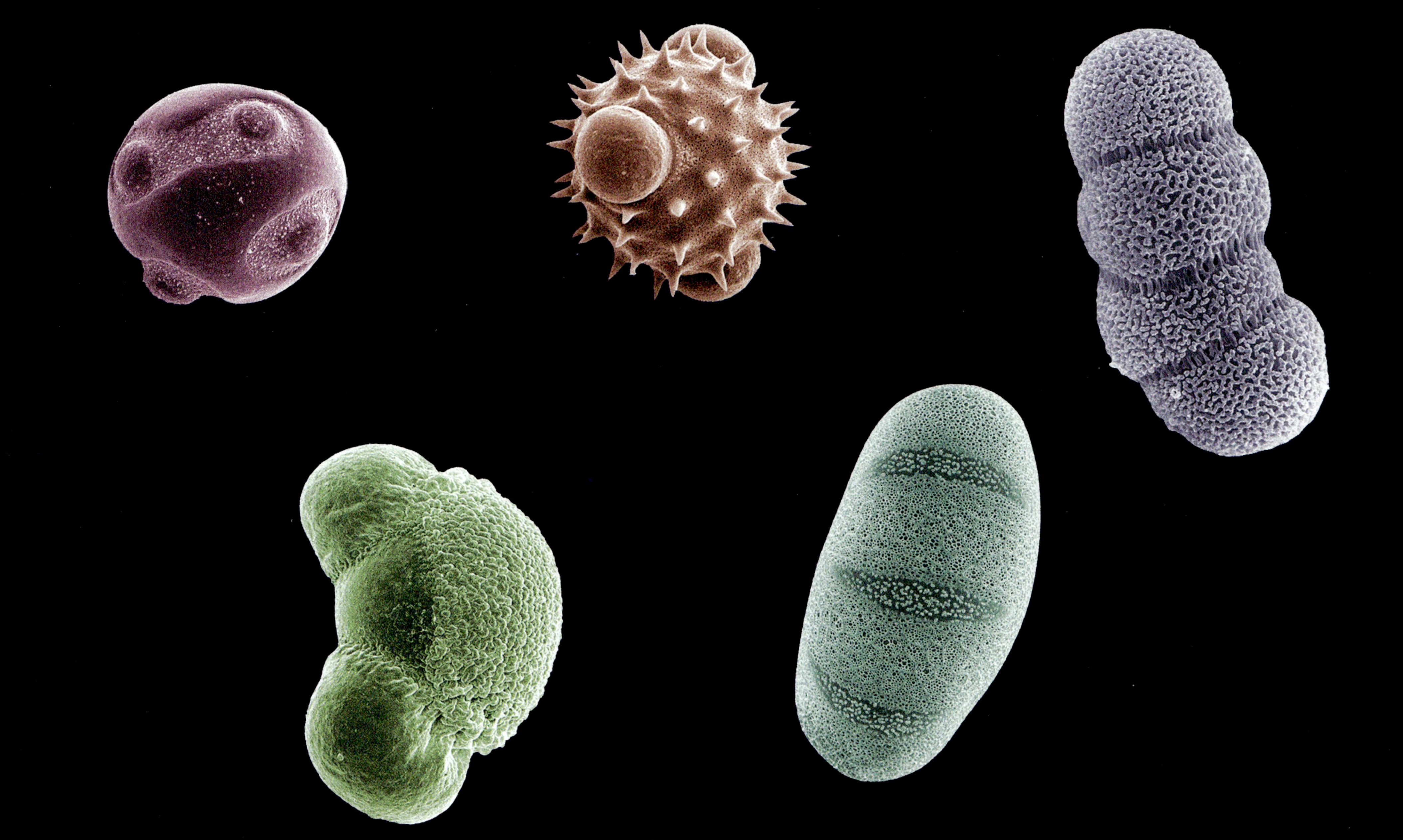

ANNEX

ANNEX

BIBLIOGRAPHY

REFERENCES AND SUGGESTIONS FOR FURTHER READING

1. **APLF** (1975) Morphologie pollinique: problèmes de terminologie, taxons-guides et pollens périporés. Bull Soc Bot France 122: 1-272
2. **BANKS H** (2003) Structure of pollen apertures in the Detarieae *sensu stricto* (Leguminosae: Caesalpinioideae), with particular reference to underlying structures (Zwischenkörper). Ann Bot 92: 425-435
3. **BANKS H, STAFFORD P, CRANE PR** (2007) Aperture variation in the pollen of *Nelumbo* (Nelumbonaceae). Grana 46: 157–163
4. **BEUG HJ** (1961) Leitfaden der Pollenbestimmung. Lief. 1. Gustav Fischer, Stuttgart
5. **BEUG HJ** (2004) Leitfaden der Pollenbestimmung für Mitteleuropa und angrenzende Gebiete. Verlag Dr. Friedrich Pfeil, München
6. **BLACKMORE S** (1992) Scanning electron microscopy in palynology. In: Nilsson S, Praglowski J (eds) Erdtman's Handbook of Palynology. 2nd ed. Munksgaard, Copenhagen, pp 403-431
7. **BLACKMORE S** (2000) The palynological compass: the contribution of palynology to systematics. In: Nordenstam B, El-Ghazaly G, Kassas M (eds) Plant Systematics for the 21st Century. Portland Press, London, pp 161-177
8. **BLACKMORE S, BARNES SH** (1995) Garside's rule and the microspore tetrads of *Grevillea rosmarinifolia* A. Cunningham and *Dryandra polycephala* Bentham (Proteaceae). Rev Palaeobot Palynol 85: 111-121
9. **BLACKMORE S, CANNON MJ** (1983/84) Palynology and systematics of Morinaceae. Rev Palaeobot Palynol 40: 207-226
10. **BLACKMORE S, CRANE PR** (1998) The evolution of apertures in the spores and pollen grains of embryophytes. In: Owens SJ, Rudall PJ (eds) Reproductive Biology. Royal Botanic Gardens, Kew
11. **BLACKMORE S, WORTLEY A, SKVARLA JJ, ROWLEY JR** (2007) Pollen wall development in flowering plants. New Phytol 174: 483-498
12. **BOR J** (1979) Pollen morphology and the bi-reticulate exine of the *Phyllanthus* species (Euphorbiaceae) from Mauritius and Réunion. Rev Palaeobot Palynol 27: 149-172
13. **BUCHNER R, WEBER M** (2000 onwards) PalDat - a palynological database: Descriptions, illustrations, identification, and information retrieval. *http://www.paldat.org/*
14. **DICKINSON HG, ELLEMAN CJ, DOUGHTY J** (2000) Pollen coatings - chimaeric genetics and new functions. Sex Pl Reprod 12: 302-309
15. **DOYLE J** (2005) Early evolution of angiosperm pollen as inferred from molecular and morphological phylogenetic analyses. Grana 44: 227-251
16. **DUCKER S, KNOX B** (1985) Pollen and pollination: a historical review. Taxon 34: 401-419
17. **EHRENBERG CG** (1838) Über die Bildung der Kreidefelsen und des Kreidemergels durch unsichtbare Organismen. Abh Kgl Akademie Wiss Berlin: 59-147
18. **ERDTMAN G** (1943) An introduction to pollen analysis. Chronica Botanica, Waltham, Massachusetts
19. **ERDTMAN G** (1945) Pollen morphology and plant taxonomy. Vol. 3 *Morina* L. Svensk Bot Tidskr 39: 187-191
20. **ERDTMAN G** (1947) Suggestions for the classification of fossil and recent pollen grains and spores. Svensk Bot Tidskr 41: 104-114
21. **ERDTMAN G** (1952) Pollen Morphology and Plant Taxonomy. Angiosperms. Almqvist & Wiksell, Stockholm

22. **ERDTMAN G** (1957) Pollen and Spore Morphology. Plant Taxonomy. Gymnospermae, Pteridophyta, Bryophyta. Almqvist & Wiksell, Stockholm
23. **ERDTMAN G** (1960) The acetolysis method. Svensk Bot Tidskr 54: 561-564
24. **ERDTMAN G** (1969) Handbook of Palynology - An Introduction to the Study of Polllen Grains and Spores. Munksgaard, Copenhagen
25. **ERDTMAN G, BERGLUND B, PRAGLOWSKI L, NILSSON S** (1961) An Introduction to a Scandinavian pollen flora. Almqvist & Wiksell, Stockholm
26. **ERDTMAN G, DUNBAR A** (1966) Notes on electron micrographs illustrating the pollen morphology in *Armeria maritima* and *Armeria sibirica*. Grana Palynol 6: 338-354
27. **ERDTMAN G, STRAKA H** (1961) Cormophyte spore classification. Geol Fören Förenhandl 83: 65-78
28. **ERDTMAN G, VISHNU-MITTRE** (1956) On terminology in pollen and spore morphology. The Palaeobotanist 5: 109-111 (reprinted in Grana Palynologica 1: 6-9, 1958)
29. **FÆGRI K** (1956) Recent trends in palynology. Bot Rev 22: 639-664
30. **FÆGRI K, IVERSEN J** (1950) Textbook of modern pollen analysis. Munksgaard, Copenhagen
31. **FÆGRI K, IVERSEN J** (1989) Textbook of Pollen analysis. 4th ed. John Wiley & Sons, Chichester
32. **FERGUSON DF, ZETTER R, PAUDAYAL KN** (2007) The need for the SEM in paleopalynology. C R Palevol 6: 423-430
33. **FEUER SM, KUIJT J** (1985) Fine structure of mistletoe pollen VI. Small flowered neotropical Loranthaceae. Ann Missouri Bot Gard 72: 187-212
34. **FISCHER H** (1890) Beiträge zur vergleichenden Morphologie der Pollenkörner. Thesis, Berlin
35. **FLYNN JJ, ROWLEY JR** (1967) Methods for direct observation and single-stage surface replication of pollen exines. Rev Palaeobot Palynol 3: 227-236
36. **FRITZSCHE J** (1837) Über den Pollen. Mém Sav Étrang Acad Sci Pétersbourg 3: 649-672
37. **FUJIKI T, ZHOU Z, YASUDA Y** (2005) Asian enviromental history 1. The Pollen Flora of Yunnan, China. Vol. I. Lustre Press/ Roli Books
38. **FURNESS CA** (2007) Why does some pollen lack apertures? A review of inaperturate pollen in eudicots. Bot J Linn Soc 155: 29-48
39. **FURNESS CA, RUDALL PJ** (1999) Microsporogenesis in Monocotyledons. Ann Bot 84: 475-499
40. **FURNESS CA, RUDALL PJ** (2003) Apertures with lids: distribution and significance of operculate pollen in Monocotyledons. Int J Plant Sci 164: 835-854
41. **FURNESS CA, RUDALL PJ, SAMPSON FB** (2002) Evolution of microsporogenesis in angiosperms. Int J Plant Sci 163: 235-260
42. **GARSIDE S** (1946) The developmental morphology of the pollen of Proteaceae. J S African Botany 11: 27-34
43. **GERLACH D** (1969) Botanische Mikrotechnik. Georg Thieme Verlag, Stuttgart
44. **GÖPPERT HR** (1837) De floribus in statu fossili, commentatio botanica. Grass, Bart & Co, Vratislaviae (Thesis of University)
45. **GRAYUM MH** (1992) Comparative external pollen ultrastructure of the Araceae and putatively related taxa. Monogr Syst Bot Missouri Bot Gard 43
46. **GUPTA A, UDAR R** (1986) Palyno-Taxonomy of Selected Indian Liverworts. J. Cramer, Berlin-Stuttgart
47. **HALBRITTER H** (1998) Preparing living pollen material for scanning electron microscopy using 2,2-dimethoxypropane (DMP) and critical-point drying. Biotech Histochem 73: 137-143
48. **HALBRITTER H, HESSE M** (1995) The convergent evolution of exine shields in Angiosperm pollen. Grana 34: 108-119
49. **HALBRITTER H, HESSE M** (2004) Principal modes of infoldings in tricolp(or)ate Angiosperm pollen. Grana 43: 1-14

50. **HALBRITTER H, HESSE M** (2005) Specific ornamentation of orbicular walls and pollen grains, as exemplified by Acanthaceae. Grana 44: 308-313
51. **HALBRITTER H, WEBER M, ZETTER R, FROSCH-RADIVO A, BUCHNER R, HESSE M** (2008) PalDat - Illustrated Handbook on Pollen Terminology. *http://www.paldat.org/*, Wien
52. **HARLEY MM** (1999) Tetrad variation: its influence on pollen form and systematics in the Palmae. In: Kurmann MH, Hemsley AR (eds) The Evolution of Plant Architecture. Royal Botanic Gardens, Kew, pp 289-304
53. **HARLEY MM** (2004) Triaperturate pollen in the monocotyledons: configuration and conjecture. Pl Syst Evol 247: 75-122
54. **HARLEY MM, FERGUSON IK** (1990) The role of the SEM in pollen morphology and plant systematics. In: Claugher D (ed) Scanning Electron Microscopy in Taxonomy and Functional Morphology. Syst Ass Special Volume 41: 45-68. Clarendon Press, Oxford
55. **HARRIS WF** (1955) A manual of the spores of New Zealand Pteridophyta. A discussion of spore morphology and dispersal with reference to the identification of the spores in surface samples and as microfossils. New Zealand Dept Sci Industr Res Bull 116: 1-186. R.E. Owen, Government Printer, Wellington.
56. **HAY WW, SANDBERG PA** (1967) The Scanning Electron Microscope, a major break-through for micropaleontology. Micropaleontology 13: 407-418
57. **HESLOP-HARRISON J, HESLOP-HARRISON Y** (1980) Cytochemistry and function of the Zwischenkörper in grass pollens. Pollen Spores 22: 5-10
58. **HESSE M, HALBRITTER H, WEBER M, BUCHNER R, FROSCH-RADIVO A, ULRICH S** (1999) Frustrate. In: Buchner R, Weber M (2000 onwards) PalDat - a palynological database: Descriptions, illustrations, identification, and information retrieval. *http://www.paldat.org/*
59. **HESSE M, ZETTER R** (2005) Ultrastructure and diversity of recent and fossil-zonaaperturate pollen grains. Pl Syst Evol 255: 145-176
60. **HUYSMANS S, EL-GHAZALY G, SMETS E** (1998) Orbicules in angiosperms: morphology, function, distribution, and relation with tapetum types. Bot Rev 64: 240-272
61. **HYDE HA** (1944) Pollen analysis and the museums. Mus J 44: 145-149
62. **HYDE HA** (1955) Oncus, a new term in pollen morphology. New Phytologist 54: 255
63. **HYDE HA, WILLIAMS DW** (1944) The right word. Pollen Analysis Circular 8: 6
64. **IVERSEN J, TROELS-SMITH J** (1950) Pollenmorphologische Definitionen und Typen. Danm Geol Unders, ser 4, 3: 1-54
65. **JACKSON DD** (1928) A glossary of botanic terms. 4th ed. Duckworth, London
66. **KESSELER R, HARLEY MM** (2004) Pollen. The hidden sexuality of flowers. Papadakis Publisher, London
67. **KLAUS W** (1987) Einführung in die Paläobotanik. Fossile Pflanzenwelt und Rohstoffbildung, Band I. Grundlagen - Kohlebildung - Arbeitsmethoden/Palynologie. Franz Deuticke, Wien
68. **KNOX RB** (1984) The pollen grain. In: Johri BM (ed) Embryology of Angiosperms. Springer, Berlin, pp 197-271
69. **KNOX RB, McCONCHIE CA** (1986) Structure and function of compound pollen. In: Blackmore S, Ferguson IK (eds) Pollen and Spores, Form and Function. Linnean Society of London, London, pp 265-282
70. **KOSANKE RM** (1950) Pennsylvanian spores of Illinois and their use in correlation. Illinois State Geol Survey Bull 74: 1-128
71. **KREMP GOW** (1968) Morphologic Encyclopedia of Palynology. 2nd ed. Arizona Press, Tucson
72. **KRESS WJ, STONE DE** (1982) Nature of the sporoderm in monocotyledons, with special reference to the pollen grains of *Canna* and *Heliconia*. Grana 21: 129-148

73. **KUPRIYANOVA LA** (1956) K voprosu o stroenii obolochki pyl'tsevykh zeren [The structure of the membrane of the pollen grains]. Bot Zhurn 41: 1212-1216
74. **LUGARDON B** (1981) Les globules des Filicinées, homologues des corps d'Ubisch des Spermatophytes. Pollen Spores 23: 93-124
75. **MOAR NT** (1993) Pollen Grains of New Zealand Dicotyledonous Plants. Lincoln, Canterbury, New Zealand
76. **MOHL H** (1834) Über den Bau und die Formen der Pollenkörner. Beiträge zur Anatomie und Physiologie der Gewächse. Erstes Heft. Chr. Fischer und Comp., Bern
77. **MOON H-K, VINCKIER S, SMETS E, HUYSMANS S** (2008) Comparative pollen morphology and ultrastructure of Mentheae subtribe Nepetinae (Lamiaceae). Rev Palaeobot Palynol: 149: 174-186
78. **MOORE PD, WEBB JA, COLLINSON ME** (1991) Pollen analysis. 2nd ed. Blackwell, Oxford
79. **NADOT S, FORCHIONI A, PENET L, SANNIER J, RESSAYRE A** (2006) Links between early pollen development and aperture pattern in monocots. Protoplasma 228: 55-64
80. **NOREM WL** (1958) Keys for the classification of spores and pollen. J Paleontol 32: 666-676
81. **PACINI E** (1997) Tapetum character states: analytical keys for tapetum types and activities. Can J Bot 75: 1448-1459
82. **PACINI E, HESSE M** (2005) Pollenkitt - its composition, forms and functions. Flora 200: 399-415
83. **POTONIÉ R** (1934) I. Zur Morphologie der fossilen Pollen und Sporen. Arb Inst Paläobotanik Petrographie Brennsteine 4: 5-24
84. **POTONIÉ R, KREMP GOW** (1954) Die Gattungen der paläozoischen Sporae dispersae und ihre Stratigraphie. Geol Jahrb 69: 111-194
85. **POTONIÉ R, KREMP GOW** (1955) Die Sporae dispersae des Ruhrkarbons, ihre Morphographie und Stratigraphie mit Ausblicken auf Arten anderer Gebiete und Zeitabschnitte. Palaeontographica Abt B 98: 1-136
86. **POZHIDAEV AE** (2000a) Pollen variety and aperture patterning. In: Harley MM, Morton CM, Blackmore S (eds) Pollen and Spores: Morphology and Biology. Royal Botanic Gardens, Kew, pp 205-225
87. **POZHIDAEV AE** (2000b) Hypothetical way of pollen aperture patterning. 2: Formation of polycolpate patterns and pseudoaperture geometry. Rev Palaeobot Palynol 109: 235-254
88. **PRAGLOWSKI J, PUNT W** (1973) An elucidation of the micro-reticulate structure of the exine. Grana 13: 45-50
89. **PUNT W, BLACKMORE S, NILSSON S, LE THOMAS A** (1994) Glossary of Pollen and Spore Terminology. LPP Contributions Series 1, LPP Foundation, Laboratory of Palaeobotany and Palynology, University of Utrecht, Utrecht
90. **PUNT W, HOEN PP, BLACKMORE S, NILSSON S, Le THOMAS A** (2007) Glossary of pollen and spore terminology. Rev Palaeobot Palynol 143: 1-81
91. **PURKINJE J** (1830) De cellulis antherarum fibrosis nec non de granorum pollinarium formis commentatio phytotomica. Accedunt Tabulae Lithographicae XVIII. Vratislaviae
92. **REILLE M** (1992) Pollen et Spores d'Europe et d'Afrique du Nord. Laboratoire de Botanique Historique et Palynologie, Marseille
93. **REILLE M** (1995) Pollen et Spores d'Europe et d'Afrique du Nord, Supplement 1. Laboratoire de Botanique Historique et Palynologie, Marseille
94. **REILLE M** (1998) Pollen et Spores d'Europe et d'Afrique du Nord, Supplement 2. Laboratoire de Botanique Historique et Palynologie, Marseille
95. **REITSMA TJ** (1969) Size modification of recent pollen grains under different treatments. Rev Palaeobot Palynol 9: 175-202

96. **REITSMA TJ** (1970) Suggestions towards unification of descriptive terminology of Angiosperm pollen grains. Rev Palaeobot Palynol 10: 39-60
97. **RESSAYRE A, DREYER L, TRIKI-TEURTROY S, FORCHIONI A, NADOT S** (2005) Post-meiotic cytokinesis and pollen aperture pattern ontogeny: comparison of development in four species differing in aperture pattern. Am J Bot 92: 576-583
98. **RODRÍGUEZ-GARCÍA MI, FERNÁNDEZ MC** (1988) A review of the terminology applied to apertural thickenings of the pollen grain: Zwischenkörper or oncus? Rev Palaeobot Palynol 54: 159-163
99. **ROWLEY JR, DAHL AO** (1977) Pollen development in *Artemisia vulgaris* with special reference to Glycocalyx material. Pollen Spores 19: 169-284
100. **ROWLEY JR, FLYNN JJ** (1966) Single-stage carbon replicas of microspores. Stain Technol 41: 287-290
101. **ROWLEY JR, SKVARLA JJ** (2000) The elasticity of the exine. Grana 37: 1-7
102. **ROWLEY JR, SKVARLA JJ** (2006) Pollen development in *Epilobium* (Onagraceae): Late microspore stages (a review). Rev Palaeobot Palynol 140: 91-112
103. **SAMPSON FB** (2000) Pollen diversity in some modern Magnoliids. Int J Plant Sci, Suppl 6: S193-S210
104. **SAMPSON FB** (2007) Variation and similarities in pollen features in some basal angiosperms, with some taxonomic implications. Pl Syst Evol 263: 59-75
105. **SCHOLS P, FURNESS CA, MERCKX V, WILKIN P, SMETS E** (2005) Comparative pollen development in Dioscoreales. Int J Plant Sci 166: 909-924
106. **SCHWENDEMANN AB, WANG G, MERTZ ML, MCWILLIAMS RT, THATCHER SL, OSBORN JM** (2007) Aerodynamics of saccate pollen and its implications for wind pollination. Am J Bot 94: 1371-1381
107. **SELLING O** (1947) Studies in Hawaiian pollen statistics. Bernice P Bishop Mus, Spec Publ 38: 1-430
108. **SKVARLA JJ, DEVORE ML, CHISSOE WF** (2004) Lophate sculpturing of Vernonieae (Compositae) pollen. Rev Palaeobot Palynol 133: 51-68
109. **SKVARLA JJ, LARSON DA** (1965) An electron microscopic study of pollen in the Compositae with special reference to the Ambrosieae. Grana Palynol 6: 210-269
110. **SPURR AR** (1969) A low-viscosity epoxy resin embedding medium for electron microscopy. J Ultrastruct Res 26: 31-43
111. **TAKAHASHI M** (1989) Pattern determination of the exine in *Caesalpinia japonica* (Leguminosae: Caesalpinioideae). Am J Bot 76: 1615-1626
112. **THANIKAIMONI G** (1980 «1978») Complexities of aperture, columella and tectum. Proc IVth Intern Palynol Conf I., Lucknow, p. 228-239
113. **THANIKAIMONI G, CARATINI C, NILSSON S, GRAFSTRÖM E** (1984) Omniaperturate Euphorbiaceae pollen with striate spines. Bull Jard Bot Belg 54: 105-125
114. **THIÉRY J-P** (1967) Mise en évidence des polysaccharides sur coupes fines en microscopie électronique. J Microscopie 6: 987-1018
115. **THOMSON PW, PFLUG H** (1953) Pollen und Sporen der mittel-europäischen Tertiärs. Palaeontographica, Abt B 94: 1-138
116. **THORNHILL JW, MATTA RK, WOOD WH** (1965) Examining three-dimensional microstructures with the scanning electron microscope. Grana Palynol 6: 3-6
117. **TRAVERSE A** (1955) Pollen analysis of the Brandon Lignite of Vermont. US Dept Interior Washington, Washington
118. **TRAVERSE A** (1988) Paleopalynology. Unwin Hyman, Boston
119. **TRAVERSE A** (2007) Paleopalynology. 2nd ed., Springer, Dordrecht
120. **TRYON AF, LUGARDON B** (1991) Spores of the Pteridophyta: Surface, wall structure and diversity based on electron microscopy studies. Springer, New York

121. **Van CAMPO M** (1958) Palynologie africaine. 2 Bull IFAN (A) 20: 753-759
122. **Van CAMPO M** (1971) Palynologie et evolution - Précisions nouvelles sur les structures comparées des pollen de Gymnospermes et d'Angiospermes. C Rend Acad Sci Paris, sér D 272: 2071-2074
123. **Van CAMPO M, GUINET P** (1961) Les pollens composés. L'exemple des Mimosacées. Pollen Spores 3: 201-218
124. **Van CAMPO M, LUGARDON B** (1973) Structure grenue infratectal de l'ectexine des pollens de quelques Gymnospermes et Angiospermes. Pollen Spores 15: 171-189
125. **Van der HAM R, GROB G, HETTERSCHEID W, STAR W, Van HEUVEN BJ** (2005) Notes on the genus *Amorphophallus* (Araceae) – 13. Evolution of pollen ornamentation and ultrastructure in *Amorphophallus* and *Pseudodracontium*. Grana 44: 252-265
126. **VINCKIER S, SMETS E** (2002) Morphological and ultrastructural diversity of orbicules in relation to evolutionary tendencies in Apocynaceae s. l. Ann Bot 90: 647-662
127. **VINCKIER S, CADOT P, SMETS E** (2005) The manifold characters of orbicules: structural diversity, systematic significance, and vectors for allergens. Grana 44: 300-307
128. **WALKER JW, DOYLE JA** (1975) The bases of angiosperm phylogeny: palynology. Ann Missouri Bot Garden 62: 664-723
129. **WALKER JW, WALKER AG** (1981) Comparative pollen morphology of the Madagascan genera of Myristicaceae (*Maloutchia, Brochneura, Haematodendron*). Grana 20: 1-17
130. **WEBER M** (1992) Nature and distribution of the exine-held material in mature pollen grains of *Apium nodiflorum* L. (Apiaceae). Grana 31:17-24
131. **WEBER M, FROSCH A** (1995) The development of the transmitting tract in the pistil of *Hacquetia epipactis* (Apiaceae). Int J Plant Sci 156: 615-621
132. **WEBER M, HALBRITTER H, HESSE M** (1999) The basic pollen wall types in Araceae. Int J Plant Sci 160: 415-423
133. **WODEHOUSE RP** (1928) The phylogenetic value of pollen grain characters. Ann Bot 42: 891-934
134. **WODEHOUSE RP** (1935) Pollen grains. Their structure, identification and significance in science and medicine. McGraw-Hill, New York
135. **WORTLEY AH, FUNK VA, ROBINSON H, SKVARLA JJ, BLACKMORE S** (2007) A search for pollen morphological synapomorphies to classify rogue genera in Compositae (Asteraceae). Rev Palaeobot Palynol 146: 169-181
136. **ZETTER R** (1989) Methodik und Bedeutung einer routinemäßig kombinierten lichtmikroskopischen und rasterelektronenmikroskopischen Untersuchung fossiler Mikrofloren. Cour Forsch-Inst Senckenberg 109: 41-50

INDEX TO PLANT TAXA

B

C

D

E

F

G

H

I

J

T

U

PICTURE CREDITS

Except the below mentioned micrographs, all pictures in "*Pollen Terminology. An illustrated Handbook*" are courtesy of PalDat (*http://www.paldat.org/*) and the Society for the Promotion of Palynological Research in Austria (*http://www.autpal.at/*), respectively.

The picture of *Apium nodiflorum* in chapter "Illustrated Terminology – aperture, aperturate" is from Weber M (1992), Fig. 1B.

The two pictures of *Cistus creticus* in chapter "General Chapter - Pollen Morphology – Harmomegathy" are from HALBRITTER and HESSE (2004), Figs. 2C, D.

The two pictures of *Iris histrioides* in chapter "Illustrated Terminology – ring-like aperture" are from HESSE and ZETTER (2005), Figs. 6B, D.

The picture of *Ruspolia seticalyx* in chapter "Illustrated Terminology – Ubisch body" is from HALBRITTER and HESSE (2005), Fig. 1A.

Drawings by Heidemarie HALBRITTER and Ralf BUCHNER.

NOTES

NOTES

NOTES

QUICK REFERENCE